Quality by Design

N. Belavendram

Prentice Hall

London   New York   Toronto   Sydney   Tokyo   Singapore

First published 1995 by
Prentice Hall International (UK) Limited
Campus 400, Maylands Avenue
Hemel Hempstead
Hertfordshire, HP2 7EZ
A division of
Simon & Schuster International Group

Printed and bound in Great Britain by
Redwood Books, Trowbridge, Wiltshire

Library of Congress Cataloging-in-Publication Data

Available from the publisher

British Library Cataloguing in Publication Data

A catalogue record for this book is available
from the British Library

ISBN 0-13-186362-2

1 2 3 4 5   99 98 97 96 95

Dominic

Contents

CHAPTER 1
INTRODUCTION TO QUALITY BY DESIGN   1

CHAPTER 4
OBJECTIVE FUNCTIONS IN ROBUST DESIGN   143

CHAPTER 5
BASIC ANALYSIS OF VARIANCE
209

CHAPTER 6
MODIFYING ORTHOGONAL ARRAYS   **317**

CHAPTER 10
CONDUCTING AN EXPERIMENT   **499**

FOREWORD

Even today, Quality is still seen by many business people, including senior executives, as a manufacturing problem to be solved by the inspector or by quality assurance. Business for them is becoming increasingly difficult as competitive pressures increase. Essentially they have failed to recognize quality as a strategic objective for management in every sector and for every business. For them quality by design is a sophisticated version of quality assurance, something that is to be added at the *END* of the process and which increases costs.

They have yet to learn, as did the Japanese, and many others since, that quality by design requires a clearly demonstrated commitment from the highest level of management. Quality cannot be added on as a package. It must be built into the culture of the business *AND* the design of the process, so it includes everyone, reflecting in every activity, at every level, in the business. It belongs to us all. Poor quality is not *what we get from others, it is the result of an inadequate culture of which we are a part.*

Design for manufacture plays a vital role in manufacturing industry. Unless engineering efforts build quality into a process or product from the design stage, by robust design, the product may not be competitive and so may eventually be forced out of the market. Any method that provides robust design is a tool for engineers and engineering quality. The design of experiments is one such important tool for designing a product for manufacture.

Design for manufacture can be classified into three stages: System Design, Parameter Design and Tolerance Design. The design process must ensure that products can be manufactured to the right quality despite variations in the process.

System Design is the brainchild of a functional expert. The System Design must ensure the functionality of the design. Examples of System Design are the thermionic valve, the transistor and integrated circuits. There is no formal way of establishing a System Design since it relates to a new concept or idea.

Parameter Design is a method of experimentation to design a product by selecting the optimum conditions of Control factors (factors which we can control) so that the product is least sensitive to uncontrollable factors (Noise factors) such as wear, aging and ambient conditions. Noise factors are difficult and usually expensive to control. The Parameter Design examines Control factors and Noise factors exposed to the full spectrum of manufacturing and customer noise through the orthogonal arrays in order to achieve robustness. It is a search for control factor levels at which the functional characteristic is relatively stable despite variations in noise. This translates to the ability

xvii

of using inexpensive components and materials and less control of the external (environmental) conditions.

The signal-to-noise (SN) ratio is an objective performance measure. The SN ratio is an evaluation of the stability of performance of an output characteristic. The SN ratio measures the level of performance and the effect of noise factors on performance. The form of the SN ratio is tied directly to the Quality Loss Function. The Quality Loss Function is a mathematical formula that can be used to evaluate the effect of the stability of the performance of an output characteristic in terms of monetary units. Better performance as measured by a higher SN ratio implies a smaller loss as measured by the corresponding Quality Loss Function.

Where mathematical models relating the functional characteristic to the control factors exist, computer-aided parameter design (CAPD) may be performed by computer simulation of internal and external noise through the Direct Product Design. When a design cannot be further improved by Parameter Design but still higher SN ratios are required, then the next step may be Tolerance Design. Again, where mathematical models exist, computer-aided tolerance design (CAPD) may be performed by computer simulation. However, it must be remembered that Tolerance Design increases the cost of the product. The question therefore is where to invest the money to be most cost effective in narrowing the input tolerance to narrow the output variability. Hence, it is very important to identify the variation caused by each element so that only the critical components are replaced. The analysis of variance (ANOVA) and the importance of rho (ρ, percent contribution) of the component can then be linked to cost justification in Tolerance Design. Such a design procedure allows the engineer to spend the money only on those components that need to be upgraded.

Summarizing, unless engineering efforts are made to build quality into a product from the design stage, by robust design, the product may not be fully competitive and may eventually be forced out of the market. Any method that provides robust design is a tool for engineers and engineering quality. The design of experiment is one such important tool for designing a product for manufacture.

The main aim of this book is to emphasize that design for manufacture must consider:

- System Design – to provide functional design,
- Parameter Design – to provide robustness by attenuation of cause,
- Tolerance Design – to identify components that need to be upgraded,
- Robustization of process and product design prior to manufacture,
- Intrinsic quality and reliability,
- Emphasis on design and not inspection.

PREFACE

As customers we all want products that will work each time, every time. Most of us have bitter experiences of products failing at the most awkward of times. All of us have experiences of repairing or returning a faulty product we bought recently.

In the past it was the function of a design engineer to draw up a plan of a product. The product engineer was faced with the problem of making the product work. The quality control engineer sorted the good from the bad. The sales engineer devised offers to sell the product. The public relations officer dealt with all the returns and the complaints.

Today, an engineer has to design a product with manufacturing in mind. If products need to be manufactured easily they have to be designed so. Equally, products need to be insensitive to the spectrum of uncontrollable factors surrounding the operational environment of the product. More importantly, the customer must see the value for money. This form of engineering builds *quality by design*, where quality is an intrinsic function of good design.

Achieving such a product is not a black art although it is magic in a sense. Of course, if there are unlimited resources then we can do anything. In reality, we have to make the best use of resources – indeed the best use of energy. And unless *improved quality* means a lower loss to the society of which we are a part, there can be no quality improvement.

ACKNOWLEDGMENTS

I thank Professors Robert Burnside and William McEwan (OBE) for enabling my Doctor of Philosophy programme of study, *Taguchi Methods for Manufacturing Systems Design*, in the University of Paisley. Through the latter, I met Dr W. Edwards Deming and Professor Yuin-Wu. I had the greatest privilege of talking to Dr Deming – about Sir Ronald Fisher, the British statistician. It gave me a sense of feeling that I had spoken to someone who had worked with Dr Fisher himself – and hence a special meaning for the analysis of variance. I also thank Professor Yuin-Wu for encouraging me to take a leadership in quality engineering.

I am greatly indebted to Dr Ewan McArthur for so many hours of discussion on various issues in this book. Without his encouragement and guidance, this book may have taken much longer to complete.

I have never met Dr Genichi Taguchi, but I believe his idea of minimizing loss-to-society, among others, is revolutionary in world class quality. Of course, new ideas are always contested for a while until people begin to see the benefit. And then they ask why it was not available sooner. I hope that time is now. And to Dr Taguchi, Arigatō gozaimashita, sensei.

Similarly, somehow I feel I should also thank Dr Madhav Phadke and Dr Nicholas Logothetis. Again I do not know them, but I have read and studied many of their works which together with a few others have extended my horizon.

I also thank Charles Murray, Ian McKeown and Kenneth McElroy my Polaroid collaborators for their cooperation in my Doctor of Philosophy programme of study. In particular I wish to thank Ian McKeown for being my industrial supervisor.

Recalling my days as a quality engineer in Texas Instruments, Malaysia, I extend my gratitude to Jerry Lee, Nazimuddin Majeed and Eric Ng Eng Wah besides many others, and particularly, Peter Hor Siew Weng, for encouraging my academic interests at a time when everything seemed impossible.

To my brothers and sisters, who believe I should be with them at least on Christmas day, I promise I will come out to play.

And to my parents, Belavendram Pillai and Souriamah, who wonder what all my calculations are about – for them, they keep saying, no number is large enough to express their love for me. I was no great religious person until I wondered how to thank my parents for everything they have sacrificed for me. Now I pray, God help me.

Last but not least, I shall try to thank Sarojani Devi, Valli Sarojah and Carolyn Frances for their unbounded love. I shall try because, I have yet to learn how to thank someone who gave her everything.

ABBREVIATIONS

$\bar{A}$	Average of factor A.
A_l	Linear component of factor A.
A_q	Quadratic component of factor A.
A_0	Average loss at the specification limit(s).
CAPD	Computer-Aided Parameter Design.
CATD	Computer-Aided Tolerance Design.
CI	Confidence interval.
cos	Cosine of an angle.
df	Degrees of freedom.
e1	Primary error.
e2	Secondary error.
Pooled e	Pooled error.
$E(X)$	Expected value (mean) of X.
$V(X)$	Variance of X.
Exp	The experiment number.
$F_{\alpha, \nu 1, \nu 2}$	F-ratio with alpha risk (α), numerator degrees of freedom ($\nu 1$), and denominator degrees of freedom ($\nu 2$).
f	Function.
f'	First differential.
f''	Second differential.
f'''	Third differential.
k	Cost coefficient.
log	Logarithm to base ten.
ln	Logarithm to base e, (e = 2.718281828).
$L(y)$	Financial loss at y.
MSD	Mean squared deviation
n_{eff}	Effective number of units.
$n!$	Factorial n, $(n \times (n - 1) \times (n - 2) \times \ldots \times 1)$.
P_{ij}	Element P for the ith control factor array and jth noise factor array.
QLF	Quality Loss Function.
sin	Sine of an angle.
SA	Sum of squares due to factor A.
SA'	Corrected (or pure) sum of squares due to factor A.
Se	Sum of squares due to error.
Sm	Sum of squares due to mean.

SN	Signal-to-noise.
VA	Variance of factor A.
y_i	The i th element of y.
α	Alpha error or risk.
β	Beta error or risk.
γ	Initial tolerance.
δ	Infinitesimal change or difference.
Δ_0	Tolerance of a quality characteristic.
ϵ	Error.
η	Signal-to-noise ratio.
λ	Upgraded tolerance.
μ	Process average for mean data.
νA	Degrees of freedom of factor A.
π	Pi (3.141592654)
ρ	The percent contribution to total sum of squares.
σ	The sample standard deviation.
σ_n	The population standard deviation.
Σ	The sum of numbers.
Ω	The Omega transformation.
$\wp$	Loss per piece
$\Re$	Loss reduction.
$\Im$	Gain in loss per piece.

SYNOPSIS

Quality by Design is based on the Taguchi methodology, which was introduced to the United States by Dr Genichi Taguchi in 1980. Since then, many companies have used this methodology and claimed great benefits. Equally, many others, particularly statisticians, have claimed that the methodology is neither accurate nor precise. The original works of Taguchi are found in his two-volume book, *System of Experimental Design*.

Many industrial engineers have applied Taguchi Methods, usually after a short course, and have claimed wonders. Others, who had only seen crude applications have completely denied any use of Taguchi's techniques.

While there are many books on this subject, most have been written by statisticians for statisticians and engineers who may already have a first degree in mathematics or statistics. As an engineer myself, with neither a mathematics nor a statistics degree, I have found these books very difficult to follow. Often, I have found numerous books which attempt to explain something in Chapter 1 by a reference to Chapter 5! I have tried to avoid that in this book. While some statistics and mathematics is inevitable, I have tried to keep the text as simple as possible.

Chapter 1 provides an **Introduction to Quality by Design.** It describes the theory and practical aspects of Quality by Design and how these aspects can be applied to practical problems. The concepts of quality loss functions and loss-to-society are also introduced. Three particular quality loss functions, namely, nominal-the-best, smaller-the-better and larger-the-better are discussed with respect to one-piece and many-pieces. The chapter explains the use of customer loss data to establish realistic manufacturing tolerances. The chapter also explains the need to quantify quality improvements in monetary units so that objective comparisons may be made.

Chapter 2 describes **The Design Process** and provides an introduction to the roles of quality assurance and the quality loss function. The quality loss function is an objective way of establishing the monetary loss borne by customers. The idea of Parameter Design and a practical example is included. Robust Design adds a new dimension to statistical experimental design – it explicitly addresses the following concerns faced by all product and process designers: (i) How to reduce economically the variation of a product's function in the customer's environment, and (ii) how to ensure that decisions that are found to be optimum during laboratory experiments will also prove to be so in manufacturing and in customer environments.

Chapter 3 describes how to design experiments using **Orthogonal Arrays and Matrix Experiments**. It also explains the effects of interactions on factor effects and how to analyze them. The selection of quality characteristics based on additivity is also explained. An explanation of fair comparison is given, together with the concept of degrees of freedom. Interaction is explained with interaction breakdown. The main difficulty found in industrial experiments is related to interactions. Selection of good quality characteristics is explained with respect to good additivity.

Chapter 4 deals with **Objective Functions in Robust Design**. The aim of this chapter is to provide the reader with methods of calculating signal-to-noise ratios for a number of quality characteristic types and performing attribute data analysis. Static problems are distinguished from dynamic problems. SN ratios for static problems are developed for smaller-the-better, nominal-the-best, larger-the-better, signed-target and percent defective quality characteristics. This chapter also introduces the relationship between loss-to-society and signal-to-noise ratio. A method for increasing the operating limits for a process is illustrated using the operating window analysis. Methods for improving additivity in percent defective data are illustrated using the Omega (Ω) transformation.

Chapter 5 provides an introduction to **Basic Analysis of Variance**. Fundamental ideas of statistics are developed from first principles. A simple idea of analysis of variance is introduced through the no-way (or no-factor) analysis of variance and built up to a one-way (or one-factor) analysis of variance and eventually to a multi-way (or multi-factor) analysis of variance using orthogonal arrays. The concept of conserving the total sum of squares and total degrees of freedom is also introduced. Analysis of variance calculations for special designs are shown. Advanced ideas of percentage contribution of sum of squares due to a source are treated. Pooling of insignificant factors is introduced. Methods of calculating confidence intervals for factor levels, mean effects and predicted process averages are also shown.

Chapter 6 introduces ways of **Modifying Orthogonal Arrays**. Industrial experimentation does not always lend itself to direct applications of standard orthogonal arrays. Sometimes it is necessary to tailor an orthogonal array for a more specific purpose. This chapter introduces five ways of modifying orthogonal array designs for specific purposes. These are the multi-level design, dummy level design, combination factor design, pseudo-factor (nested) design and the pseudo-factor (idle-column) design. This chapter also introduces the use of distributed interaction designs, which although does not require any modification, nevertheless, provides a special method of factor assignment. Each of these techniques has its own uniqueness and merit for application.

Chapter 7 introduces **Computer-Aided Parameter Design**. A relatively simple idea of a trajectory is used to illustrate the importance of parameter design. Initially the problem is solved using a conventional approach and later using the parameter design. The results of both techniques are compared in terms of the normal distributions of the trajectory for a controlled noise environment.

Chapter 8 introduces **Computer-Aided Tolerance Design**. This is a method of trade-off between reductions in quality loss due to performance variation and increases in manufacturing costs. Tolerance design identifies components (or subsystems) of a system that contribute significantly to the functional variation of the system. Only such significant components need to be replaced using higher grade components. Of course, the unselective replacement of components using higher grade components will lead to unnecessary manufacturing costs.

Chapter 9 introduces **Managing the Design of Experiments**. This chapter introduces a few principles of total quality management, including quality awareness of a company, quality perception and the cost of quality. A number of management issues concerning the implementation of robust design within a company and guidelines for successful experimentation are also given. Some common criticisms of robust design are also addressed, with explanations of how to overcome these criticisms.

Chapter 10 introduces **Conducting an Experiment** and how to report the experiment. The chapter gives a method of approaching experimentation particularly with respect to forming a multi-disciplinary team for brainstorming and team activities. The relevant data that need to be collected before and after experimentation are highlighted. The communication of quality improvements for production personnel and cost-savings calculations for management are demonstrated through graphical representations.

The reader may note that many experimental results in the text have been simplified so that the reader can readily grasp the underlying principles and not wrestle with the mathematics or statistics. The reader should also note that most calculations have been rounded-up to two decimal places although actual calculations have been carried to much higher precision on the spreedsheet used for these calculations. The reader is urged to complete the self-assessment question(s) at the end of each section before proceeding to the next. Answers are given immediately after each section for the benefit of the reader. The reader who completes this text should be able to perform a range of design of experiments and for the interested reader, to look forward to advanced quality by design.

Finally, while great care has been taken to ensure the correctness of all calculations, tables and figures, any errors, omissions, oversights or misinterpretations especially those of Taguchi's philosophy are the author's responsibility.

CHAPTER 1

INTRODUCTION TO QUALITY BY DESIGN

AIMS:
The aim of this chapter is to introduce the concept of *Quality by Design* using the Taguchi methodology. It describes the theoretical and practical aspects of Quality by Design and how these aspects can be developed for industrial application.

OBJECTIVES:
When you have completed studying this chapter you should be able to:
- explain customer perception of quality,
- explain the interpretation of loss,
- describe loss-to-society,
- derive the quality loss functions,
- calculate the quality loss functions,
- explain why process capability is not a good measure of quality,
- determine loss due to defects,
- determine loss due to dispersion.

OVERVIEW:
This chapter explains the importance of making products to target values rather than to specifications. The concepts of quality loss functions and loss-to-society are also introduced. Three particular loss functions, namely, nominal-the-best, smaller-the-better and larger-the-better, are discussed with respect to one piece (an individual product) and many pieces (a sample of products). The chapter explains the use of customer loss data to establish realistic manufacturing tolerances and explains the need to quantify quality improvements in monetary units so that objective comparisons may be made. The chapter forms the basis for the development of subsequent chapters.

1.1 Basics of Quality by Design

1.1.1 Introduction

The objective of engineering design, a major part of research and development (R&D), is to produce drawings, specifications, and other relevant information, needed to manufacture products that meet customer requirements. A large amount of engineering effort is consumed in conducting experiments (either hardware, software or simulation) to produce the information required to make decisions. Efficiency in generating such information is the key to making market decisions, keeping development and manufacturing costs low, as well as producing high-quality products. Quality by Design is one such engineering methodology.

1.1.2 Customer's perception of quality

In April 17, 1979, an article appeared in the *Asahi*[1], a Japanese newspaper, discussing the preference of American consumers for television sets built by Sony in Japan rather than those built by Sony in America. Apparently identical subsidiary Sony plants had been built in San Diego, California and in Tokyo, Japan. Both facilities were intended to manufacture television sets for the American market and did so using identical blueprints. Despite these similarities, when television sets were put on display, the American consumer displayed a preference for the sets that had been made in Japan.

The reason for this preference was traced to the colour density of the television sets. The voltage specification for the colour density circuit was 115 ± 20 volts and American sets were made-to-specifications. Before shipping, every set was inspected, resulting in no defective televisions being shipped out. This ensured that no set with the colour density out-of-specification was shipped from the American plant. Why then were the Japanese sets preferred?

When the television sets were put on display, the Japanese sets were perceived as better quality. What then is Quality? A careful study of the problem revealed that Sony America and Sony Japan had different characteristic distributions of colour density in their television sets. From Figure 1:1.1 it can be seen that:

- area A corresponds to the proportion of television sets that are common to both the American and Japanese manufacturing plants,
- area B shows the American television sets that are poor in performance that do not have similar Japanese counterparts,
- area C shows the Japanese television sets that have a better performance that do not have similar American counterparts,

---

1   Genichi Taguchi and Don Clausing, *Robust Quality*, Harvard Business Review, January-February 1990;
Thomas B Barker, *Quality Engineering by Design: Taguchi's Philosophy*, Quality Progress, December 1986.

- area D shows that all the effort in 100 % screening by Sony America can affect only this minute and indeed insignificant proportion of television sets.

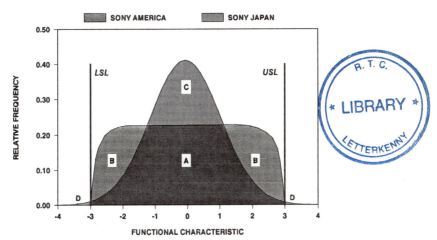

Figure 1:1.1 Colour density distributions of Sony America and Japan sets. *LSL* and *USL* are the lower and upper specification limits, repectively.

This study showed clearly that the performance of the Japanese television sets was indeed superior. Thus, meeting specifications alone is not enough. Although they dictate the limits of product functionality, specifications (and hence tolerances) themselves do not ensure quality. Measurments of conformance (or nonconformance) and the associated fraction defective, percent defective and parts per million, etc., as did the Sony America plant, are not good measures of quality. Rather, measurements of quality must focus on concentration about the target value as did the Sony Japan plant. Indeed, as concentration around the target value increases, other wasteful losses such as the need for 100 % inspection as a tool for ensuring quality also decreases.

1.1.3 Taguchi's definition of quality

Referring to the case of Sony America and Japan, according to the conventional method, products that met specifications were deemed to be perfect in the American plant. Therefore, a set that had a 10 volt deviation from target (115 volts) conformed to specifications and so was not regarded as a quality problem. Taguchi, however, proposes a different view of quality that relates quality to cost and loss in monetary units, not only to the manufacturer at the time of production, but also to the consumer and society as a whole.

> *The quality of a product is the (minimum) loss imparted by the product to society from the time the product is shipped.*

With Taguchi's definition of quality, the goal of the manufacturer should be to provide products and services that meet customer needs and expectations over the life of the product or service, at a cost that represents customer value.

1.1.4 What is loss – and who pays for it?

We usually think of loss as additional manufacturing costs incurred up to the point a product is shipped. After that, it is the customer who bears the cost of quality loss. When the consumer refuses to continue paying for the costs of poor quality, the manufacturer will go out of business. When a product is under warranty, the manufacturer pays in warranty costs. When the warranty expires, the customer may pay for repair or rework on a product. But indirectly, it is the manufacturer who will ultimately pay for the quality loss as a result of negative customer reaction and costs which are difficult to capture and account for, such as:

- returns,
- warranty costs,
- customer complaints and dissatisfaction,
- time and money spent by the customer,
- eventual loss of market share and growth.

1.1.5 Interpretation of loss

Historically, the objective of quality control has been to control functional variation and its related problems. However, since no method of quantitative evaluation of quality or quality loss had been established, the problems of quality control and their resolutions have usually been treated subjectively. The objective of Taguchi's *Quality Loss Function* (QLF) is quantitative evaluation of quality loss due to functional variation.

A quality characteristic is the object of interest of a product or process. It may be also called a functional characteristic. Generally, any quality characteristic will have a target. There are three types of targets:

- nominal-the-best,
- smaller-the-better,
- larger-the-better.

1.   Nominal-the-best

A nominal-the-best characteristic is a measurable characteristic with a specific user-defined target value. An example is the colour television output voltage supply studied in the case of Sony America and Sony Japan. The target is a nominal value of 115 ± 20 volts.

2.   Smaller-the-better

A smaller-the-better characteristic is a non-negative measurable characteristic that has an ideal state or target of 0 (zero). An example is tyre wear. The smaller the tyre wear, the better the tyre. The target is zero.

3.   Larger-the-better

A larger-the-better characteristic is a non-negative measurable characteristic that has an ideal state or target of ∞ (infinity). An example is fuel efficiency. The more kilometres per litre of fuel, the better the fuel efficiency. The target is infinity.

For each quality characteristic there exists some function which uniquely defines the relationship between economic loss and the deviation of the quality characteristic from its target value. The time and resources required to obtain such a relationship for each quality characteristic represents a considerable investment. Taguchi found the quadratic representation of the quality loss function to be an efficient and effective way to assess the loss due to deviation of a quality characteristic from its target value, i.e. due to poor quality. A premise of the Taguchi methodology is that useful results must be obtained quickly and at low cost. The use of the quadratic approximation for the quality loss function is consistent with this philosophy.

From the study of perceived quality based on the Sony America and Japan colour television sets, we will see later in this chapter that:

● conformance to specification limits is an inadequate measure of quality loss,
● quality loss is caused by customer dissatisfaction,
● quality loss can be related to product characteristics,
● quality loss is a financial loss,
● the quality loss function is an efficient tool for evaluating quality loss at the earliest stage of product or process development.

1.1.6 QLF – Nominal-the-best: One piece

If y is some variable value of a nominal-the-best quality characteristic the quality loss function for y can be written as $L(y)$. The loss $L(y)$ can be expanded by a Taylor's series about the target value m. Hence,

$$L(y) = L(m + (y - m))$$
$$= L(m) + \frac{L'(m)(y - m)}{1!} + \frac{L''(m)(y - m)^2}{2!} + \dots \qquad (1:1.1)$$

$L(m)$ is always a constant and may be ignored since its effect is to uniformly raise or lower the value of $L(y)$ at all values of y. Also, since $L(y)$ is a minimum at $y = m$, $L'(m) = 0$. The $(y - m)^2$ term is the dominant term in Equation (1:1.1), larger powers being neglected. Therefore,

$$L(y) = \frac{L''(m)(y - m)^2}{2!}$$
$$L(y) \approx k (y - m)^2$$

Expressed in this way, $L(y)$ is the average quality loss or the *loss-per-piece* ($\wp$). The quadratic loss function of a nominal-the-best quality characteristic y around the target value m is therefore given by Equation (1:1.2). This is the quadratic form for the quality loss function for nominal-the-best.

$$L(y) = k\,(y - m)^2 \qquad\qquad (1:1.2)$$

where

y	=	the value of the quality characteristic (e.g. length, concentration, force, flatness, etc.),
$L(y)$	=	the loss in dollars ($\$$) per piece of product when the quality characteristic is equal to y,
m	=	the target value of y,
k	=	the proportionality constant or *cost coefficient*, which depends on the financial criticality of y.

Let us develop the quality loss function for a colour television power supply circuit. The power supply voltage is a quality characteristic, denoted by y. Further, suppose that the manufacturing engineer knows that the ideal voltage m, for the colour television set is 115 volts. This value m, is referred as the target value of y. Since this is a nominal value, the quality characteristic is nominal-the-best. Suppose the voltage at which half the customers view the television sets as poor quality is 115 ± 20 volts. The *customer tolerance* Δ is thus $\pm$ 20 volts. These limiting voltages are called the customer specification limits. The *lower specification limit (LSL)* is 95 volts ($115 - 20 = 95$). The *upper specification limit (USL)* is 135 volts ($115 + 20 = 135$). To say that the customer specification is 115 ± 20 volts implies that if the output voltage were gradually reduced, about half the customers would bring back their sets for repair or replacement at 95 volts. Similarly, if the output voltage were gradually increased, about half the customers would bring back the sets for repair or replacement at 135 volts. These values represent an average viewpoint and are referred to as the *LD-50*[2]. If the average cost of fixing (i.e. long-term repairing, resetting or replacing) a colour television that is out-of-specification at Δ is A_0, then,

$$L(y) = k\,(y - m)^2$$

$$k = \frac{L(y)}{(y - m)^2}$$

and at the specification,

---

2   The notation *LD-50* represents the lethal dose of a drug at which 50 % of the test individuals are killed.

$$k = \frac{A_0}{\Delta^2}$$

$$\therefore L(y) = \frac{A_0}{\Delta^2} (y - m)^2$$

If the average cost A_0 is $ 200.00 at a deviation Δ of 20 volts, then, the cost coefficient k is 0.50 $ volt^{-2}, as shown below:

$$k = \frac{A_0}{\Delta^2}$$

$$= \frac{200 \ \$}{(20 \ \text{V})^2}$$

$$= 0.50 \ \$ \ \text{V}^{-2}$$

and the quality loss function is:

$$L(y) = 0.50 (y - 115)^2$$

This quadratic quality loss function for the colour television voltage is shown graphically in Figure 1:1.2.

At $y = m$, i.e. in this case $y = m = 115$ volts, the loss is a minimum (zero). This is appropriate since m is the best value for y. Hence, if a recently manufactured television has a voltage of 115 volts, i.e. a 0 volt deviation from target, the quality loss is:

$$L(115 \ \text{V}) = 0.50 (115 - 115)^2$$
$$= 0.50 \times 0^2$$
$$= \$ \ 0.00$$

If a recently manufactured television has a voltage of 125 volts, i.e. a 10 volt deviation from target, the quality loss is:

$$L(125 \ \text{V}) = 0.50 (125 - 115)^2$$
$$= 0.50 \times 10^2$$
$$= \$ \ 50.00$$

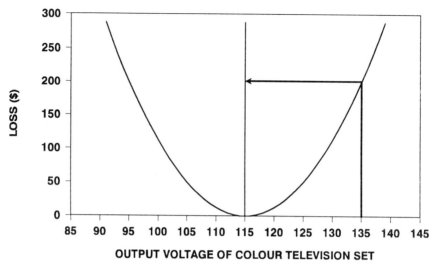

Figure 1:1.2 Nominal-the-best characteristic.

Thus, on average, the cost is $ 50.00. If the product develops a fault while still under warranty, the producer will pay this cost; otherwise, the customer will have to pay. In either case, this is a loss-to-society.

One may claim that 125 volts is within specification. True. But the point is, as long as the set is not on target, some customers will be dissatisfied. At a deviation of 10 volts (125 − 115 volts) that dissatisfaction equates to $ 50.00. When the voltage deviation increases, so does the quality loss and hence the dissatisfaction. Eventually, at the specification limits, half the customers will no longer accept their sets and demand repair, replacement, etc., and the quality loss-to-society at that point is $ 200.00.

1.1.7 QLF – Nominal-the-best: Many pieces

The quality loss function as described in Section 1.1.6 is used for assessing the quality loss for one piece of product. More often, we need to evaluate the average quality loss of a product based on a sample of products. The average quality loss is called the loss-per-piece ($\wp$). Hence, if we take a sample of a product, there will be many values of y, represented as y_i. The average of such a sample may be denoted $\bar{y}$. Of course $\bar{y}$ may or may not be equal to the target value, m. The average quality loss can be formulated from Equation (1:1.2):

$$L(y) = k (y - m)^2$$

$$L(y) = \frac{k (y_1 - m)^2 + k (y_2 - m)^2 + ... + k (y_n - m)^2}{n}$$

$$= k \left[\frac{(y_1 - m)^2 + (y_2 - m)^2 + ... + (y_n - m)^2}{n} \right] \qquad (1:1.3)$$

The term in brackets is the average of all the values of $(y_i - m)^2$ and is called the *mean squared deviation* (*MSD*). More specifically it may be referred to as the mean squared deviation from target and is mathematically equivalent to:

$$MSD = \frac{(y_1 - m)^2 + (y_2 - m)^2 + ... + (y_n - m)^2}{n}$$

$$= \frac{1}{n} \sum_{i=1}^{n} (y_i - m)^2$$

$$= \frac{1}{n} \sum_{i=1}^{n} \left(y_i^2 - 2 y_i m + m^2 \right)$$

$$= \frac{\sum y_i^2 - 2 m \sum y_i + \sum m^2}{n}$$

Substituting $\sum y_i = n \bar{y}$

and $\sum m^2 = n m^2$,

$$MSD = \frac{\sum y_i^2}{n} - \frac{2 m n \bar{y}}{n} + \frac{n m^2}{n}$$

$$= \frac{\sum y_i^2}{n} - 2 m \bar{y} + m^2$$

To simplify this equation further, we add $(-\bar{y}^2 + \bar{y}^2) = 0$ into the last equation and regroup the terms as follows:

$$MSD = \frac{\sum y_i^2}{n} - 2\,m\,\bar{y} + m^2$$

$$= \frac{\sum y_i^2}{n} - \bar{y}^2 + \bar{y}^2 - 2\,m\,\bar{y} + m^2$$

$$= \left[\frac{\sum y_i^2}{n} - \bar{y}^2\right] + \left[\bar{y}^2 - 2\,m\,\bar{y} + m^2\right]$$

$$= \left[\frac{\sum y_i^2}{n} - \bar{y}^2\right] + [\bar{y} - m]^2$$

$$= \sigma_n^2 + [\bar{y} - m]^2 \qquad\qquad (1{:}1.4)$$

where σ_n is usually called the *population standard deviation* and $(\bar{y} - m)$ is the *bias* of the sample from the target. For practical reasons however, we use σ_{n-1} which is usually called the *sample standard deviation*[3]. Since,

$$L(y) = k\,[MSD]$$

substituting $[MSD] = [\sigma^2 + (\bar{y} - m)^2]$,

$$L(y) = k\,[\sigma^2 + (\bar{y} - m)^2] \qquad\qquad (1{:}1.5)$$

Notice that the quality loss function always determines the loss-per-piece, $\wp$. Therefore, the total loss can be calculated simply by multiplying by the total number of products.

1.1.8 QLF – Smaller-the-better: One piece

Having developed the nominal-the-best, the smaller-the-better quality loss function can be developed rather easily. In fact, the smaller-the-better quality loss function can be extended directly from the nominal-the-best quality loss function. Referring to Equation (1:1.5), the formula for the nominal-the-best quality loss function was given as

$$L(y) = k\,[\sigma^2 + (\bar{y} - m)^2]$$

---

3   On most scientific calculators this is given as the σ_{n-1} or S button or function. On a spreadsheet, the reader should check for the appropriate function.

In the case of the smaller-the-better quality loss function, e.g. percent shrinkage of a speedometer casing, the target value m can be assumed to be zero (0). By substituting the target $m = 0$, the equation reduces to:

$$L(y) = k\,(y - m)^2$$
$$L(y) = k\,(y - 0)^2$$
$$\therefore L(y) = k\,y^2 \tag{1:1.6}$$

If the *LD-50* occurs at Δ corresponding to a loss of A_0, then,

$$L(y) = k\,y^2$$
$$\therefore k = \frac{L(y)}{y^2}$$

At the specification,

$$k = \frac{A_0}{\Delta^2}$$

Therefore,

$$L(y) = \frac{A_0}{\Delta^2}\,y^2 \tag{1:1.7}$$

If the average loss $A_0 = \$\,80$ when the average product shrinkage is 1.5 % then the quality loss function is calculated as follows:

$$k = \frac{A_0}{\Delta^2}$$
$$= \frac{80.00\ \$}{(1.5\ \%)^2}$$
$$= 35.56\ \$\ \%^{-2}$$
$$\therefore L(y) = 35.56\,y^2 \tag{1:1.8}$$

An example of a quality loss function for a smaller-the-better type characteristic is shown in Figure 1:1.3. Here, as the percent shrinkage of a speedometer casing increases, so does the quality loss. A smaller-the-better characteristic can never take a negative value. Thus, it is a one-sided quality loss function because *y* cannot take a negative value. The ideal value is equal to zero and as this value increases, the quality loss becomes larger.

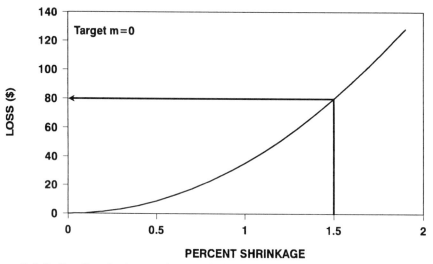

Figure 1:1.3 Smaller-the-better characteristic.

1.1.9 QLF – Smaller-the-better: Many pieces
For the smaller-the-better characteristic, the quality loss function for many pieces can be derived from Equation (1:1.5).

$$L(y) = k \, [MSD]$$

$$= k \left[\sigma^2 + (\bar{y} - m)^2 \right]$$

Since the target for smaller-the-better characteristic is $m = 0$, the quality loss function becomes:

$$L(y) = k \left[\sigma^2 + (\bar{y} - m)^2 \right]$$

$$= k \left[\sigma^2 + (\bar{y} - 0)^2 \right]$$

$$= k \left(\sigma^2 + \bar{y}^2 \right) \tag{1:1.9}$$

Again, it must be remembered that $L(y)$ represents the average loss-per-piece of product, estimated from a sample of products with a sample standard deviation σ and mean $\bar{y}$.

1.1.10   QLF – Larger-the-better: One piece

The quality loss function for a larger-the-better characteristic for one piece can be derived from the quality loss function for smaller-the-better. Mathematically, a larger-the-better characteristic, e.g. weld strength, can be regarded as the inverse of a smaller-the-better characteristic. Thus, from Equation (1:1.6),

$$L(y) = k \left(\frac{1}{y}\right)^2$$

$$\therefore k = L(y) \ y^2 \qquad\qquad (1:1.10)$$

If the *LD-50* occurs at Δ corresponding to a loss of A_0, then,

$$k = L(y) \ y^2$$

$$= A_0 \ \Delta^2 \qquad\qquad (1:1.11)$$

If A_0 = \$ 100.00 at an average weld strength Δ = 0.50 kg cm^{-2} then the quality loss function is calculated as follows:

$$k = A_0 \ \Delta^2$$

$$= 100 \times 0.50^2$$

$$= 25.00 \ \$ \ \text{kg cm}^{-2}$$

$$\therefore L(y) = 25.00 \left(\frac{1}{y}\right)^2$$

An example of the quality loss function for a larger-the-better type characteristic is shown in Figure 1:1.4. Here, as the weld strength increases the quality loss decreases. Ideally, the target m is infinity.

A larger-the-better quality characteristic cannot take a negative value. Zero is the worst value and as the characteristic value becomes larger, the performance becomes better – that is, the quality loss becomes smaller. The ideal value is infinity, at which point the loss is zero. A special case of the larger-the-better characteristic exists where a characteristic can take a range of values such as 0 % to 100 %. Such a characteristic is a pseudo-larger-the-better characteristic and may be regarded as a larger-the-better characteristic for most practical purposes.

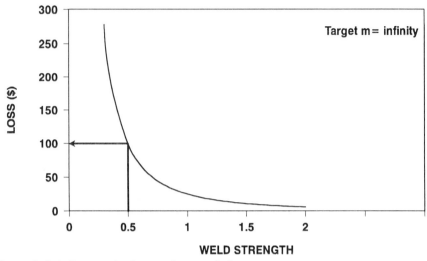

Figure 1:1.4 Larger-the-better characteristic.

1.1.11   QLF – Larger-the-better: Many pieces

The larger-the-better quality loss function for many pieces can be found by averaging the quality loss for one piece. Using the equation for larger-the-better, the loss per piece averaged from many pieces is:

$$L(y) = k \times \frac{1}{n} \left(\frac{1}{y_1^2} + \frac{1}{y_2^2} + ... + \frac{1}{y_n^2} \right)$$

$$= k \times \frac{1}{n} \sum_{i=1}^{n} \left(\frac{1}{y_i^2} \right) \qquad (1:1.12)$$

$$= k \, [MSD]$$

Notice that [*MSD*] is defined variously depending on the target value of the quality characteristic. While [*MSD*] for nominal-the-best and smaller-the-better characteristics can be calculated relatively easily, calculation of [*MSD*] for larger-the-better characteristic requires the use of Taylor's theorem. The reader who is unfamiliar with Taylor's theorem may proceed to Equation (1:1.19) without any loss of continuity. In general,

$$f(y) = f(\mu + (y - \mu))$$

can be expanded by Taylor's expansion:

$$f(y) = \frac{f(\mu)}{0!} + \frac{f'(\mu)\ (y - \mu)}{1!} + \frac{f''(\mu)\ (y - \mu)^2}{2!} + \frac{f'''(\mu)\ (y - \mu)^3}{3!} + \dots$$

$$(1:1.13)$$

where f', f'' and f''' are the first, second and third derivatives of $f(y)$. Since the third and higher derivatives have very little significance, we rewrite Equation (1:1.13) as:

$$f(y) \approx f(\mu) + f'(\mu)\ (y - \mu) + \frac{f''(\mu)\ (y - \mu)^2}{2} \qquad (1:1.14)$$

If we let

$$f(y) = \frac{1}{y^2} \qquad (1:1.15)$$

and differentiating with respect to y,

$$f'(y) = \frac{-2}{y^3}$$

$$f''(y) = \frac{6}{y^4}$$

so that:

$$f(y) \approx f(\mu) + f'(\mu)\ (y - \mu) + \frac{f''(\mu)\ (y - \mu)^2}{2}$$

$$= \frac{1}{\mu^2} + \frac{-2}{\mu^3}\ (y - \mu) + \frac{6}{\mu^4}\ \frac{(y - \mu)^2}{2}$$

$$= \frac{1}{\mu^2} + \frac{-2}{\mu^3}\ (y - \mu) + \frac{3}{\mu^4}\ (y - \mu)^2 \qquad (1:1.16)$$

The expected value of this equation is:

$$E(f(y)) \approx E\left(\frac{1}{\mu^2}\right) + \frac{-2}{\mu^3} E(y - \mu) + \frac{3}{\mu^4} E(y - \mu)^2$$

$$= \frac{1}{\mu^2} + \frac{-2}{\mu^3} \times 0 + \frac{3}{\mu^4} \sigma^2$$

$$= \frac{1}{\mu^2} + \frac{3\sigma^2}{\mu^4}$$

$$= \frac{1}{\mu^2}\left[1 + \frac{3\sigma^2}{\mu^2}\right] \tag{1:1.17}$$

since the expected value of $(y - \mu)$ is zero and the expected value of $(y - \mu)^2$ is the variance σ^2. Thus we may rewrite:

$$MSD = \frac{1}{n} \sum_{i=1}^{n} \left(\frac{1}{y_i}\right)^2$$

$$\approx \frac{1}{n} \times \frac{n}{\mu^2}\left[1 + \frac{3\sigma^2}{\mu^2}\right]$$

$$\approx \frac{1}{\mu^2}\left[1 + \frac{3\sigma^2}{\mu^2}\right] \tag{1:1.18}$$

And the quality loss function can be written as:

$$L(y) = k\,[MSD]$$

$$L(y) \approx \frac{k}{\mu^2}\left[1 + \frac{3\sigma^2}{\mu^2}\right] \tag{1:1.19}$$

and we can use Equation (1:1.19) to calculate the quality loss for a larger-the-better characteristic. Again, it must be remembered that $L(y)$ represents the average loss-per-piece of product. The population parameters (μ and σ) can be estimated from a sample of products with a mean $\bar{y}$ and sample standard deviation σ_{n-1} or S.

1.1.12   Asymmetric loss function

In certain situations, deviations from the quality characteristic for a nominal-the-best characteristic may be different on either side of its tolerance. Figure 1:1.5 shows an example of an asymmetric nominal-the-best quality loss function. Here, the quality loss in one direction may be greater than in the other direction. In such cases, we use different cost coefficients, k_1 and k_2, for the two directions. Thus, the quality loss would be approximated by the following asymmetric quality loss function:

$$L(y) = k_1 \ (y - m)^2 \ ; \quad y \le m$$
$$L(y) = k_2 \ (y - m)^2 \ ; \quad y > m \qquad\qquad (1{:}1.20)$$

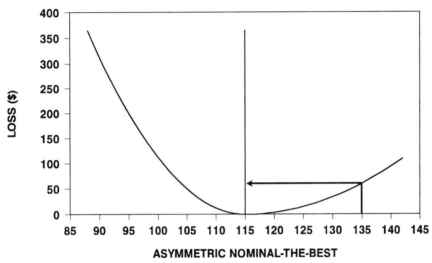

Figure 1:1.5 Asymmetric nominal-the-best characteristic.

1.1.13   Tolerancing using the quality loss function

The quality loss function is also an ideal function to establish realistic manufacturing tolerances[4]. Suppose the output voltage of the colour television sets in Section 1.1.6 could be recalibrated at the end of the production line at a cost of \$ 2.00. What is the realistic manufacturing tolerance? In other words, at what output voltage should the manufacturer spend \$ 2.00 to recalibrate a set?

---

4   While the quality loss function can be used to establish realistic manufacturing tolerances, it is not
         to be confused with tolerance design. Tolerancing merely establishes the tolerance at which the
         producer's loss equals the customer's loss. Tolerance design, however, aims to reduce variation by
         reducing the tolerances of variables.

Using the quality loss function for nominal-the-best from Section 1.1.6,

$$L(y) = 0.50\ (y - 115)^2$$

$$\$\ 2.00 = 0.50\ (y - 115)^2$$

$$0.50\ (y - 115)^2 = \$\ 2.00$$

$$(y - 115)^2 = \frac{\$\ 2.00}{0.50\ \$\ V^{-2}}$$

$$y - 115 = \pm\ \sqrt{4\ V^2}$$

$$\therefore\ y = 115 \pm 2\ V$$

This *realistic tolerance* ($\pm$ 2 volts) represents the point at which the manufacturer and consumer break even. This can be shown as follows:

$$\frac{L_{Producer}}{y_{Producer} - m} = \frac{L_{Customer}}{y_{Customer} - m}$$

$$\frac{\$\ 2.00}{(117 - 115)^2\ V^2} = \frac{\$\ 200}{(135 - 115)^2\ V^2}$$

$$0.50\ \$\ V^{-2} = 0.50\ \$\ V^{-2}$$

Thus, suppose a colour television set is shipped out at 125 volts, the manufacturer saves $ 2.00 while the consumer loses $ 50.00. Therefore, there is a net $ 48.00 loss-to-society. If, however, the manufacturer spent $ 2.00 to recalibrate the set in the factory to within $\pm$ 2 volts from the target value of 115 volts, the customer's loss would be:

$$L(y) = 0.50\ (y - 115)^2$$

$$L(117\ V) = 0.50\ (117 - 115)^2$$

$$= \$\ 2.00$$

Beyond this realistic tolerance of $\pm$ 2 volts from the target value of 115 volts, either the producer or the customer may pay for improved quality.

1.1.14   Self-assessment questions

1. Explain how a customer perceives quality and why customer loss and producer loss are both related to loss-to-society.

2. Calculation of the nominal-the-best quality loss function.
   Television sets are made with a desired target value for the output voltage of $m = 115$ volts. When the output voltage lies beyond the range of 115 ± 20 volts and the set is sold to a customer, the average cost of repairing or replacing the set is $ 200.00.
   1. Find the cost coefficient k and establish the quality loss function.
   2. The repair cost at the end of the production line is $ 4.00 per set. What are the realistic production tolerances for the output voltage?

3. Two different processes are being considered for producing these television sets. A sample of output voltages from each of the two processes is given below. Which process has the smaller loss? (Hint: compute the loss associated with each process).

Process	Data							
A	113	116	115	113	117	115	115	114
B	113	112	113	112	113	113	112	114

4. Calculation of the smaller-the-better quality loss function.
   A manufacturer who makes speedometer cable casings for cars knows that the shrinkage of the casing can be problem. Ideally, the shrinkage must be zero. When the shrinkage exceeds 1.50 %, about 50 % of customers complain. When a customer does complain the average cost of replacing the casing is $ 80.00.
   1. Establish the quality loss function.
   2. The rework cost at the end of the production line is $ 8.00 per piece. What are the realistic production tolerances for the shrinkage?
   2. In the data below, materials A and B have different shrinkage properties. If both materials cost the same, which of them is the better material?

Material	Data							
A	0.28	0.24	0.33	0.30	0.18	0.26	0.24	0.33
B	0.08	0.12	0.07	0.03	0.09	0.06	0.05	0.03

5.   Calculation of the larger-the-better quality loss function.
The weld strength of motor protector terminals needs to be maximized.
When the weld strength is 0.50 kg cm^{-2}, some welds have been known
to break, resulting in an average replacement cost of $ 100.00 to the
customer.
1.   Determine the quality loss function.
2.   The rework cost at the end of the production line is $ 10.00 per
weld. What are the realistic production tolerances for the weld
strength?
3.   An experiment was conducted to compare the existing semi-
manual welding processes A and B. Which of the processes is the
better?

Material	Data							
A	2.3	2.0	1.9	1.7	2.1	2.2	1.4	2.2
B	2.1	2.9	2.4	2.5	2.4	2.8	2.1	2.6

1.1.15   Answers to self-assessment questions

1.   Explain how a customer perceives quality and why customer loss and producer loss are both related to loss-to-society.

Answer

Any repair or adjustment cost incurred by a customer after buying a product is a loss to the customer. Customers prefer products that incur no loss. When a product is under warranty, a producer may pay for the repair or adjustment cost, but after the warranty expires, the customer has to pay for any repair or adjustment. In either case, the cost is borne by society. The objective in manufacturing quality products is to minimize this loss to society.

2.   Calculation of the nominal-the-best quality loss function.
Television sets are made with a desired target value for the output voltage of $m = 115$ volts. When the output voltage lies beyond the range of 115 ± 20 volts and the set is sold to a customer, the average cost of repairing or replacing the set is $ 200.00.
1.   Find the cost coefficient k and establish the quality loss function.
2.   The repair cost at the end of the production line is $ 4.00 per set. What are the realistic production tolerances for the output voltage?

Answer

1.

$$L(y) = k\,(y - m)^2$$

$$k = \frac{L(y)}{(y - m)^2}$$

$$= \frac{A_0}{\Delta^2}$$

$$= \frac{200\ \$}{(20\ V)^2}$$

$$= 0.50\ \$\ V^{-2}$$

$$\therefore\ L(y) = 0.50\,(y - 115)^2$$

2.

$$L(y) = 0.50 \ (y - 115)^2$$

$$4.00 = 0.50 \ (y - 115)^2$$

$$\therefore \ y = 115 \pm \sqrt{\frac{4.00}{0.50}}$$

$$= 115 \pm \sqrt{8} \ \text{V}$$

$$= 115 \pm 2.83 \ \text{V}$$

$$\approx 115 \pm 3 \ \text{V}$$

Hence, the realistic production tolerances for the output voltage is $\pm$ 3 volts.

3.   Two different processes are being considered for producing these television sets. A sample of output voltages from each of the two processes is given below. Which process has the smaller loss? (Hint: compute the loss associated with each process).

Process	Data							
A	113	116	115	113	117	115	115	114
B	113	112	113	112	113	113	112	114

Answer

For Process A,

$$\sigma_A^2 = 1.9286$$

$$(\bar{y}_A - m)^2 = (114.75 - 115.00)^2$$

$$= 0.0625$$

$$L(y) = k \left[\sigma_A^2 + (\bar{y}_A - m)^2 \right]$$

$$= 0.50 \times [1.9286 + 0.0625]$$

$$= \$ \ 1.00$$

For Process B,

$$\sigma_B^2 = 0.5000$$

$$(\bar{y}_B - m)^2 = (112.75 - 115.00)^2$$

$$= 5.0625$$

$$L(y) = k \left[\sigma_B^2 + (\bar{y}_B - m)^2 \right]$$

$$= 0.50 \times [0.5000 + 5.0625]$$

$$= \$ 2.78$$

Thus, Process A has the smaller quality loss. Notice however, that if Process B can be centred on target, then:

$$L(y) = k \left[\sigma_B^2 + (\bar{y}_B - m)^2 \right]$$

$$= 0.50 \times [0.5000 + 0.0000]$$

$$= \$ 0.25$$

and would therefore be the potentially better process.

4.   Calculation of the smaller-the-better quality loss function.
A manufacturer who makes speedometer cable casings for cars knows that the shrinkage of the casing can be problem. Ideally, the shrinkage must be zero. When the shrinkage exceeds 1.50 %, about 50 % of customers complain. When a customer does complain the average cost of replacing the casing is $ 80.00.
    1.   Establish the quality loss function.
    2.   The rework cost at the end of the production line is $ 8.00 per piece. What are the realistic production tolerances for the shrinkage?
    2.   In the data below, materials A and B have different shrinkage properties. If both materials cost the same, which of them is the better material?

Material	Data							
A	0.28	0.24	0.33	0.30	0.18	0.26	0.24	0.33
B	0.08	0.12	0.07	0.03	0.09	0.06	0.05	0.03

Answer

1.

$$L(y) = k\, y^2$$

$$k = \frac{L(y)}{y^2}$$

$$= \frac{A_0}{\Delta^2}$$

$$= \frac{80.00\ \$}{(1.50\ \%)^2}$$

$$= 35.56\ \$\ \%^{-2}$$

$$\therefore\ L(y) = 35.56\ y^2$$

2.

$$L(y) = k\, y^2$$

$$\therefore\ y = \sqrt{\frac{L(y)}{k}}$$

$$= \sqrt{\frac{8.00}{35.56}}$$

$$= 0.47\ \%$$

Hence, the realistic production tolerance is 0.47 % shrinkage.

3.

For Material A,

$$\sigma_A^2 = 0.0026$$

$$\bar{y}_A^2 = 0.0729$$

$$L(y) = k\left[\sigma_A^2 + \bar{y}_A^2\right]$$

$$= 35.56 \times [0.0026 + 0.0729]$$

$$= \$\ 2.68$$

For Material B,

$$\sigma_B^2 = 0.0009$$

$$\bar{y}_B^2 = 0.0044$$

$$L(y) = k\left[\sigma_B^2 + \bar{y}_B^2\right]$$

$$= 35.56 \times [0.0009 + 0.0044]$$

$$= \$\ 0.19$$

Material B is better since this has a smaller quality loss.

5.   Calculation of the larger-the-better quality loss function.
The weld strength of motor protector terminals needs to be maximized.
When the weld strength is 0.50 kg cm^{-2}, some welds have been known
to break, resulting in an average replacement cost of $ 100.00 to the
customer.
1.   Determine the quality loss function.
2.   The rework cost at the end of the production line is $ 10.00 per
weld. What are the realistic production tolerances for the weld
strength?
3.   An experiment was conducted to compare the existing semi-
manual welding processes A and B. Which of the processes is the
better?

Material	Data							
A	2.3	2.0	1.9	1.7	2.1	2.2	1.4	2.2
B	2.1	2.9	2.4	2.5	2.4	2.8	2.1	2.6

Answer

1.

$$L(y) = k\left(\frac{1}{y}\right)^2$$

$$k = L(y) \times y^2$$

$$= \$\ 100.00 \times (0.50\ \text{kg cm}^{-2})^2$$

$$= 25.00\ \$\ \text{kg}^2\ \text{cm}^{-4}$$

$$\therefore L(y) = 25.00\ y^2$$

2.

$$L(y) = k \left(\frac{1}{y} \right)^2$$

$$\therefore y = \sqrt{\frac{k}{L(y)}}$$

$$= \sqrt{\frac{25.00}{10.00}}$$

$$\approx 1.58 \text{ kg cm}^{-2}$$

Hence, the realistic production tolerance is 1.58 kg cm^{-2}.

3.

For Process A,

$$\sigma_A^2 = 0.0907$$

$$\bar{y}_A^2 = 3.9006$$

$$L(y) \approx \frac{k}{\bar{y}_A^2} \left[1 + \frac{3 \sigma_A^2}{\bar{y}_A^2} \right]$$

$$= \frac{25.00}{3.9006} \left[1 + \frac{3 \times 0.0907}{3.9006} \right]$$

$$= \$ \, 6.86$$

For Process B,

$$\sigma_B^2 = 0.0850$$

$$\bar{y}_B^2 = 6.1256$$

$$L(y) \approx \frac{k}{\bar{y}_B^2} \left[1 + \frac{3 \sigma_B^2}{\bar{y}_B^2} \right]$$

$$= \frac{25.00}{6.1256} \left[1 + \frac{3 \times 0.0850}{6.1256} \right]$$

$$= \$ \, 4.25$$

Process B is better since this has a smaller quality loss.

1.2   Quality and Process Capability

1.2.1 Introduction
In most manufacturing companies the department responsible for product quality is usually the quality assurance department. The manufacturing department itself is seen to be responsible for production and delivery. In such organizations, the manufacturing department is required to produce products to a quality level acceptable to the quality control department. Since the quality control department is concerned with the defect ratio, i.e. the ratio of defects to total production, whenever this ratio is almost zero, products are deemed to have passed inspection. In order to attain a low defect ratio, the manufacturing departments turn to the quality department to shut down processes that produce defective products. This leads the quality department to determine whether a production process is capable of meeting the specifications required. If it is, then the process is capable; if not, then it is the engineering department that must do something! In this section, however, we will see why process capability is not a sufficient measure of quality by comparing processes with different process capabilities and the related quality loss.

1.2.2 Process capability and quality
Specification limits are the fixed engineering limits for the product dimensions. They are set independently of the inherent process variation. The specification limits can be one-sided or two-sided, with or without target values. Only once a process is in a state of statistical control, is it sensible to assess whether or not it is capable of meeting the pre-determined specification. Where no target is specified with two-sided limits, the usual value is taken to be halfway between the two limits to minimize the potential wastage from the product dimensions falling outside the specification limits.

A unitless measure of the potential of the process to meet a two-sided specification is the C_p index:

$$C_p = \frac{USL - LSL}{6\sigma} \qquad (1:2.1)$$

where *USL* and *LSL* are the upper and lower specification limits, respectively. The range $(USL - LSL)$ is the allowable specification spread. If the process can be adequately represented by a normal distribution, then the denominator is the process spread and is taken to be six standard deviations of the process.

A value of $C_p = 1$ suggests that the specification spread equals the process spread. This means the process is just capable of meeting the specifications. Referring to a table of normal distribution, only a small proportion (0.0027) of products will fall outside the specification limits assuming the process is centred between the specification limits. If $C_p \gg 1$, then the specification spread is much larger than the process spread and we say that the process is potentially capable of meeting the specification. If $C_p \ll 1$, then the specification spread is much less than the process spread and we say that the process is not capable of meeting the specification.

When the process mean is not centred or the specification is one-sided, then a more informative index is the C_{pk} index. C_{pk} is calculated as the smaller of the following two values:

● the upper specification limit minus the process average, divided by three standard deviations of the process,

● the process average minus the lower specification limit, divided by three standard deviations of the process.

Mathematically, C_{pk} is equal to:

$$C_{pk} = \text{Minimum} \left[\frac{(USL - \mu)}{3\sigma} , \frac{(\mu - LSL)}{3\sigma} \right] \qquad (1:2.2)$$

provided that $LSL < \mu < USL$. This index compares the minimum deviation of the process mean, μ, from either specification, to half the process spread. If the process mean is centred between the specification limits, then the deviation from either specification limit is half the allowable spread and so C_p and C_{pk} will be equal. As the mean moves further from the centre of the specification limits, C_{pk} will decrease in relation to C_p. The mean will become critical if C_{pk} is less than 1, i.e. when a significant proportion, more than 1 product per thousand, is expected to fall outside a specification limit. Any action to be taken on a process will depend on the comparison of these two indices, bearing in mind that reducing the inherent variation will generally require fundamental changes. It will also be more difficult than any action undertaken to centre a potentially capable process which may require only a small adjustment to machine settings.

1.2.3 Assessing the capability of a process

When the value of C_p is appreciably greater than 1, i.e. $C_p \gg 1$, this suggests that the process is potentially very capable. If C_{pk} is considerably less than C_p and in particular less than 1, the process mean will still require centring to avoid significant number of rejected products. When the value of C_p is approximately equal to one, i.e. $C_p \approx 1$, it suggests a capable process. Centring of the process mean will be of importance particularly when C_{pk} is less than 1. When the value of C_p is appreciably less than 1, i.e. $C_p \ll 1$, this suggests the process is not capable and fundamental changes will be required to reduce the inherent variation. A review of specifications will also be beneficial if they are tighter than is necessary. The process mean should still be centred to minimize the amount of nonconforming products. However, 100 % final inspection of the product will be needed to intercept any nonconforming products.

C_{pk} is an important measure of process or product quality characteristic because it assesses the mean centring and the spread, both of which affect the quality loss function. However, the absolute values of C_p or C_{pk} are based on specification limits and have little significance. The C_{pk} process or product history provides information relative to quality improvement. These data, when used in the quality loss function, quantify the decrease in quality loss as C_{pk} increases. Should this process of reducing variation, i.e. increasing C_{pk} be ongoing, never ending, continuous, etc? No! When the added cost of further reducing the variation exceeds the quality loss that would be avoided, we should do nothing.

1.2.4 C_{pk} as a measure of quality

To visualize more clearly what C_{pk} does not reflect, let us study four processes. Processes 1, 2 and 3 all have a C_p index of 1.0. Process 1 is mean centred. That is, its process average coincides with the target value, *m*, and has a standard deviation $\sigma = 1.33$. Processes 2 and 3 are not target value centred but have reduced standard deviations compared with Process 1. Finally, Process 4 has a C_p of 0.471, is mean centred and is contrived for demonstration purposes.

In the following, we compare quality loss based on:
- the conventional method,
- the quality loss function method.

1.   The conventional method

The conventional method uses *loss-by-defect* to calculate quality loss due to defective products. All products within specification are assumed to have no quality loss. Thus, if the proportion of defective products is known, it is relatively easy to establish this quality loss.

Loss-by-defect = *number of defects × cost of product*

= proportion out of specification × total number × cost of product

2.   The quality loss function method

The quality loss function method uses *loss-by-dispersion* to calculate quality loss due to both variation and bias. This method is based on the quality loss function.

$$Loss\text{-}by\text{-}dispersion = \frac{A_0}{\Delta^2} \left[\sigma^2 + (\bar{y} - m)^2\right] \times number\ of\ products$$

The examples emphasize the differences in assessment, conclusions and prioritization that result from loss-by-defects based on the number of defectives and loss-by-dispersion based on the quality loss function. We perform calculations based on an annual production of 100,000 units. The specification is 20 ± 4 (i.e. $m = 20$ and

Δ = 4). The cost of repairing or reseting a product out-of-specification is $ 32 (i.e. A_0 = $ 32). We can now develop the cost calculations for:

- Process 1,
- Process 2,
- Process 3,
- Process 4.

1.   Process 1

Figure 1:2.1 shows Process 1. Given that the specification is 20 ± 4, the upper specification limit (*USL*) is 24 and the lower specification limit (*LSL*) is 16. The process average ($\bar{y}$) is centred at the target m = 20 with a standard deviation (σ) of 1.33.

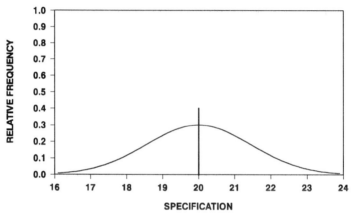

Figure 1:2.1 Process 1.

To calculate C_{pk}, we need to find the following:
$USL - \bar{y} = 24 - 20 = 4$
$\bar{y} - LSL = 20 - 16 = 4$

Since ($USL - \bar{y}$) and ($\bar{y} - LSL$) are equal we can use either of them for the minimum of ($USL - \bar{y}$) and ($\bar{y} - LSL$). Now,

$$C_{pk} = \frac{[\bar{y} - LSL]}{3\sigma}$$

$$= \frac{4}{3 \times 1.33}$$

$$= 1$$

Loss-by-defect:
Loss = number of defects × cost of product
      = (proportion out-of-specification[5] × total number) × cost of product
      = (0.0027 × 100,000) × \$ 32
      = \$ 8,640.

Loss-by-dispersion:
Loss = loss-per-piece × number of products
      = $k\,[\sigma^2 + (\bar{y} - m)^2]$ × 100,000
      = (A_0/Δ^2) × $[1.33^2 + (20 - 20)^2]$ × 100,000
      = \$ 32/4^2 × 1.7689 × 100,000
      = \$ 353,780.

Note that loss-by-defect only predicts a loss of \$ 8,640 while loss-by-dispersion predicts a much larger loss of \$ 353,780.

2.   Process 2
Figure 1:2.2 shows Process 2. The process average is observed to be centred at 18 with a standard deviation of 0.66.

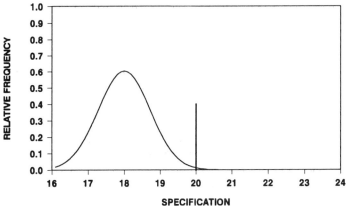

Figure 1:2.2 Process 2.

To calculate C_{pk}, we need to find:
$USL - \bar{y} = 24 - 18 = 6$
$\bar{y} - LSL = 18 - 16 = 2$

---

5   The proportion out-of-specification is obtained from Murdoch and Barnes, *Statistical Tables for Science, Engineering, Management and Business Studies*, MacMillan Education Ltd. Using the table for normal distribution, the proportion of area beyond ± 3 standard deviations (4/1.33) is 0.0027.

The minimum of $(USL - \bar{y})$ and $(\bar{y} - LSL)$ is $(\bar{y} - LSL)$. So we use $(\bar{y} - LSL)$ in the C_{pk} calculation. Now,

$$C_{pk} = \frac{[\bar{y} - LSL]}{3\sigma}$$

$$= \frac{2}{3 \times 0.66}$$

$$= 1$$

Loss-by-defect:

Loss   = number of defects × cost of product
        = (proportion out-of-specification[6] × total number) × cost of product
        = (0.00135 × 100,000) × \$ 32
        = \$ 4,320.

Loss-by-dispersion:

Loss   = loss-per-piece × number of products
        = $k [\sigma^2 + (\bar{y} - m)^2]$ × 100,000
        = (A_0/Δ^2) × $[0.66^2 + (18 - 20)^2]$ × 100,000
        = $\$ 32/4^2$ × 4.4356 × 100,000
        = \$ 887,120.

Note, again, that the loss-by-defect is \$ 4,320, while the loss-by-dispersion is \$ 887,120.

3.   Process 3

Figure 1:2.3 shows Process 3. The process average is observed to be centred at 17.2 with a standard deviation of 0.40.

To calculate C_{pk}, we need to find:
$USL - \bar{y} = 24 - 17.2 = 6.8$
$\bar{y} - LSL = 17.2 - 16 = 1.2$

The minimum of $(USL - \bar{y})$ and $(\bar{y} - LSL)$ is $(\bar{y} - LSL)$. So we use $(\bar{y} - LSL)$ in the C_{pk} calculation. Now,

---

6   The proportion out-of-specification is obtained from Murdoch and Barnes, *Statistical Tables for Science, Engineering, Management and Business Studies*, MacMillan Education Ltd. Using this table for normal distribution, the area beyond −3.00 standard deviations (since only one side of the distribution will have out-of-specification parts) is 0.00135.

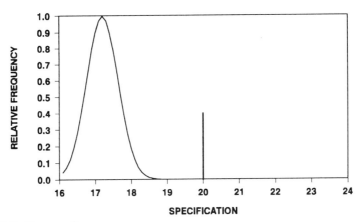

Figure 1:2.3 Process 3.

$$C_{pk} = \frac{[\bar{y} - LSL]}{3\sigma}$$

$$= \frac{1.2}{3 \times 0.40}$$

$$= 1$$

Loss-by-defect:
Loss = number of defects × cost of product
      = (proportion out-of-specification × total number) × cost of product
      = (0.00135 × 100,000) × $ 32
      = $ 4,320.

Loss-by-dispersion:
Loss = loss-per-piece × number of products
      = $k [\sigma^2 + (\bar{y} - m)^2] \times 100,000$
      = $(A_0/\Delta^2) \times [0.40^2 + (17.2 - 20)^2] \times 100,000$
      = $ $32/4^2 \times 8.0000 \times 100,000$
      = $ 1,600,000.

4.   Process 4
Figure 1:2.4 shows Process 4. The process average is observed to be centred
at 20 with a standard deviation of 2.8284.

To calculate C_{pk}, we need to find:
$USL - \bar{y} = 24 - 20 = 4$
$\bar{y} - LSL = 20 - 16 = 4$

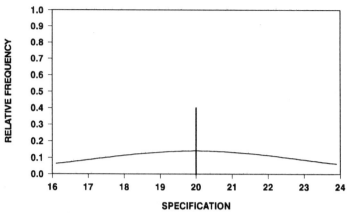

Figure 1:2.4 Process 4.

Since $(USL - \bar{y})$ and $(\bar{y} - LSL)$ are both the same, we can use either of them for the minimum of $(USL - \bar{y})$ and $(\bar{y} - LSL)$. Now,

$$C_{pk} = \frac{[\bar{y} - LSL]}{3\sigma}$$

$$= \frac{4}{3 \times 2.8284}$$

$$= 0.471$$

Loss-by-defect:
Loss = number of defects × cost of product
     = (proportion out-of-specification[7] × total number) × cost of product
     = (0.16 × 100,000) × $ 32
     = $ 512,000.

Loss-by-dispersion:
Loss = loss-per-piece × number of product
     = $k [\sigma^2 + (\bar{y} - m)^2] \times 100,000$
     = $(A_0/\Delta^2) \times [2.8284^2 + (20 - 20)^2] \times 100,000$
     = $ 32/4² × 8.0000 × 100,000
     = $ 1,600,000.

---

7   The proportion out-of-specification is obtained from Murdoch and Barnes, *Statistical Tables for Science, Engineering, Management and Business Studies*, MacMillan Education Ltd. Using this table for normal distribution, the area beyond ±1.41 standard deviations is 0.16.

For ease of comparison, the results of the four processes are tabulated in Figure 1:2.5 from which it can be seen that:

- although Processes 1 and 2 have the same C_{pk} value their losses are very different,
- using loss-by-defect, Processes 2 and 3 have the same loss despite the fact that their distributions are very dissimilar,
- the results obtained by the quality loss function, i.e. loss-by-dispersion, indicate that Processes 3 and 4 have the same loss even though Process 4 has a C_{pk} of only 0.471.

	Process 1	Process 2	Process 3	Process 4
C_{pk}	1.00	1.00	1.00	0.471
Loss-by-defect ($)	8,640	4,320	4,320	512,000
Loss-by-dispersion ($)	353,780	887,120	1,600,000	1,600,000

Figure 1:2.5 Comparison of loss-by-defect against loss-by-dispersion.

Hence, we can conclude that:

- C_{pk} by itself is not a sufficient measure of quality loss,
- loss-by-defect does not consider all losses,
- the quality loss function is a better measure of quality loss.

1.2.5 Self-assessment questions

1.   What is the role of specifications in manufacturing?

2.   Using the quality function, compare the loss-per-piece for the Sony television sets made in the USA and Japan. The sample standard deviation for American sets was $10/\sqrt{12}$ with mean m. The sample standard deviation for Japanese sets was $10/6$ with mean m.

3.   Explain why process capability is not a good measure of quality.

4.   A manufacturer buys resistors from two suppliers, A and B. An inspection of recent deliveries reveals that the resistors from Supplier A have a mean resistance value of 100 ohms with a standard deviation of 3.33 ohms, while those from Supplier B has mean resistance value of 104 ohms with a standard deviation of 2 ohms. Compare the two suppliers with regard to their process capabilities and quality loss given that the resistor specification is 100 ± 10 ohms. The repair or rework cost when a resistor fails is $ 10. The manufacturer buys 100,000 resistors from each supplier per year.

1.2.6 Answers to self-assessment questions

1.   What is the role of specifications in manufacturing?

Answer
   Specification values were originally conceptualized as criteria for determining acceptance or rejection. As quality level increased, the role of a total inspection to screen out defects diminished. Total inspection must shift to process control, and quality must be built-in through design. Nevertheless, it is a fact that less total inspection is necessary when rational process control is adopted. In that case, the role of specification values is merely to evaluate the target value.

2.   Using the quality function, compare the loss-per-piece for the Sony television sets made in the USA and Japan. The sample standard deviation for American sets was $10/\sqrt{12}$ with mean m. The sample standard deviation for Japanese sets was $10/6$ with mean m.

Answer

$$L(y) = k\left[\sigma^2 + (\bar{y} - m)^2\right]$$

$$L(USA) = k\left[\left(\frac{10}{\sqrt{12}}\right)^2 + (m - m)^2\right]$$

$$= 8.33\ k$$

$$L(Japan) = k\left[\left(\frac{10}{6}\right)^2 + (m - m)^2\right]$$

$$= 2.78\ k$$

$$\frac{L(USA)}{L(Japan)} = \frac{8.33\ k}{2.78\ k}$$

$$= 3.00$$

Thus, the quality loss associated with the American television sets was 3.00 times higher than that of the Japanese television sets.

3.   Explain why process capability is not a good measure of quality.

Answer

C_{pk} is not a good measure of quality since it does not relate directly to quality loss in a product. Hence, although a process may have a C_{pk} of 1.33 representing a capable process, it is still possible for that process to have a larger quality loss as compared to another process having a C_{pk} of, say, 1.0.

4.   A manufacturer buys resistors from two suppliers, A and B. An inspection of recent deliveries reveals that the resistors from Supplier A have a mean resistance value of 100 ohms with a standard deviation of 3.33 ohms, while those from Supplier B has mean resistance value of 104 ohms with a standard deviation of 2 ohms. Compare the two suppliers with regard to their process capabilities and quality loss given that the resistor specification is 100 ± 10 ohms. The repair or rework cost when a resistor fails is $ 10. The manufacturer buys 100,000 resistors from each supplier per year.

Answer

For Supplier A:

$$C_{pk} = \frac{110 - 100}{3 \times 3.33} = 1.00$$

$$Loss\text{-}by\text{-}defect = 0.0027 \times 100,000 \times \$ \ 10$$

$$= \$ \ 2,700$$

$$Loss\text{-}by\text{-}dispersion = L(y)$$

$$L(y) = \frac{A_0}{\Delta^2} \left[\sigma^2 + (\bar{y} - m)^2\right]$$

$$= \frac{10}{10^2} \left[3.33^2 + (100 - 100)^2\right]$$

$$= \$ \ 1.10889$$

$$Annual \ loss = \$ \ 1.10889 \times 100,000$$

$$= \$ \ 110,889$$

For Supplier B:

$$C_{pk} = \frac{110 - 104}{3 \times 2.00} = 1.00$$

$$Loss\text{-}by\text{-}defect = 0.00135 \times 100{,}000 \times \$\,10$$

$$= \$\,1{,}350$$

$$Loss\text{-}by\text{-}dispersion = L(y)$$

$$L(y) = \frac{A_0}{\Delta^2} \left[\sigma^2 + (\bar{y} - m)^2\right]$$

$$= \frac{10}{10^2} \left[2.00^2 + (104 - 100)^2\right]$$

$$= \$\,2.00$$

$$Annual\ loss = \$\,2.00 \times 100{,}000$$

$$= \$\,200{,}000$$

The two suppliers can now be compared as follows: Both suppliers have a C_{pk} of 1.00. The loss-by-defect for Suppliers A and B are $ 2,700 and $ 1,350 respectively. Thus, despite their C_{pk} being equal, their losses are different. This implies that C_{pk} is not a good measure of quality. The loss-by-dispersion for Suppliers A and B are $ 110,889 and $ 200,000, respectively. Loss-by-defect shows incorrectly, that Supplier B has a lower loss. Loss-by-dispersion shows that Supplier A has a lower loss. However, if the process used by Supplier B can be centred on target, the loss-by-dispersion would be $ 40,000 as shown below:

$$Loss\text{-}by\text{-}dispersion = \frac{A_0}{\Delta^2} \left[\sigma^2 + (\bar{y} - m)^2\right]$$

$$= \frac{10}{10^2} \left[2.00^2 + (0)^2\right]$$

$$= \$\,0.40$$

$$Annual\ loss = \$\,0.40 \times 100{,}000$$

$$= \$\,40{,}000$$

Thus, Supplier B has the potentially better process.

CHAPTER 2

THE DESIGN PROCESS

AIMS:
The aim of this chapter is to introduce the roles of quality assurance and the need for a robust design process. The principle of robustization and the theory required for parameter design are also discussed, leading to an example of parameter design.

OBJECTIVES:
When you have completed studying this chapter you should be able to:
- describe the role of quality assurance,
- describe the principle of robust design in the production process,
- explain the robust design process in relation to:
  - System Design,
  - Parameter Design,
  - Tolerance Design,
- distinguish between external, internal and unit-to-unit variation, and
- explain the use of parameter design to improve quality.

OVERVIEW:
Chapter 1 provided a method of measuring quality on an objective basis using the quality loss function. This chapter introduces the role of the various types of Quality Control activities and their applicability at various stages of product development. The emphasis is placed on the principle of robust design for manufacture. The basic theory of parameter design is discussed, leading to an early example of a parameter design experiment.

2.1 The Design Process

2.1.1 Introduction

The goal of experimentation in manufacturing is to devise ways of minimizing the deviation of a quality characteristic from its target value. This can be done only by identifying those factors which impact the quality characteristic in question and by changing the appropriate factor levels so that the deviations are minimized and the quality characteristic is on target. In other words, from a quality perspective, experimentation seeks to determine the best material, the best pressure, the best temperature, chemical formulation, cycle time, etc., which will operate together within a process to produce a desired quality characteristic such as length, durability, etc., taking cost into account.

2.1.2 Comparison of the classical and Taguchi's approach

The classical methods for design of experiments were developed by R. A. Fisher in England, in the early part of the 20th century. They include a full variety of statistical design techniques based on Latin squares[1] and developed for the agricultural industry. While rigorous, a major problem with applying Fisher's method in manufacturing industry is the time and cost required to learn and use it. Hence, Fisher's classical experimentation has been around for some time, but it has been used relatively little in industry. There has been little training of engineers in applied probability and statistics. Further, Fisher's methods are often cumbersome to implement in manufacturing industrial experimentation because of certain assumptions and procedural emphases.

Taguchi's approach to the design of experiments utilizes Robust Design, which can be applied to a wide variety of problems. The application of the method in electronics, automotive products, photography and many other Japanese industries has been an important factor in their rapid growth and their subsequent domination of international markets. Robust Design adds a new dimension to Fisher's statistical experimental design by explicitly addressing the concerns faced by all process and product designers, such as:

- how to reduce economically the variation of a product's function in the customer's environment, and
- how to ensure that decisions found to be optimum during laboratory experiments will prove to be so in manufacturing and in customer environments.

In addressing these concerns, Robust Design uses the mathematical formalism of Fisher's statistical experimental design, but the concept behind the mathematics is fundamentally different from Fisher's classical methods in many ways. The answers provided by Robust Design to the two concerns listed above make it a valuable tool for improving the productivity of research and development (R&D) activities.

---

1   Latin squares are balanced square arrangements required for unbiased statistical experimentation.

2.1.3 The objective of engineering design

The objective of engineering design, a major part of research and development, is to produce drawings, specifications, and other relevant information needed to manufacture products that consistently meet customer requirements. Knowledge of scientific phenomena and past engineering experience with similar product designs and manufacturing processes form the basis of the engineering design activity. However, when a number of new decisions related to a product must be made regarding process and product architecture, parameters of the manufacturing process and the functional characteristics of a product, it becomes necessary to engineer the whole research and development in a concurrent way. Additionally, a large amount of engineering effort is always expended in conducting research and development (either with hardware or software, by experimentation or simulation) to generate the information needed to guide these decisions. Therefore, efficiency in generating such information is the key to meeting market requirements, keeping product development and manufacturing costs low while attaining high-quality products. Robust Design is an engineering methodology for improving productivity during research and development so that high-quality products can be produced quickly and at low cost.

2.1.4 Variability due to noise factors

The factors that cause variability in a product's proper functioning are called *noise factors*. Such factors cause, for example, the brightness of a fluorescent lamp to vary with the power supply voltage, to deteriorate over time, as well as to vary between different lamps. There are three main types of noise:

- external noise,
- internal noise, and
- unit-to-unit noise.

1.   External noise (Ambient noise)

External noise refers to factors in the environment or conditions of use that influence the ideal functioning of a product. Examples of environmental noise factors are ambient temperature, humidity, dust, supply voltage, electromagnetic interference, vibrations and human error in operating a product.

2.   Internal noise (Deterioration noise)

Internal noise refers to factors that cause a product to deteriorate during storage or to wear out during use so that it can no longer achieve its target functions. Examples of internal noise factors are the wear of parts and the deterioration of components with age.

3.   Unit-to-unit noise (Variational noise)

Unit-to-unit noise refers to factors that cause differences between individual products that have been manufactured to the same specifications. This variation is inevitable in a manufacturing process and leads to variations in the product parameters from unit to unit. For example, the value of a resistor may be specified to be 100 ohms, but the resistance value may be 101 ohms in one particular unit and 98 ohms in another.

2.1.5 Examples of noise
Examples of three common products and their associated noise factors include:
- colour television power circuit,
- refrigerator,
- automobile.

1.   Colour television power circuit
The function of a power circuit in a colour television set is to convert alternating current (AC) input into direct current (DC) output. If the power circuits in all sets manufactured maintained a constant direct current output under all conditions, their voltage would be perfect. However, it is likely that the following noise factors may cause the output to deviate from its target voltage.
- external noise,
- internal noise,
- unit-to-unit noise.

1.   External noise
All variations in environmental conditions such as temperature, humidity, dust, and input voltage.

2.   Internal noise
Changes in the component and material characteristics. For example, after 10 years the resistance of a resistor may have increased by 10 %.

3.   Unit-to-unit noise
Differences between individual manufactured units, causing different output voltages from the same input voltage.

2.   Refrigerator
Some of the important noise factors related to the temperature control inside a refrigerator are given below:
- external noise,
- internal noise,
- unit-to-unit noise.

1.   External noise
The number of times the door is opened and closed, the amount of food kept and the initial temperature of the food, variation in the ambient temperature, and the fluctuation in power supply voltage.

2.   Internal noise
The leakage of refrigerant and mechanical wear of compressor parts and door seals.

3.   Unit-to-unit noise
The tightness of the door closure and the amount of refrigerant used.

3.   Automobile
The following noise factors are important for the braking distance of an automobile.
- external noise,
- internal noise,
- unit-to-unit noise.

1.   External noise
Wet or dry roads, concrete or asphalt surfaces and the number of passengers in the car.

2.   Internal noise
The wear of the drums and brake pads, and leakage of brake fluid.

3.   Unit-to-unit noise
Variations in the friction coefficient of the pads and drums, and the amount of brake fluid.

2.1.6 Role of various quality control activities

Once a decision to make a product has been made, the life cycle of that product has four major stages:
- Product Design,
- Production Process Design,
- Manufacturing, and
- Customer Usage.

The quality control activities in each of these stages are listed in Figure 2:1.1 and explained in the following sections. Quality control activities in process and product design prior to manufacturing are often called *off-line quality control*, whereas the quality control activities in manufacturing are often called *on-line quality control*. Quality control activities during customer usage involve warranty and service.

Product realisation	Quality control activity	Ability to reduce effect of noise factors		
		External	Internal	Unit-to-unit
Product Design	System Parameter Tolerance	✓ ✓ ✓	✓ ✓ ✓	✓ ✓ ✓
Production Process Design	System Parameter Tolerance	X X X	X X X	✓ ✓ ✓
Manufacturing	Detection / Correction Feedforward Control Screening	X X X	X X X	✓ ✓ ✓
Customer usage	Warranty and Repair	X	X	X

Figure 2:1.1 Quality control activities during various product stages.

2.1.7 Product Design (Off-line quality control)

A firm decides on a set of target functions, prepares the necessary specifications and drawings, and begins to manufacture the product. Some of the manufactured units may achieve the target functions while others do not. This is due to unit-to-unit noise. The product may break down after prolonged use, its functions having been impaired by deterioration due to internal noise. The product may function well under normal conditions but not under high temperature or humidity, or when the power supply voltage is 20 % off nominal. These variations are due to external noise.

Good product quality means there is little functional variation due to any of the above types of noise. The product functions as intended under a wide range of conditions for the duration of its design life. The aim in product quality is for the product functions to remain on target despite fluctuations in temperature, humidity, supply voltage, and other environmental (external) noise, even when components and materials degrade or wear down during prolonged use (internal noise) as well as to consistency (unit-to-unit noise). Quality with respect to an objective function can be measured as the amount of deviation from the functional target value determined from the specifications.

Noise factors cause variations in the external, internal and unit-to-unit factors of products resulting in variations in the product's objective functions. If products maintain on-target performance under a wide range of conditions (external noise), over a long period of time (internal noise) and consistently (unit-to-unit noise), the manufacturer earns a reputation for quality and reliability.

2.1.8 Production Process Design (Off-line quality control)

Assuring functional quality requires finding ways to reduce the effects of all three types of noise. The most important means is through a three-step design process, an aspect of off-line quality control involving:

- system design (or primary design),
- parameter design (or secondary design),
- tolerance design (or tertiary design).

1. System design (Primary design)

System design (or primary design) is the functional design stage focused on the pertinent technology. System design requires technical knowledge and extensive experience in an area of specialization to initially *design* or specify the process or product. For example, a person familiar with internal combustion engines may be selected to design a prototype of a radically new type of car engine. System design does not utilize design optimization methods such as the design of experiments. In this step the design engineer surveys the relevant technology and asks, for example, what kinds of circuits could be used to convert alternating current to direct current, or what reaction processes could be used to produce a desired combustion rate. It is a search for the best available technology.

As another example, an automatic control system can be considered in this step. In the case of the television power circuit discussed in Section 2.1.5, one might suggest a system design that measures the instantaneous output voltage and automatically corrects the deviation from the 115 V target. This could be done by controlling a circuit parameter or perhaps by changing the value of a variable resistance element. It would still be difficult, however, to control deterioration and other changes in the automatic control system itself and of course would add to the cost.

2. Parameter design (Secondary design)

Parameter design (or secondary design) provides a means of both reducing cost and improving quality by making effective use of experimental design methods. This involves the determination of parameter values that are least sensitive to noise. When the goal is to design a process or product with high stability and reliability, parameter design is the most important step in which functional non-linearity is used to best advantage. This is also the step in which we seek the combination of parameter levels that reduces the effect of noise. It is the central step in robust design and the answer to the requirement to design a product or process that exhibits high reliability under a wide range of conditions, despite the use of inexpensive, highly variable materials and parts that will easily deteriorate.

Indeed, this is the aim of design of experiments. Studying a large number of factors and selecting the optimum combination of factor levels is an application of non-linearity, even when it is not recognized as such. The key to success in the design of experiments is to select objective characteristics and to conduct studies of the above factors with product quality in mind.

Where raw materials and components tend to be highly variable, parameter design is the most important quality control measure. If the product can be designed so that its output characteristic is resistant to external, internal and unit-to-unit noise, then it will function on target despite variability in its component parts while also affording a low cost. This is the ideal way to deal with all three sources of noise. The greatest advantage of design of experiments is in this secondary design process of finding the optimum combination of parameter levels.

3.   Tolerance design (Tertiary design)

Tolerance design (or tertiary design) is a means of controlling factors that affect the target value by using higher grade components and inevitably increasing the cost. After the system has been designed (through system design) and the nominal mid-values of its parameters determined (through parameter design), the next step is to set the tolerances of the parameters (through tolerance design). Noise factors as well as system parameters are considered in a design of experiment to determine the extent of their impact on the output characteristics. Narrower tolerances must be given to noise factors that will have the greatest influence on the output characteristics.

The methodology of tolerance design is different from that of parameter design. In parameter design we do not attempt to control the effect of noise factors; rather, we attempt to identify factor levels that make the quality characteristic insensitive to noise factors. At the tolerance design stage, however, we attempt to control the noise factors by keeping them within narrow tolerances. Often this will increase the product cost as well. This is why every possible effort should be made to incorporate quality design measures in the system and parameter design steps. Narrow tolerances should be the last resort used only when parameter design gives inadequate results, and never without careful evaluation of the loss due to variability. Cost calculations, including those through the quality loss function, must be used to determine the tolerances.

The results of system, parameter and tolerance design obtained by the production process design department can now be passed to the production department in the form of specifications. The production department then designs a manufacturing process that will satisfy these specifications adequately.

2.1.9 Production (On-line quality control)

After a production process has been determined, variability in a product may be due to variability in materials, differences in purchased components, process drift, tool wear, machine failure and human error. These sources of variability are dealt with by on-line quality control during normal production. There are three forms of on-line quality control:

- process diagnosis and adjustment,
- prediction and correction, and
- measurement and action.

1.   Process diagnosis and adjustment

This is also known as process control. The process is diagnosed at regular intervals and if it is normal, production is continued. If it is not, the production process is stopped temporarily (shut down) until the cause is found. The process is then restored to its normal state and production is restarted. This is usually the stage where many companies have traditionally used Statistical Process Control (SPC). Common forms of SPC employed are variable control charts such as the X-bar Sigma $(\bar{X} - S)$ and X-bar Range $(\bar{X} - R)$ charts, as well as attribute control charts such as percent defective control charts, familiarly called P-charts or D-charts. Specific control limits are determined from process capability and are used to control the upper and lower limits of variation in a process. Advanced methods using pre-control charts or cumulative sum (Cusum) charts allow some preventive adjustments to be made when imminent failure can be diagnosed.

2.   Prediction and correction

A quantitative characteristic to be controlled is measured at regular intervals, and the measured value is used to predict the (mean) characteristic value of the product if production is continued without adjustment. If the predicted value differs from the target value, the level of a corrective signal factor is modified to reduce the difference. This method may be used to feed forward or feed backward control. However, it depends heavily on rational system design.

3.   Measurement and action

This is also called inspection. Each unit manufactured is measured, and if it is out of specification, it is reworked or scrapped. This method of quality control deals only with the product, while the first two methods deal mainly with the process.

2.1.10   Customer Usage (On-line quality control)

With all the quality control efforts in product design, process design, and manufacturing, some defective products may still be shipped to the customer. The only way to prevent further damage to the manufacturer's reputation for quality is to provide field service and to compensate the customer for the loss caused by the defective product.

2.1.11 **Self-assessment questions**

1. What is the role of quality engineering in product design?

2. Distinguish between off-line and on-line quality control and their relevance.

3. What is a robust design process?

4. Describe the quality assurance a manufacturer can offer after the sale of a product.

5. Distinguish between external, internal and unit-to-unit noise factors.

6. Identify and explain three noise factors that may affect the functioning of a laser printer.

2.1.12   Answers to self-assessment questions

1.   What is the role of quality engineering in product design?

Answer

The role of quality engineering in product design is to ensure that the product is designed to be robust against noise factors. A robust product or process is one whose response is least sensitive to all noise factors. A product's response depends on the values of the control and noise factors through a non-linear function. We exploit the non-linearity to achieve robustness. The three major steps in designing a product or a process are:
1.   System design: selection of the production technology.
2.   Parameter design: selection of the optimum levels of the control factors to maximize robustness.
3.   Tolerance design: selection of the optimum values of the tolerance factors (material type, tolerance limits) to balance the improvement in quality loss against the increase in the unit cost.

2.   Distinguish between off-line and on-line quality control and their relevance.

Answer

Design quality efforts are referred to as off-line quality control and production quality efforts are referred to as on-line quality control. Off-line quality control includes any of the design and development activities that may take place before products are manufactured and made available to customers. When products are manufactured to be available for customers, the on-line activities begin. Of course, the on-line activity must be well planned before the start of production. The loss function reflects the off-line and on-line quality control efforts in the equation:

$$L(y) = k \left[\sigma^2 + (\bar{y} - m)^2 \right]$$

The variance σ^2 is reduced and the target value m determined during the off-line phase. The average value produced $\bar{y}$ is controlled during the on-line phase. Both of these phases will reduce the loss associated with the product. The real purpose of a quality control chart is to provide the proper, close to the target, average value from a process. Incidental (special) causes of variation may be identified and controlled or eliminated, but this is the tolerance design approach. If a large amount of variation is present at the introduction of a process, this is an indication of the lack of off-line quality control efforts. The more off-line quality control work is done, the more robust a process or product will be against disturbances (external and internal noise) in the environment during the life of the process or product. Again, to reduce the loss-to-society as calculated in the quality loss function, there must be a reduced variance (off-line quality control work) and an average near the target value (on-line quality control work).

3.   What is a robust design process?

Answer

Taguchi views the robust design of a product or process as a three-step process:

- system design,
- parameter design,
- tolerance design.

1.   System design

System design is the phase when new concepts, ideas, methods, etc., are generated to provide new or improved products to customers. One way to remain competitive in the world economy is to be a leader in system design.

2.   Parameter design

The parameter design phase is crucial to improving the uniformity of a product and can be done at no cost or even at a saving. This means that certain parameters of a product or process design are set to make the performance less sensitive to the causes of variation.

3.   Tolerance design

The tolerance design phase improves quality by narrowing tolerances on the process or product parameters to reduce the performance variation. When tolerances are narrowed, variation can be reduced and quality improved. However, narrowing tolerances is usually expensive and may be unnecessary if parameter design were used first. One serious mistake a designer can make is to use expensive materials, components, or processes for a product when lower cost of-the-shelf items could be used if a parameter design approach is applied first.

4.   Describe the quality assurance a manufacturer can offer after the sale of a product.

Answer

     None. However, a warranty and service guarantee may be regarded as quality control measures to maintain customer satisfaction and goodwill.

5.   Distinguish between external, internal and unit-to-unit noise factors.

Answer

     Noise factors are those factors over which the manufacturer has no direct control but which vary with the customer's environment and usage. In general, noise factors are those which the manufacturer desires not to have to control at all. Noise factors can be classified into:

- external noise,
- internal noise,
- unit-to-unit noise.

1.   External noise

External noise refers to environmental factors such as ambient temperature, humidity, pressure or people. External noise produce variation from outside the product.

2.   Internal noise

Internal noise produces variations from inside or within the product. Internal noise is function and time-related, such as deterioration, wear, colour fading, shrinkage and drying out.

3.   Unit-to-unit noise

Unit-to-unit noise manifests itself in part-to-part variation. Unit-to-unit noise can also be due to differences between batches of materials used in a production process.

     Products may be sensitive to all three forms of noise simultaneously. Quality control at the design stage of a product or process provides less functional variation due to external or internal noise. Quality control at the production stage provides less functional variation from one part to another.

6.   Identify and explain three noise factors that may affect the functioning of a laser printer.

Answer

The three noise factors are:
- external noise,
- internal noise,
- unit-to-unit noise.

1.   External noise

The type of paper (surface texture, moisture content, etc.) used is usually variable and hence may cause jams in the feeder mechanism.

2.   Internal noise

Feed rollers that wear out due to constant friction can become smooth and hence fail to roll the paper forward.

3.   Unit-to-unit noise

Variations in the elasticity of the rubber rollers used and friction on spindles can cause a poor traction force and hence cause paper jams.

2.2   Principle of Robustization

2.2.1 Introduction
Noise factors cannot be eliminated; they are inherent in any system. Since noise factors cannot be eliminated, the quality characteristic of a product will not attain its intended target value. Of course, any deviation of a quality characteristic from the intended ideal will cause a loss. The principle of robustization attempts to reduce this loss by considering control and signal factors against noise factors in order to identify product specifications that make the quality characteristic insensitive to noise.

2.2.2 Sample distribution and the quality loss function
Suppose the distribution of a quality characteristic y resulting from all sources of noise is normally distributed as shown in Figure 2:2.1, where LSL and USL are the lower and upper specification limits, respectively. From Chapter 1 the loss-to-society for one piece of product at y for a nominal-the-best characteristic is given by:

$$L(y) = k (y - m)^2 \qquad\qquad (2{:}2.1)$$

where k is the cost coefficient, m is the target value and y is the specific value of y. For a sample of products this equation has been shown to be:

$$L(y) = k \left[\sigma^2 + (\bar{y} - m)^2\right] \qquad\qquad (2{:}2.2)$$

Thus, the quality loss function has three components:
* cost coeficient, k,
* variance, σ^2,
* bias squared, $(\bar{y} - m)^2$.

1.   Cost coeficient, k,
The cost coeficient k indicates the financial criticality of the quality loss function in relation to the quality characteristic.

2.   Variance, σ^2
The quantity σ is the standard deviation of the sample. Note that the square of the standard deviation is called the variance, σ^2. The variance results from the mean squared deviations of y around its own mean value $\bar{y}$.

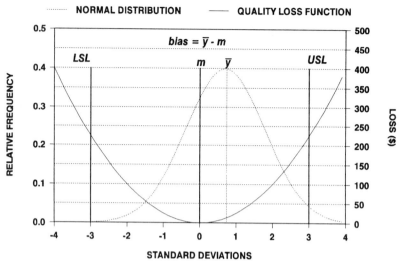

Figure 2:2.1 Distribution of a quality characteristic and the associated quality loss function for nominal-the-best.

3.   **Bias squared, $(\bar{y} - m)^2$**
The quantity $(\bar{y} - m)$ is called the bias or offset. The bias squared $(\bar{y} - m)^2$, results from the deviation of the average value of the sample $(\bar{y})$ from the target value m.

Between these three components of the quality loss function, little can be done about the cost coeficient k. The bias $(\bar{y} - m)$ can usually be easily reduced or eliminated by adjustment or calibration. Reducing the variance (σ^2) requires decreasing the standard deviation of the sample. Reducing the variance is usually the more difficult step. Robust design attempts to minimize both σ^2 and $(\bar{y} - m)^2$ simultaneously.

2.2.3 Reducing loss through reducing variance
There are four approaches to reduce variation:
● screening out bad products,
● discovering and eliminating the cause of discrepancy,
● narrowing the tolerance,
● application of Robust Design.

1.   **Screening out bad products**
Here, products that fall outside the lower specification limit (*LSL*), $m - \Delta_0$, and the upper specification limit (*USL*), $m + \Delta_0$, are rejected as defective. The rejected products are either reworked or scrapped. Because inspection, scrapping

and reworking are expensive, this method of reducing the variance leads to a higher cost per unit of accepted product.

2.   Discovering and eliminating the causes of discrepancy
The variance can also be reduced by discovering the cause of a discrepancy and eliminating it. For example, if the cause of paper jams in a photocopier is due to the type of paper used, then the customer could be asked to use certain *high quality* paper or to feed the paper in a particular direction marked on the packaging. Such methods of reducing the variance are frequently used due to need or urgency to get things done.

3.   Narrowing the tolerance
Another method of reducing the variance is to reduce the tolerance. If the tolerance on a particular component is identified as a major contributor to system performance, then a narrower tolerance can be specified for that component. For example, a 1 % tolerance resistor may be used instead of a 10 % tolerance resistor to control the voltage in a circuit. Usually a producer would not want to use a narrower tolerance unless it is necessary because components with narrower tolerances are often more expensive as well.

4.   Application of Robust Design
The best method of reducing the variance is to apply Robust Design. This method consists of making the performance of a product insensitive to noise factors. In many cases the manufacturing tolerances do not have to be tightened, the product's usage environment does not have to be controlled tightly, and cheaper material or components can be used, all of which make Robust Design the most economical of the four methods. Robust Design accomplishes this by taking advantage of product or process parameters to exploit the inherent non-linearity of the relationships among the various factors and the quality characteristic.

2.2.4 Classification of data types
Data used in robust design can be of three types:
- variable,
- attribute,
- digital.

1.   Variable
Variable data are amenable to measurement on a continuous scale.

2.   **Attribute**

Attribute data are not continuously scalable but can be classified on a discretely graded scale (two or more classes), often based on subjective judgments such as good, fair or bad.

3.   **Digital**

Digital data can only take values of 0 or 1.

2.2.5 Classification of quality characteristics

A quality characteristic is the object of interest of a product or process. It is also called the functional characteristic or response variable. Examples are tool wear, tyre wear, weld strength, room temperature, fuel economy, engine power, etc. A quality characteristic can also be classified according to its target value:

- nominal-the-best,
- smaller-the-better,
- larger-the-better,
- signed-target,
- classified attribute.

1.   **Nominal-the-best**

A nominal-the-best characteristic is a measurable characteristic with a specific user-defined target value. Values may be positive or negative. Examples of nominal-the-best characteristics include clearance, pressure, viscosity and weight. See Figure 2:2.2.

Characteristic type	Target	Example
Nominal-the-best	centred on given value	television voltage, turntable revolution per second
Smaller-the-better	as small as possible (0, zero)	tool wear, surface roughness
Larger-the-better	as large as possible (∞, infinity)	fuel economy, weld strength
Signed-target	0, zero	residual current
Classified attribute	-	low-medium-high, good-bad

Figure 2:2.2 Quality characteristics.

2.   Smaller-the-better
A smaller-the-better characteristic is a non-negative measurable characteristic that has an ideal state or target value of 0 (zero). Examples of smaller-the-better characteristics include deterioration, shrinkage and wear. See Figure 2:2.2.

3.   Larger-the-better
A larger-the-better characteristic is a non-negative measurable characteristic that has an ideal state or target value of infinity. Examples of larger-the-better characteristics include fuel efficiency, product life, mean time between failures and material strength. See Figure 2:2.2.

4.   Signed-target
A signed-target characteristic is a measurable characteristic that has an ideal state or target value of 0 (zero). It is different from smaller-the-better in that signed-target characteristic can have negative values. Examples of signed-target characteristics include current flow in a bridge circuit and colour match in a print. See Figure 2:2.2.

5.   Classified attribute
Attribute characteristics are not continuous variables, but can be classified on a discretely graded scale. They are often based on subjective judgments such as Good/Bad. The simplest type of attribute characteristic is the Go/No-go characteristic which has only two levels. Typical levels are Pass/Fail, Accept/Reject and Good/Bad. A second type of attribute characteristic is a fundamentally continuous characteristic that may be divided into a number of classes, (usually at least three classes) and provides more information than the fundamental Go/No-go characteristic. Typical classes in an experiment to study paper jams in a photocopier transport mechanism are I (0 or failed paper feed), II (1 – 150 paper feed), III (151 – 300 paper feed) and IV (301 – ∞ paper feed). Other examples of classified attribute characteristics include appearance, cracking and porosity. See Figure 2:2.2.

2.2.6 Classification of parameters
A number of other factors can influence the quality characteristic (response variable) of a product, as shown in Figure 2:2.3. These factors can be classified into:
- noise factors,
- control factors,
- signal factors,
- scaling factors.

1.   Noise factors
A parameter that causes the deviation of a quality characteristic from the target value is called a noise factor. Noise factors can have uncontrollable and

unpredictable influences on the quality characteristic. Noise factors are usually difficult, expensive or not intended to be controlled. However, for experimental purposes, they may need to be controlled on a small scale.

2.   Control factors

These are parameters whose values are controlled by the design engineer. A control factor may take one or more values called levels. At the end of an experiment, a suitable level of the control factors will be selected. One aspect of robust design is to search for the optimal level settings for the control factors so as to make the quality characteristic insensitive to noise. Examples of control factors are welding current, material type and preheat temperature.

3.   Signal factors

These are factors which change the true values of the quality characteristic to be measured. A quality characteristic in a design of experiment in which the signal factor takes a constant value (in which case it is not included as a factor) is called a static characteristic. When the signal factor takes a number of values, the quality characteristic is called a dynamic characteristic. A signal factor is not set by a design engineer but by the user based on the intended result.

4.   Scaling factors

These are factors which are used to shift the mean level of a quality characteristic to achieve the required functional relationship between a signal factor and the quality characteristic. Scaling factors are also called adjustment factors. An example is the gearing ratio in the steering mechanism of a car.

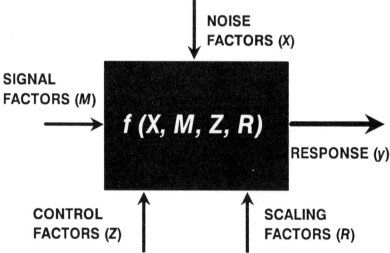

Figure 2:2.3 Factors affecting a quality characteristic.

2.2.7 Example of a classification of factors

Consider a truck being studied for manoeuvrability. Two control factors could be the front springs (soft or standard) and the steering geometry (Ackerman or parallel). At the end of the experiment the design engineer will want to choose one type of spring (soft or standard) and one type of steering geometry (Ackerman or parallel). Noise factors could be the turning direction (left or right), road conditions (wet or dry) and type of tyre (steel or radial). At the end of the experiment the engineer will not want to choose any of these factor levels. Signal factors could be steering angles of 20°, 25° and 30°. The engineer may wish to study the manoeuvrability at these angles but will not want to set them as these will be continuously changed during use. A scaling factor could be the gear box system. It may be necessary for the customer to change the scaling factor according to the condition of use.

2.2.8 Parameter diagram and modelling

A diagrammatic representation of the manoeuvrability experiment with the various factors affecting the quality characteristic is shown in Figure 2:2.3. The quality characteristic y (response variable) is a function of X (noise), M (signal), Z (control) and R (scaling). That is $y = f(X, M, Z, R)$. There are two approaches for optimising such a complex system:

- micro-modelling,
- macro-modelling.

1.   Micro-modelling

As the name suggests, micro-modelling is based on an in-depth understanding of the system. It begins with the development of a mathematical model of the system, which, in industrial experimentation, would be very complex. When systems are complex one must make assumptions that simplify the operation, as well as put considerable effort into developing the model. Furthermore, the more simplification one does, the less realistic the model will be, and hence the less adequate it will be for precise optimization.

2.   Macro-modelling

In macro-modelling, the step of building a mathematical model of the system is bypassed. The primarily concern is to obtain the optimum system configuration, and not to obtain a detailed understanding of the system itself. As such, macro-modelling gives faster and more efficient results. It gives the specific information needed for optimization with a minimum expenditure of experimental resources.

2.2.9 Choice of factors

There are a number of properties of factors that need to be considered:

- factor levels,
- number of factor levels,
- range of factor levels,
- feasibility of factor levels.

1.   Factor levels

These are levels of values or attributes assigned to a factor, which may be a control, noise, signal or scaling factor. A control factor such as force, temperature, or current may be represented as in Figure 2:2.4. In order to make a comparison we need to obtain at least two measurements. Therefore a factor must have at least two levels. In Figure 2:2.4 the factor *Force* can be regarded as being studied at two levels, coded A1 and A2 corresponding to Levels 1 and 2. Their respective forces are 10 and 20 newtons. That is, in this experiment the effect of *Force* is studied at two settings, namely, 10 and 20 newtons. Similarly, *Temperature* is studied at three levels, coded B1, B2 and B3 corresponding to Levels 1, 2 and 3. Their respective temperatures are 100, 120 and 140 °C. How many measurements are done at each setting depends of course on the engineer. Often the control factors are coded A, B, C, D, etc. while the noise factors are coded P, Q, R, etc.

Response	Code	Level 1	Level 2	Level 3	Units
Force	A	10	20	-	newton
Temperature	B	100	120	140	°C
Pressure	C	low	medium	high	-
Texture	D	first	second	third	grade

Figure 2:2.4 Factor levels.

2.   Number of factor levels

Deciding on the levels to use for each factor selected for the experiment is an important phase in planning. Determining the levels for the qualitative factors will usually be clear from the nature of the problem being investigated. However, choosing the appropriate levels for the quantitative factors is a more difficult task. The number of levels selected and level settings will depend on how much is known about the process or product. If a new process or product is being investigated, it may be desirable to run three levels for some of the variables to evaluate non-linearity over the range of the variables. If more is known about the effect of certain variables, then 2-level factors may be sufficient to derive the desired information from the analysis of the experimental results.

When uncertainty exists about the number of levels to choose for a given variable, then three levels might provide sufficient information. A lot depends on the cost of experimentation and how much the scope of the experiment is increased by going from two to three levels. Thus, when two levels are used, we can fit only a linear function (or straight line). When three levels are used, we can fit a quadratic function (or a curve). Similarly, when four levels are used, we can fit a cubic function (also a curve). See Figure 2:2.5.

3.   Effect of the number of factor levels on the number of experiments
If a factor is studied at two levels there must be at least one measurement at each level. If an experiment studies two factors each at two levels, then there must be four measurements. In general, if there are m factors at 2-levels, there will be 2^m measurements. Similarly, if there are n factors at 3-levels, there will be 3^n measurements. When there is a mixture of m 2-level and n 3-level factors, the will be $2^m \times 3^n$ measurements. Since cost is usually a constraint in studying many levels of a factor, it would be more economical to study 2-level factors. If, however, the quality characteristic is a very important one, such as in the medical field, then it may be necessary to study 3-level factors.

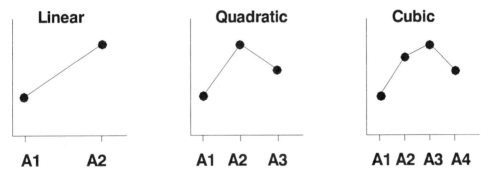

Figure 2:2.5 Effect of the number of factor levels studied.

4.   Range of factor levels
Another question that must be addressed is the range of values to be covered for a given variable. Assuming a linear relationship, the wider the range used in the experiment, the better will be the chance of discovering the real effect of that variable on the quality characteristic. However, the wider the range, the less reasonable is the assumption of a linear effect for the variable. The selection of the range depends in part on whether the purpose of the experiment is to explore over a broad region or to fine tune to achieve optimum conditions. If the levels of the factor are chosen such that levels 1 and 2 are narrow, corresponding to

A in Figure 2:2.6, then the factor effects may not appear to be significant. If the levels of the factor are chosen such that levels 1 and 2 are too wide, corresponding to C in Figure 2:2.6, then again the factor effects may not appear to be significant. Therefore some engineering knowledge is necessary to select the range of factor levels.

5.   Feasibility of levels

Another consideration in choosing the levels of the factors is whether or not it is possible to run the various treatment combinations chosen for the experiment. This is especially true in chemical processes. The experimenter should not go beyond the range of physically or practically useful conditions. Choosing too narrow a range for the levels of a factor results in little useful information other than increasing the sensitivity of the experiment. It is sometimes necessary to change levels and repeat an experiment due to lack of knowledge of these operating boundaries.

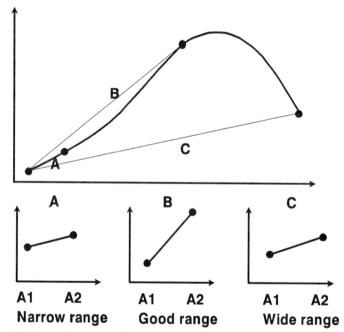

Figure 2:2.6 Effect of range of factors.

6.   Sliding levels

Sliding levels involve the effects of two control factors. Consider the results of a soldering process based on two control factors, as shown in Figure 2:2.7.

		Temperature	
		low	high
Current	10 A	150 °C : bad	170 °C : good
	20 A	150 °C : good	170 °C : bad

Figure 2:2.7 Factor assignment of two *related* factors.

When the solder bath temperature is low (150 °C) and the current is low (10 A) the solder is not fully melted and poor results are seen. When the solder bath temperature is high (170 °C) and the current is low (10 A) the results are good. The same happens when the solder bath temperature is low and the current is high. However, the results are again poor when the solder bath temperature is high and the current is high. Obviously, the factors cannot be both low or both high. When investigating such control factors it is advisable to offset the factors as shown in Figure 2:2.8. Thus, when the current is low, the low temperature is set at 160 °C whereas when the current is high, the low temperature is set at 150 °C. In this way, known combinations of factors that are infeasible will not be studied and hence the efficiency of experimentation is increased. The use of this technique of factor level assignment is called sliding levels.

		Temperature	
		low	high
Current	10 A	160 °C	170 °C
	20 A	150 °C	160 °C

Figure 2:2.8 Factor assignment of two *related* factors using sliding levels.

7.   Safety of plant and personnel
The experimental level settings may be critical to personnel or plant safety. The experimenter must accept all responsibility for the choice of factors and test extremes. The experimenter must ensure that all experimental test combinations result in safe operating conditions if applied to the actual system level use.

2.2.10   Effects of factors
Although control factors are studied so that the ideal levels of the factors can be established, a control factor may:

- affect the mean only,
- affect the variance only,
- affect the mean and variance, or
- affect neither.

1.   Affect the mean only
A factor that affects the mean but not the variance can be used to adjust the mean value, of a process or product, to the target value. A factor that affects the mean only is usually called an adjustment factor. See Figure 2:2.9, A.

2.   Affect the variance only
A factor that affects the variance but not the mean can be used to reduce variation of a process or product. See Figure 2:2.9, B.

3.   Affect the mean and variance
A factor that affects both the mean and the variance must be used very carefully. However, such a factor allows some flexibility in balancing target requirements. See Figure 2:2.9, C.

4.   Affect neither
A factor that does not affect the mean or variance is not a useless factor. Indeed, the *better level* of such a factor can be used advantageously depending on other factors such as cost, convenience, etc. See Figure 2:2.9, D.

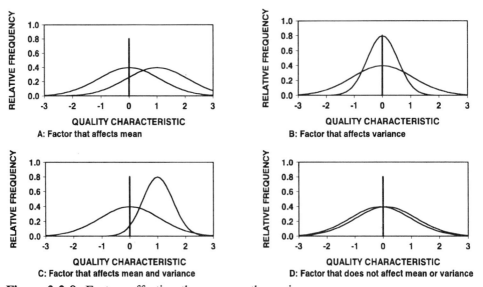

Figure 2:2.9 Factors affecting the mean or the variance.

In industrial experimentation, it is very seldom that a factor affects the mean only or the variance only. Generally, it is more likely that a factor that affects the mean also affects the variance to some extent, and vice versa. What we mean when we say a factor affects the mean is that, to a large extent, the factor affects the mean only. Similarly, when we say that a factor affects neither the mean nor the variance, we mean that the factor does not affect the mean or the variance appreciably.

2.2.11   Self-assessment questions

1.   Describe four ways of reducing the variation in a production sample.

2.   Distinguish between the three target characteristic types.

3.   Discuss the choice of the number and the range of factor levels in an experiment. Why would you need precautionary measures?

4.   An experimenter wishes to increase the fuel efficiency and the power output of an automobile engine. Is this feasible? What constraints is the engineer likely to face?

2.2.12   Answers to self-assessment questions

1.   Describe four ways of reducing the variation in a production sample.

Answer
Four ways of reducing the variation in a production sample are product screening, elimination of the cause, tightening of the tolerance and Robust Design. In product screening, products that are smaller or larger than the allowable specifications are discarded. Hence the sample has a smaller variation. This however leads to a high cost of production. Elimination of the cause of the variation reduces variation in a production sample. However some effort is necessary to monitor the causes of variation. The method also increases the production costs. Tightening of the tolerance can also be used to reduce variation in a production sample. However, this will cause more non-conforming products or will result in higher unit cost of the product. Robust Design methods of reducing variation enables robust products at the lowest cost.

2.   Distinguish between the three target characteristic types.

Answer
Nominal-the-best has a specified value and is usually associated with a tolerance. An example of nominal-the-best is colour television voltage, e.g. 115 ± 20 volts. The objective is to be as close as possible to 115 volts. The smaller-the-better characteristic has a target value of zero (0). An example of smaller-the-better is tyre wear. Customers prefer tyres that wear out slowly. The larger-the-better characteristic has a target value of infinity (∞). An example is automobile fuel efficiency. The more kilometres per litre, the better. Although in such a case it is not possible to attain infinity, the objective is to improve fuel efficiency as much as possible.

3.   Discuss the choice of the number and the range of factor levels in an experiment. Why would you need precautionary measures?

Answer
By studying a factor at two levels it is only possible to make a linear inference. By studying a factor at three levels it is possible to make a quadratic inference. Thus it would appear to be better to study a factor at three levels. However, the number of measurements in an experiment increases as 3^n, where n is the number of factors. Two-level factors do not give quadratic effects but the measurements increase only as 2^n where n is the number of factors. Therefore, engineering caution need to be exercised when selecting the number of factor levels. The range of

factor levels is another important issue in studying the effect of a factor. If the range of a factor level is too small or too large, this can give an erroneous non-significant effect. Again, engineering caution need to be exercised when selecting the range of factor levels.

4.   An experimenter wishes to increase the fuel efficiency and the power output of an automobile engine. Is this feasible? What constraints is the engineer likely to face?

Answer

From a quality viewpoint this is infeasible. This is a case of contrasting quality characteristic requirements. Since the laws of energy cannot be violated, there has to be a compromise. Therefore the engineer is likely to face opposing factor or level requirements. What is to be compromised depends on what minimizes the loss to society. The best that can be done is to increase the efficiency of the engine. This is usually the case in the design of modern engines.

2.3 Examples of Parameter Design

2.3.1 Introduction
A great deal of engineering time is spent generating information about how different design parameters affect performance under different usage conditions. Parameter Design serves as a methodology that enables an engineer to generate information needed for decision-making with the minimum experimental effort. In this section we consider a simple experiment that serves a starting point for Parameter Design. The experiment is not even an idealistic one. Nevertheless, it demonstrates some common pitfalls associated with the design of experiments. Methods of overcoming some of these pitfalls are treated in subsequent chapters.

2.3.2 Ina Seito tile experiment
During the late 1950s, Ina Seito Tile Company[2] of Japan faced the problem of high variability in the dimensions of the tiles it produced. The company had purchased an expensive new tunnel kiln in order to increase productivity, but it showed a high variability in the tile dimensions, and more than half of the tiles were outside specifications. Since rejecting those tiles outside the specified dimensions was an expensive answer to the problem, the company assigned a team of expert engineers to investigate the cause. The team's analysis showed that the tiles at the centre of the pile inside the kiln experienced lower temperatures than those at the edge. Since redesigning the tunnel kiln was an expensive countermeasure, the Tile Company used parameter design to overcome the effects of uneven temperature. In order to understand this experiment we need to know something about orthogonal arrays (See Figure 2:3.1).

The $L_8(2^7)$ orthogonal array shown in Figure 2:3.1 is commonly used in the design of experiments. In this array, the first column Exp represents the experiment numbers. There are eight experiments in the $L_8(2^7)$ orthogonal array. The next seven columns labelled A to G represent the seven factors that can be assigned to this orthogonal array. Each of the columns A to G has a number of ones and twos which represent the levels of a factor assigned to the column. Orthogonal arrays have a number of unique properties:
- equal proportions of experiments,
- equal proportions of remaining factor levels,
- equal proportions of combinations of factor levels.

---

2   Genichi Taguchi, *Introduction to Quality Engineering – Designing Quality into Products and Processes*, 1986, Asian Productivity Organisation.

Exp	A	B	C	D	E	F	G
1	1	1	1	1	1	1	1
2	1	1	1	2	2	2	2
3	1	2	2	1	1	2	2
4	1	2	2	2	2	1	1
5	2	1	2	1	2	1	2
6	2	1	2	2	1	2	1
7	2	2	1	1	2	2	1
8	2	2	1	2	1	1	2

Figure 2:3.1 The $L_8(2^7)$ orthogonal array. Exp denotes the experiment number.

1.   Equal proportions of experiments

Consider column A. There are four ones (Experiments 1, 2, 3 and 4) and four twos (Experiments 5, 6, 7 and 8) in this column. Similarly, in column B, there are also four ones (Experiments 1, 2, 5 and 6) and four twos (Experiments 3, 4, 7 and 8). This is also true for all the remaining columns.

2.   Equal proportions of remaining factor levels

Consider column A. The factor level 1 (denoted A1) appears in Experiments 1, 2, 3 and 4. While factor A is in level 1, half of factor B is in level 1 and the other half is in level 2. Similarly, while factor A is in level 1, half of factor C in level 1 and the other half is in level 2. This is true for all the remaining factors. Indeed, this is also true if we started with any other factor.

3.   Equal proportions of combinations of factor levels

Consider columns A and B. In Experiments 1 and 2, factor A is in level 1 and factor B is in level 1. We denote this as (1, 1). In Experiments 3 and 4, factor A is in level 1 and factor B is in level 2. We denote this as (1, 2). Similarly, there are (2, 1) and (2, 2) in Experiments 5, 6 and 7, 8 respectively. Notice that these combinations (1, 1), (1, 2), (2, 1) and (2, 2) occur an equal proportion of times. That is, one-quarter of the Experiment has (1, 1), another one-quarter has (1, 2), another one-quarter has (2, 1) and one-quarter has (2, 2) combinations. Indeed this is true for any two pairs of the columns A to G. It is this balanced property for which the array is called orthogonal. These unique properties of the $L_8(2^7)$ orthogonal array thus enable seven fair comparisons of factors, namely, A1 against A2, B1 against B2, and so on, to G1 against G2.

2.3.3 Ina Seito tile experiment, continued

Suppose we were studying seven[3] control factors (A, B, C, D, E, F, and G), each of which has two levels, as shown in Figure 2:3.2. The seven factors can be assigned to columns A through G of the $L_8(2^7)$ orthogonal array in Figure 2:3.1. Eight tile mixtures can then be prepared according to Figure 2:3.1, moulded into tiles and fired in the tunnel kiln.

Factor	Level 1	Level 2
A Lime content	A1 5 %	A2 1 %
B Granularity	B1 coarse	B2 fine
C Agalmatolite	C1 43 %	C2 53 %
D Agalmatolite type	D1 current mixture	D2 cheaper mixture
E Charge quantity	E1 1300 kg	E2 1200 kg
F Waste return	F1 0 %	F2 4 %
G Feldspar content	G1 0 %	G2 55

Figure 2:3.2 Factors and levels. Current levels are shown shaded.

Referring to Experiment 1 in Figure 2:3.1, the factors levels are A1, B1, C1, D1, E1, F1 and G1. Specifically, the clay preparation used was a 5 % admixture (level A1) of the lime additive in its coarse-grained form (level B1), with a 43 % (level C1) agalmatolite content of the present type (level D1) with no waste return (level F1) and no feldspar (level G1) in a 1300 kg charge (level E1). The $L_8(2^7)$ orthogonal array can thus be read as a set of instructions for eight particular experiments that will allow fair comparisons of A1 against A2, B1 against B2, and so on, to G1 against G2.

2.3.4 Results of experiments

A sample of 100 tiles was taken under each of the eight sets of experimental conditions. The numbers of tiles that were unacceptable are given in the last column (Per 100) of Figure 2:3.3. Since the quality characteristic is the percentage of unacceptable tiles, this implies that the quality characteristic type is smaller-the-better.

---

3   Modifications of the technique can be used when there are fewer than seven factors or if some of them have more than two levels.

Exp	A	B	C	D	E	F	G	Per 100
1	1	1	1	1	1	1	1	16
2	1	1	1	2	2	2	2	17
3	1	2	2	1	1	2	2	12
4	1	2	2	2	2	1	1	6
5	2	1	2	1	2	1	2	6
6	2	1	2	2	1	2	1	68
7	2	2	1	1	2	2	1	42
8	2	2	1	2	1	1	2	26

Figure 2:3.3 The $L_8(2^7)$ orthogonal array with results of experiment.

The grand average $\bar{y}$, of unacceptable tiles is the overall average of the experiments. This is calculated as:

$$\bar{y} = \frac{(16 + 17 + 12 + 6 + 6 + 68 + 42 + 26)}{800} \times 100 \% = 24.125 \%$$

The effects of A1 (5 % lime content) and A2 (1 % lime content) can be compared by taking the average of defectives in those experiments using level A1 (Experiment numbers 1, 2, 3 and 4) with the average for the experiments using level A2 (Experiment numbers 5, 6, 7, and 8). These averages are:

$$\overline{A1} = \frac{16 + 17 + 12 + 6}{400} \times 100 \% = 12.75 \%$$

$$\overline{A2} = \frac{6 + 68 + 42 + 26}{400} \times 100 \% = 35.50 \%$$

This indicates that if the amount of lime content (factor A) is increased from A2 (1 %) to A1 (5 %), the percent of defectives would fall from 35.50 % to 12.75 % under experimental conditions.

Similarly, B1 and B2 can be compared by comparing the average for the experiments using level B1 (Experiment numbers 1, 2, 5, and 6) with the average for the experiments using level B2 (Experiment numbers 3, 4, 7, and 8). The same can be

done for all the remaining factors and the results are shown in the response table
(Figure 2:3.4) and response graph (Figure 2:3.5).

   The difference between the highest and the lowest values for each factor is also
calculated and entered into Figure 2:3.4. Since this difference represents the
significance of a factor, a ranking is done on all the factors. The ranking is established
to be: A, F, G, E, D, B and C. This implies that factors A, F and G are more
important than the rest of the factors E, D, B and C. Here, we note that while ranking
gives the order of importance it does not indicate the relative magnitude of importance.

	A	B	C	D	E	F	G
Level 1	12.75	26.75	25.25	19.00	30.50	13.50	33.00
Level 2	35.50	21.50	23.00	29.25	17.75	34.75	15.25
Difference	22.75	5.25	2.25	10.25	12.75	21.25	17.75
Rank	1	6	7	5	4	2	3

Figure 2:3.4 Response table of factor effects.

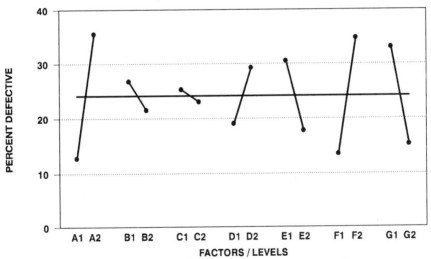

Figure 2:3.5 Response graph of factor effects for the tile experiment.

   From the response table (Figure 2:3.4) the optimum set of conditions is then
selected by choosing all factor levels with the lowest percentage of defectives since the
percentage of defectives is a smaller-the-better quality characteristic. In order of
ranking, the optimum condition is therefore A1, F1, G2, E2, D1, B2 and C2.

2.3.5 Prediction of the process average under optimum condition

Having determined the optimum condition from the orthogonal array experiment, we should be able to predict the process average $\mu_{Predicted}$ under the optimum condition. This is done by summing the effects of the higher ranking factors. The effect of a significant factor is its effect about the experimental average. From above, the mean effects of factor A at levels 1 and 2 are 12.75 % and 35.50 %, respectively. Since, the experimental average is 24.125 %, the effect of factor level A1 is to change the mean effect by (12.75 − 24.125), that is −11.375 %. In effect this is $(\overline{A1} - \overline{y})$. The total effect of the higher ranking factors about the experimental average would be:

$$\mu_{Predicted} = \overline{y} + (\overline{A1} - \overline{y}) + (\overline{F1} - \overline{y}) + (\overline{G2} - \overline{y})$$

$$= \overline{A1} + \overline{F1} + \overline{G2} - 2 \times \overline{y}$$

$$= 12.75 + 13.50 + 15.25 - 2 \times 24.125$$

$$= -6.75 \ \%$$

Notice that only the three highest ranking factor effects have been used to estimate the predicted process average. As a rule of thumb, we may take about half the number of factors in an experiment. This is to ensure we allow for experimental error. Notice also that the predicted average is a negative quantity. At this introductory stage, we shall take this to mean approximately zero percent (0 %) defective. All of these points are covered in later chapters.

2.3.6 Conclusions

From the response graph (Figure 2:3.5) it can be seen that factor A has a large effect on the percent defectives. Factor C has a relatively small effect on the percent defectives. That does not mean that factor C is unimportant. In fact, since it does not make much difference (only 2.25 %) whether C1 or C2 is used, the engineer may choose the factor level based on other criteria. In this case, C1 (43 %) could be chosen over C2 (53 %) since C1 represents less Agalmatolite used, the most expensive component of the mix. Similarly, since lime content (factor A1) is the cheapest material in the process, increasing the amount of limestone from 1 % to 5 % contributes a cost saving.

Thus, the problem of high variation in tile dimensions was solved by minimizing the effect of the cause of the variation (non-uniform temperature distribution) without controlling the cause (the tunnel kiln design) itself. As illustrated by this example, the fundamental principle of Robust Design is to improve the quality of a product by minimizing the effect of the cause of variation without eliminating that cause. This is achieved by optimizing the product and process designs through parameter design where the quality characteristic is made least sensitive to the cause of variation.

Sometimes, however, parameter design alone does not always lead to sufficiently high quality. Further improvement can be obtained by controlling the causes of variation (tolerance design) where it is economically justifiable, typically by using more expensive equipment, higher grade components, better environmental controls, etc. all of which lead to higher product cost, or operating cost, or both. The benefits of improved quality must justify the added product cost.

2.3.7 Self-assessment questions

1. An engineer performed an experiment on the weld strength of wires. The control factors are A, B, C, D, E, F and G. The results of the experiment in kilonewtons (kN) are:

Exp	A	B	C	D	E	F	G	Results	
1	1	1	1	1	1	1	1	40	37
2	1	1	1	2	2	2	2	42	48
3	1	2	2	1	1	2	2	59	54
4	1	2	2	2	2	1	1	47	57
5	2	1	2	1	2	1	2	43	50
6	2	1	2	2	1	2	1	30	26
7	2	2	1	1	2	2	1	28	27
8	2	2	1	2	1	1	2	37	43

1. Explain why an orthogonal array is used.
2. Determine the average response for each factor level, i.e. A, B, C, D, E, F and G.
3. Using a response table, comment on the significance of the factor effects.
4. Plot the factor effects on a response graph and indicate the recommended factor levels for optimization.
5. What is the predicted weld strength at the optimum condition?

2.3.8 Answers to self-assessment questions

1.   An engineer performed an experiment on the weld strength of wires. The control factors are A, B, C, D, E, F and G. The results of the experiment in kilonewtons (kN) are:

Exp	A	B	C	D	E	F	G	Results	
1	1	1	1	1	1	1	1	40	37
2	1	1	1	2	2	2	2	42	48
3	1	2	2	1	1	2	2	59	54
4	1	2	2	2	2	1	1	47	57
5	2	1	2	1	2	1	2	43	50
6	2	1	2	2	1	2	1	30	26
7	2	2	1	1	2	2	1	28	27
8	2	2	1	2	1	1	2	37	43

1.   Explain why an orthogonal array is used.
2.   Determine the average response for each factor level, i.e. A, B, C, D, E, F and G and draw a response table.
3.   Using the response table, comment on the significance of the factor effects.
4.   Plot the factor effects on a response graph and indicate the recommended factor levels for optimization.
5.   What is the predicted weld strength at the optimum condition?

Answers

1.   An orthogonal array is a balanced array of many factors. It thus enables an engineer to evaluate the significance of many factors simultaneously. The orthogonal property of these arrays ensures fair comparisons of all factors.

2.   Response table of factors.
      Using the given data, the average result for each experiment is calculated. These are shown below:

Experiment No.	1	2	3	4	5	6	7	8
Average Result	38.5	45.0	56.5	52.0	46.5	28.0	27.5	40.0

From the results of each experiment, the response for each level can be calculated.

For factor A at level 1:
Factor A is at level 1 in Experiments 1, 2, 3 and 4. Hence the average response of factor A is the average of the results for which factor A is in level 1. That is:

$$\overline{A1} = \frac{38.5 + 45.0 + 56.5 + 52.0}{4} = 48.0$$

For factor A at level 2:
Factor A is at level 2 in Experiments 5, 6, 7 and 8. Hence the average response of factor A is the average of the results for which factor A is in level 2. That is:

$$\overline{A2} = \frac{46.5 + 28.0 + 27.5 + 40.0}{4} = 35.5$$

This procedure is repeated for the remaining factors and levels. The results are shown in the response table below:

	A	B	C	D	E	F	G
Level 1	48.00	39.50	37.75	42.25	40.75	44.25	36.50
Level 2	35.50	44.00	45.75	41.25	42.75	39.25	47.00
Difference	12.50	4.50	8.00	1.00	2.00	5.00	10.50
Rank	1	5	3	7	6	4	2

3. From the differences of the factor level responses, it can be seen that the significant factors are A, G and C.

4.   Response graph of factor levels.

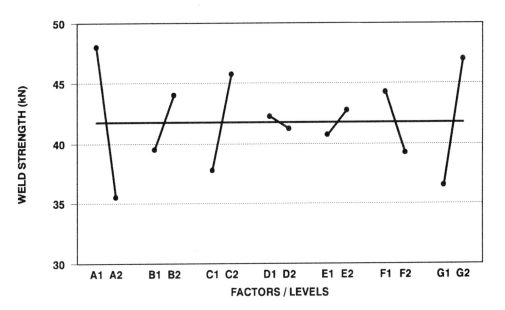

The optimum condition is A1, B2, C2, D1, E2, F1 and G2. In this case the higher levels are chosen because the quality characteristic, weld strength, has a larger-the-better target.

5.   What is the predicted weld strength at the optimum condition? Using only the three most significant factor effects A1, G2 and C2,

$$\mu_{Predicted} = \bar{y} + (\overline{A1} - \bar{y}) + (\overline{G2} - \bar{y}) + (\overline{C2} - \bar{y})$$

$$= \overline{A1} + \overline{G2} + \overline{C2} - 2 \times \bar{y}$$

$$= 48.00 + 47.00 + 45.75 - 2 \times 41.75$$

$$= 57.25 \text{ kN}$$

CHAPTER 3

ORTHOGONAL ARRAYS AND MATRIX EXPERIMENTS

AIMS:
The aim of this chapter is to describe how to design experiments using orthogonal arrays and related criteria. It also describes the effect of interactions on control factors and explains how to analyze interactions. The importance of selecting a quality characteristic based on additivity is also explained.

OBJECTIVES:
When you have completed studying this chapter you should be able to:
- explain how to choose an orthogonal array for an experiment based on the required degrees of freedom,
- explain interactions and draw graphs of interaction breakdowns,
- analyze the effects of interactions and the consequences of confounding effects,
- calculate the process average for an experiment involving interaction,
- distinguish between good and poor quality characteristics in terms of additivity,
- translate a poor quality characteristic into a good quality characteristic.

OVERVIEW:
This chapter introduces a number of concepts that are necessary in order to understand the use of an orthogonal array experiment. First, an explanation of fair comparison is given, together with the concept of degrees of freedom. The degrees of freedom associated with 2- and 3-level factors are explained. Second, interaction is introduced with examples. The degrees of freedom of interactions are explained, together with a graphical representation of interaction. Interaction breakdown in also discussed in detail. An example of interaction breakdown and the prediction of the process average when interactions are significant is also discussed. The difficulty in industrial experimentation is related to interactions in experiments. Finally, the selection of good quality characteristics is explained with respect to good additivity and reproducibility of laboratory experiments under industrial conditions.

3.1　Matrix Experiments

3.1.1 Introduction

A matrix is an array of numbers that have a special meaning. A matrix experiment consists of a set of experiments where the factors and levels are changed according to the matrix. These factors and levels are typically the settings of the various process or product parameters we wish to study. Conducting matrix experiments using special matrices, called orthogonal arrays, allows the effects of several parameters to be determined efficiently and is an important technique in Robust Design. Orthogonal arrays were described briefly in Chapter 2 in the Ina Seito example. This chapter explains more fully how to use orthogonal arrays for experiments.

3.1.2 Degrees of freedom for factors and levels

The degrees of freedom is the number of independent measurements available to estimate sources of information. The number of degrees of freedom indicates the number of independent (fair) comparisons that may be made within a set of data. In general, the number of degrees of freedom associated with a factor (v_{f1}) is equal to one less than the number of levels for that factor.

$$v_{f1} = number\ of\ levels\ -\ 1$$

Consider the heights of Allan, Brian and Collin. If we measure the height of Allan to be A1 and the height of Brian to be A2, then we can only make one fair comparison; (A1 − A2) as shown in Figure 3:1.1. If we measure the height of Collin to be A3, then we can only make two fair comparisons from the three height measurements, namely, (A1 − A2) and (A1 − A3), as shown in Figure 3:1.1. The other comparison (A2 − A3) can be found from (A1 − A3) − (A1 − A2), since:

$$\begin{aligned}(A1 − A3) − (A1 − A2)\ &=\ A1 − A3 − A1 + A2\\ &=\ A2 − A3\end{aligned}$$

and is therefore not a fair comparison. This is also frequently referred to as being not independent. Let us consider three further examples:

Example 1
Consider a factor such as temperature, studied at 2 levels, say 100 °C and 120 °C. How many degrees of freedom are there for temperature?

　　　　The number of degrees of freedom for a factor is one less than the number of levels. Therefore:

v_{f1}　　= number of levels −1

v_{f1}　　= 2 − 1 = 1.

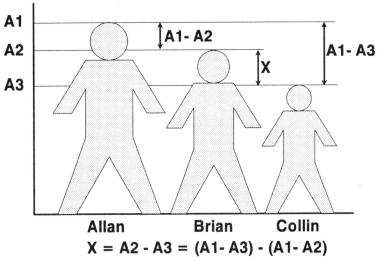

$$X = A2 - A3 = (A1 - A3) - (A1 - A2)$$

Figure 3:1.1 Explanation of degrees of freedom.

Example 2
Consider a factor such as voltage, studied at 3 levels, say 1 V, 5 V and 9 V. How many degrees of freedom are there for voltage?

The number of degrees of freedom for a factor is one less than the number of levels. Therefore:

ν_{fl} = number of levels -1

ν_{fl} $= 3 - 1 = 2.$

Example 3
An engineer wants to study three 2-level factors and one 4-level factor. How many degrees of freedom does that require?

A 2-level factor has 1 degree of freedom. Since, each 2-level factor has 1 degree of freedom, three 2-level factors require 3 degrees of freedom. A 4-level factor has 3 degrees of freedom. Therefore, the engineer requires at least 6 degrees of freedom.

ν_{fl} $= 3 \times (2 - 1) + 1 \times (4 - 1)$

ν_{fl} $= 6$

3.1.3 Orthogonal arrays
An orthogonal array is a matrix of numbers arranged in columns and rows. Each column represents a specific factor or condition that can be changed from experiment to experiment. Each row represents the state of the factors in a given experiment. The array is called orthogonal because the levels of the various factors are balanced and can be separated from the effects of the other factors within the experiment. That is, an

orthogonal array is a balanced matrix of factors and levels, such that the effect of any factor or level is not confounded with the effect of any other factor or level.

Exp	Columns						
	1	2	3	4	5	6	7
1	1	1	1	1	1	1	1
2	1	1	1	2	2	2	2
3	1	2	2	1	1	2	2
4	1	2	2	2	2	1	1
5	2	1	2	1	2	1	2
6	2	1	2	2	1	2	1
7	2	2	1	1	2	2	1
8	2	2	1	2	1	1	2

Figure 3:1.2 The $L_8(2^7)$ orthogonal array.

3.1.4 Degrees of freedom of orthogonal arrays

Like measurements made with factor levels, experiments made with orthogonal arrays can only give a certain number of fair comparisons for factors. This is the degrees of freedom in an orthogonal array (ν_{OA}), which is always 1 less than the number of experiments because one degree of freedom is taken up by the overall mean.

$$\nu_{OA} = \textit{number of experiments} - 1$$

Information about an orthogonal array can usually be found from the way the orthogonal array is named. With some exceptions, the number of degrees of freedom for an orthogonal array (ν_{OA}) equals the number of degrees of freedom (ν_{fl}) for the factor levels. An orthogonal array is usually denoted as shown in Figure 3:1.3 and includes information on the following:

- the L notation,
- the number of rows,
- the number of columns,
- the number of levels,

1.   The L notation
The L notation indicates that the information is based on the Latin square arrangement of factors. A Latin square arrangement is a square matrix arrangement of factors with separable factor effects. Thus, the L notation indicates that the information is an orthogonal array information.

2.   The number of rows
The number of rows indicates the number of experiments required when using that orthogonal array.

3.   The number of columns
The number of columns indicates the number of factors that can be studied in the orthogonal array.

4.   The number of levels
The number of levels indicates the number of factor levels.

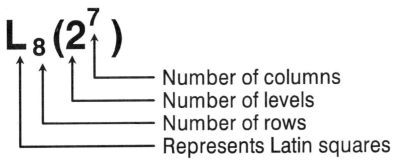

Figure 3:1.3 Notation of orthogonal arrays.

Thus, the information in an $L_8(2^7)$ orthogonal array can be read to mean: The $L_8(2^7)$ orthogonal array allows seven 2-level factors to be studied in eight experiments.

As another example, consider the $L_{36}(2^{11} \times 3^{12})$ orthogonal array. Clearly, we can expect eleven 2-level columns, twelve 3-level columns and a total of 36 rows in the orthogonal array. This would then allow up to eleven 2-level factors and up to twelve 3-level factors to be studied in a total of 36 experiments.

3.1.5 Comparing degrees of freedom
In most considerations of orthogonal arrays, the number of degrees of freedom of an orthogonal array is equal to the degrees of freedom of the factor levels in that orthogonal array. We illustrate this with a few examples.

Example 1

The $L_4(2^3)$ orthogonal array requires four experiments to study three 2-level factors.

ν_{OA} = Degrees of freedom of orthogonal array
    = number of experiments $-$ 1
    = 4 $-$ 1 = 3
ν_{fl} = Degrees of freedom of factor levels
    = 3 $\times$ (2 $-$ 1) = 3

since there are three factors; each with 1 degree of freedom.
Hence, $\nu_{OA} = \nu_{fl}$

Example 2

The $L_{36}(2^{11} \times 3^{12})$ orthogonal array requires 36 experiments to study eleven 2-level factors and twelve 3-level factors.

ν_{OA} = Degrees of freedom of orthogonal array
    = number of experiments $-$ 1
    = 36 $-$ 1 = 35
ν_{fl} = Degrees of freedom of factor levels
    = [11 $\times$ (2 $-$ 1)] + [12 $\times$ (3 $-$ 1)] = 35.

since there are eleven factors with 1 degree of freedom and twelve factors with 2 degrees of freedom.
Again, $\nu_{OA} = \nu_{fl}$.
An apparent exception to this rule is shown in the next example.

Example 3

The $L_{18}(2^1 \times 3^7)$ orthogonal array requires 18 experiments to study one 2-level factor and seven 3-level factors.

ν_{OA} = Degrees of freedom of the orthogonal array
    = number of experiments $-$ 1
    = 18 $-$ 1 = 17
ν_{fl} = Degrees of freedom of factor levels
    = [1 $\times$ (2 $-$ 1)] + [7 $\times$ (3 $-$ 1)] = 15.

since there are one factor with 1 degree of freedom and seven factors with 2 degrees of freedom.

Here, $\nu_{OA} \neq \nu_{fl}$ and it appears that the degrees of freedom calculated by the number of factor levels is less than the degrees of freedom calculated for the orthogonal array. However, this is because the $L_{18}(2^1 \times 3^7)$ orthogonal array has a special property where two degrees of freedom are taken up between a 2-level factor and a 3-level factor. If these degrees of freedom are included in the degrees of freedom for factor levels, the degrees of freedom for the orthogonal array is equal to the degrees of freedom for the factor levels.

3.1.6 Matching degrees of freedom for experimentation
In order to use a standard orthogonal array the degrees of freedom of factors and levels must be matched with the degrees of freedom for that orthogonal array. For the $L_8(2^7)$ orthogonal array:

ν_{OA} = Degrees of freedom of orthogonal array
    = number of experiments − 1
    = 8 − 1 = 7 and
ν_{fl} = Number of factors × Degrees of freedom of factor levels
    = Number of factors × (number of levels − 1)
    = 7 × (2 − 1) = 7

    In effect, this means that we cannot study more than seven 2-level factors using the $L_8(2^7)$ orthogonal array. When selecting a suitable orthogonal array for experimentation, we must calculate the number of degrees of freedom for the factors and levels (ν_{fl}) and compare it to the number of degrees of freedom for the orthogonal array (ν_{OA}). In any case, ν_{OA} must be equal to or greater than ν_{fl}. Mathematically, $\nu_{OA} \geq \nu_{fl}$. If however, an experiment does not use up all the degrees of freedom of an orthogonal array, that is, not all the columns in the orthogonal array are used, orthogonality is still preserved and the experiment can be continued. Indeed, it is not necessary to fill up all the columns of an orthogonal array before an experiment can be conducted. If each column of an orthogonal array is assigned to a factor, then the experiment is said to be using a *saturated orthogonal array*. For simplicity, an orthogonal array is usually referred to by the number of experiments. Thus, an $L_8(2^7)$ orthogonal array is usually called an L_8 orthogonal array and an $L_{18}(2^1 \times 3^7)$ orthogonal array is usually called an L_{18} orthogonal array.

3.1.7 Selecting an orthogonal array
Some common orthogonal arrays are shown in Figure 3:1.4. Taguchi has tabulated 18 orthogonal arrays which are called *standard orthogonal arrays*[1]. In many cases, one of these arrays can be used directly to plan a matrix experiment, as shown in the following examples.

Example 1
An experimenter wishes to study seven 2-level control factors. What is a suitable orthogonal array?
    Since a 2-level factor has one degree of freedom, seven 2-level factors require seven degrees of freedom ($\nu_{fl} = 7 \times (2 − 1) = 7$). Thus, the smallest orthogonal array that can be used must have at least seven degrees of freedom. Figure 3:1.4 shows that the smallest common orthogonal array with seven

---

1   Taguchi and Konishi, *Orthogonal Arrays and Linear Graphs: Tools for Quality Engineering*, 1987, ASI Press.

degrees of freedom is the $L_8(2^7)$ orthogonal array. The name of this orthogonal array indicates there will be eight experiments accommodating up to seven 2-level factors.

2 levels	3 levels	4 levels	5 levels	mixed levels
$L_4(2^3)$	$L_9(3^4)$	$L_{16}(4^5)$	$L_{25}(5^6)$	$L_{18}(2^1 \times 3^7)$
$L_8(2^7)$	$L_{27}(3^{13})$	$L_{64}(4^{21})$	-	$L_{32}(2^1 \times 4^9)$
$L_{12}(2^{11})$	$L_{81}(3^{40})$	-	-	$L_{36}(2^{11} \times 3^{12})$
$L_{16}(2^{15})$	-	-	-	$L_{36}(2^3 \times 3^{13})$
$L_{32}(2^{31})$	-	-	-	$L_{54}(2^1 \times 3^{25})$
$L_{64}(2^{63})$	-	-	-	$L_{50}(2^1 \times 5^{11})$

Figure 3:1.4 Common orthogonal arrays.

Example 2

An experimenter wishes to study one 2-level factor and six 3-level control factors. What is a suitable orthogonal array?

Since a 2-level factor has one degree of freedom and a 3-level factor has two degrees of freedom, one 2-level factor and six 3-level control factors require 13 degrees of freedom. Mathematically,

$$\nu_{fl} \quad = 1 \times (2 - 1) + 6 \times (3 - 1) = 13$$

From Figure 3:1.4, the smallest orthogonal array with at least 13 degrees of freedom is the $L_{16}(2^{15})$ orthogonal array. Although this orthogonal array can accommodate fifteen 2-level factors, it cannot accommodate any 3-level factors directly. Therefore, we cannot use the standard $L_{16}(2^{15})$ orthogonal array. The next possibility is the $L_{18}(2^1 \times 3^7)$ orthogonal array. This can accommodate one 2-level and seven 3-level factors. It is possible to assign a 2-level factor to the 2-level column and six 3-level factors to six of the seven 3-level columns, leaving one 3-level column unassigned. The orthogonality of an orthogonal array experiment is not lost by keeping one or more empty column. So, an $L_{18}(2^1 \times 3^7)$ orthogonal array is a good choice for this experiment.

By a similar reasoning, it would be possible to suggest an $L_{36}(2^{11} \times 3^{12})$ orthogonal array for the experiment. However, to do so would result in inefficient experimentation since 36 experiments would have to be conducted. In general, the experimenter should seek the smallest orthogonal array for an experiment.

3.1.8 Self-assessment questions

1. Explain what is meant by the term degrees of freedom.

2. Illustrate by an example that for three measurements, X, Y and Z, we can only make two fair comparisons.

3. An experimenter wishes to study one 2-level factor (A), and five 3-level factors (B, C, D, E and F). Calculate the total degrees of freedom required for this experiment.

4. Describe the orthogonal array $L_{36}(2^3 \times 3^{13})$. Calculate the number of factors and levels you can study with this orthogonal array.

3.1.9 Answers to self-assessment questions

1.   Explain what is meant by the term degrees of freedom.

Answer

    The degree of freedom is the number of fair comparisons that can be made from an experiment. The more degrees of freedom in an experiment, the more information there is in the experiment.

2.   Illustrate by an example that for three measurements, X, Y and Z, we can only make two fair comparisons.

Answer

    Suppose the first fair comparison is $(X - Y)$. The second fair comparison is $(X - Z)$. $Z - Y$ (or $-[Y - Z]$) is not a fair comparison as it can be obtained by comparing the first and second fair comparisons as follows:

$(X - Y) - (X - Z) = X - Y - X + Z = Z - Y.$

Hence, only two fair comparisons can be made from three measurements.

3.   An experimenter wishes to study one 2-level factor (A), and five 3-level factors (B, C, D, E and F). Calculate the total degrees of freedom required for this experiment.

Answer

    The answer may be tabulated as follows:

Factor	Number of factors and levels	Degrees of freedom
A	one 2-level factor	$1 \times (2 - 1)$
B, C, D, E and F	five 3-level factors	$5 \times (3 - 1)$
Total degrees of freedom		11

4.   Describe the orthogonal array $L_{36}(2^3 \times 3^{13})$. Calculate the number of factors and levels you can study with this orthogonal array.

Answer

    There are 3 factors at 2 levels; degrees of freedom   =   3
    There are 13 factors at 3 levels; degrees of freedom   = 26
    Total degrees of freedom   = 29

The $L_{36}(2^3 \times 3^{13})$ has 35 degrees of freedom. Three of these are taken up by three 2-level factors and 26 of these are taken up by thirteen 3-level factors. The remaining six $(35 - 3 - 26)$ are distributed between the 2-level and 3-level columns. The total number of experiments is 36. Therefore, the $L_{36}(2^3 \times 3^{13})$ orthogonal array can be used to study a maximum of three 2-level factors and thirteen 3-level factors in a total of 36 experiments.

3.2   Interactions

3.2.1 Introduction
When the effect of one factor depends on the level of another factor, an *interaction* is said to exist. In other words, an interaction occurs when the collective effect of two (or more) factors taken together is different from the sum of each of the factors taken individually. When such an effect is strong, it becomes difficult for an engineer to predict the effect of a factor selection. This is a disadvantage in industrial experimentation. This chapter will show one method of treating an interaction by assigning the interaction to an orthogonal array column.

3.2.2 Interactions
Consider the following experiment in which a person takes a combination of medicines as shown in Figure 3:2.1. There are two factors, Medicine A and Medicine B. Suppose each of them can be administered at two levels: level 1, *do not take medicine*, or level 2, *take medicine*:

	Level 1	Level 2
Medicine A	Do not take	Take
Medicine B	Do not take	Take

Figure 3:2.1 Factor assignment of Medicines A and B.

The assignment of the experiment (Exp) is shown in Figure 3:2.2 and the results are based on the person's feeling of well-being, on a scale of 0 (feeling ill) to 10 (feeling good).

Medicine A	Level 1		Level 2	
Medicine B	Level 1	Level 2	Level 1	Level 2
Exp	1	2	3	4
Feeling	0	5	5	10

Figure 3:2.2 Interaction experiment.

This experiment can be represented in the $L_4(2^3)$ orthogonal array shown in Figure 3:2.3. The rows numbered 1 to 4 represent experiments. The columns labelled

1, 2 and 3 represent columns which can be assigned to factors. In this case column 1 is assigned to Medicine A and column 2 is assigned to Medicine B. Column 3 is left unassigned. However, this column is used to estimate the interaction effect between columns 1 and 2, i.e. the interaction between A and B. This is represented as $A \times B$ and simply means the $A \times B$ *interaction*. The $\times$ sign here, does *not* mean *A times B*.

Exp	Columns		
	1	2	3
1	1	1	1
2	1	2	2
3	2	1	2
4	2	2	1

Figure 3:2.3 The $L_4(2^3)$ orthogonal array.

      Notice that column 3 (Figure 3:2.3) has a unique relationship to columns 1 and 2. That is, when the factor levels of columns 1 and 2 are the same, then column 3 has a level 1 otherwise a level 2. This is represented as:

1, 1 → 1
1, 2 → 2
2, 1 → 2
2, 2 → 1

      Figure 3:2.4 shows the factor assignment and includes the results of the experiment. It also shows that the effect of taking both medicines together is the sum (10) of the effects of taking the medicines individually (5 and 5). The reader should note that $A \times B$ at level 1 is denoted $A \times B1$ and implies $(A \times B)1$ and not $A \times (B1)$, etc. Similarly, $A \times B$ at level 2 is denoted $A \times B2$.

      The effects of factors A and B and the interaction $A \times B$ (Figure 3:2.4) are shown in the response table in Figure 3:2.5. An explanation on the calculation of the response table is as follows:

- effects of factor A,
- effects of factor B,
- effects of factor $A \times B$.

Exp	1	2	3	Results
	A	B	A×B	
1	1	1	1	0.0
2	1	2	2	5.0
3	2	1	2	5.0
4	2	2	1	10.0

Figure 3:2.4 Interaction experiment.

	A	B	A×B		B1	B2
Level 1	2.50	2.50	5.00	A1	0.00	5.00
Level 2	7.50	7.50	5.00	A2	5.00	10.00
Difference	5.00	5.00	0.00	-	-	-
Rank	1	1	3	-	-	-

Figure 3:2.5 Response table with interaction breakdown.

1.   **Effects of factor A**

Mean effect of A at level 1

$\overline{A1}$ = mean effect of Experiment numbers 1 and 2
         = (0 + 5) / 2 = 2.50

Mean effect of A at level 2

$\overline{A2}$ = mean effect of Experiment numbers 3 and 4
         = (5 + 10) / 2 = 7.50

2.   **Effects of factor B**

Mean effect of B at level 1

$\overline{B1}$ = mean effect of Experiment numbers 1 and 3
         = (0 + 5) / 2 = 2.50

Mean effect of B at level 2

$\overline{B2}$ = mean effect of Experiment numbers 2 and 4
         = (5 + 10) / 2 = 7.50

3.   Effects of factor $A \times B$

Mean effect of $A \times B$ at level 1

$\overline{A \times B1}$ = mean effect of Experiment numbers 1 and 4
                 = (0 + 10) / 2 = 5.00

Mean effect of $A \times B$ at level 2

$\overline{A \times B2}$ = mean effect of Experiment numbers 2 and 3
                 = (5 + 5) / 2 = 5.00

The corresponding response graph is shown in Figure 3:2.6.

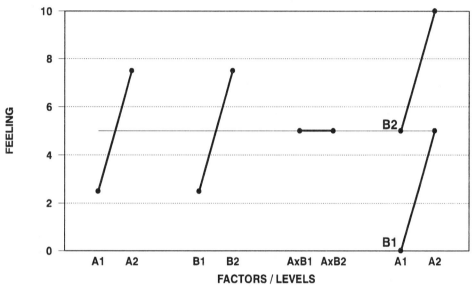

Figure 3:2.6   Response graph of the medicine experiment.

From this figure, we can see that the average effect of both $A \times B$ level 1 and $A \times B$ level 2 is 5. In other words, column 3 has no effect. Since column 3 from the $L_4(2^3)$ represents the interaction of columns 1 and 2, it is clear that the interaction effect is zero. When the interaction effect is zero, we say that factors A and B are *additive*. After all, we did start by saying that the individual effect of each medicine was a feeling of 5 and that the combined effect of taking both medicines together was a feeling of 10.

Another way of showing interaction is the *interaction breakdown* included in the right side of Figure 3:2.6 where the average effects of one factor level (e.g. B1 and B2) are drawn across the average effects of another factor level (e.g. A1 and A2). Here, we want to look at the average effect of B1 across A1 and A2 and the average effect of B2 across A1 and A2. Since the average effect of A1B1 = 0 and A2B1 = 5, we draw a corresponding line and mark it B1. Similarly, since the average effect of A1B2 = 5 and A2B2 = 10, we draw a corresponding line and mark it B2. Notice that when there is no interaction (i.e. average effect of A×B1 = 5 and A×B2 = 5) then the *interaction breakdown is parallel.* This is an important feature of interaction. Using this concept of interaction breakdown we can extend the idea of interaction.

If the interaction breakdown merely shows the interaction (e.g. A×B) effect in another way, why do we need it? Is not the interaction effect enough? The answer to this question is that the interaction effect does not show the breakdown of each factor level combination (e.g. A1B1, A2B2, etc.). While the interaction effect shows if an interaction effect is significant or not, it still takes the interaction breakdown to show which factor level combination has the highest or lowest value. Often when interaction is studied, we need the interaction breakdown to establish the highest or lowest factor level effect.

Suppose we performed the same experiment but obtained the results shown in Figure 3:2.7. The corresponding response table and graph are shown in Figures 3:2.8 and 3:2.9, respectively. Referring to Figure 3:2.8, the mean effect of factor A is 2.5 and there is no difference between A1 and A2. Similarly, the mean effect of factor B is 2.5 and there is no difference between B1 and B2. The mean effect of A×B, however, shows an important effect.

Exp	A	B	A×B	Results
1	1	1	1	0.0
2	1	2	2	5.0
3	2	1	2	5.0
4	2	2	1	0.0

Figure 3:2.7 Interaction experiment.

	A	B	A×B		B1	B2
Level 1	2.50	2.50	0.00	A1	0.00	5.00
Level 2	2.50	2.50	5.00	A2	5.00	0.00
Difference	0.00	0.00	5.00	-	-	-
Rank	3	3	1	-	-	-

Figure 3:2.8 Response table with interaction breakdown.

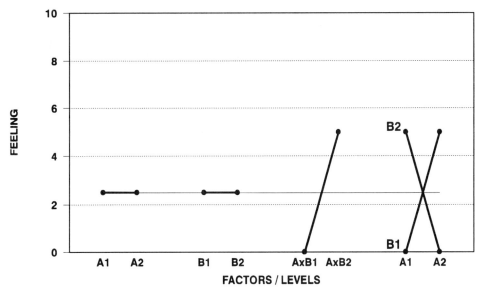

Figure 3:2.9 Response graph of the medicine experiment.

Remembering that:
A×B1 → A1B1 (take neither) and A2B2 (take both),
A×B2 → A1B2 (take B alone) and A2B1 (take A alone),
and noting that A×B2 represents A1B2 and A2B1, we see that the average effect of
A×B2 is 5. This is acceptable since the effect of either medicine alone was stated to
be a feeling of 5.

Noting that A×B1 represents A1B1 and A2B2, we can see that an effect of zero
for A1B1 is acceptable since neither medicine is taken. However, A2B2 shows that if
both medicines are taken together then the effect is also 0. In other words, the effects
of the two medicines have cancelled each other out. This is the direct consequence of
interaction. Interaction causes non-additivity, i.e., we cannot add the effect of Medicine
A to the effect of Medicine B. Non-additivity means that we cannot know the combined
effect of two factors together although we know the individual effects.

Another way of showing interaction is the interaction breakdown, as included
on the right of Figure 3:2.9, where the effects of both factor levels (e.g. B1 and B2)
are drawn across the effect of another factor level (e.g. A1 and A2). Here, we want to
look at the effect of B1 across A1 and A2. Since A1B1 = 0 and A2B1 = 5 we draw
a corresponding line and mark it B1. Similarly, since A1B2 = 5 and A2B2 = 0 we
draw a corresponding line and mark it B2. Notice that when there is full interaction,
the *interaction breakdown is diagonal*. This is an important feature of interaction.

In the medicine experiment above, we have chosen to show a clear example of
additivity against a clear example of non-additivity. In reality, factor effects are seldom
purely additive or purely non-additive. When an interaction effect is found to be zero,

this would suggest that that column could have been used to represent another factor, say C, in the first place. Doing so would enable us to study three factors for the same number of experiments and thus improve experimental efficiency. But the inherent difficulty is that there is no way of knowing whether a third factor can be substituted into an interaction column until the experiment has been conducted with only two factors. And then again, how would an engineer know if the factor C itself did not have an interaction with A or B or both?

So how does an engineer overcome this problem?

Normally, we would substitute an interaction column (column 3 in this example) with a factor such as C. An orthogonal array in which all columns are substituted with factors is called a *saturated orthogonal array design*. If, as in this case, an interaction effect existed between factors A and B then that effect (A×B) would be *confounded* with the effect of factor C. Confounded means that the effects of factors A and B would be *mixed up inseparably* with that of factor C. Thus, if the interaction effect is not zero, we cannot estimate the main effect due to C. To overcome this problem, we do three things:

- predict the quality characteristic,
- conduct a confirmation experiment,
- compare prediction versus confirmation.

1.   Predict the quality characteristic

Using the results of the orthogonal array experiment, we calculate the predicted value of the quality characteristic $\mu_{Predicted}$ based on the mean effects of factor levels.

2.   Conduct a confirmation experiment

The confirmation experiment is of the utmost importance in the design of an experiment. The confirmation experiment is conducted at the optimum factor level settings and the quality characteristic $\mu_{Confirmation}$ at the optimum settings is obtained.

3.   Compare prediction versus confirmation

The crux of a design of experiment is in the comparison of $\mu_{Predicted}$ against $\mu_{Confirmation}$. When comparing $\mu_{Predicted}$ against $\mu_{Confirmation}$ good agreement may be assumed if the confirmation value is within ± 5 % of the predicted value. If there is a good agreement between the predicted value and the confirmation experiment then additivity is present and interaction effects cannot be dominant. Hence, there will be good reproducibility of small scale experiments to large scale production and the experimenter may implement the optimum condition on a large scale. If there is a poor agreement between the predicted value and the confirmation experiment then additivity is not present, in which case, interaction effects may be dominant. Hence, there will not be good reproducibility of small scale experiments to large scale production and the experimenter should not implement the predicted optimum conditions on a large scale.

3.2.3 The purpose of a confirmation experiment

The primary purpose of a confirmation experiment is to warn when there are strong interaction effects. When interactions effects are strong, additivity is poor and the reproducibility of experimental results is consequently also poor. If the predicted optimum conditions deduced from orthogonal array experimental results are not confirmed by a confirmation experiment, it is possible that laboratory optimization is inadequate for industrial or customer usage (downstream) conditions. Thus, it is imperative that a confirmation experiment be conducted to ensure reproducibility and thereby prevent faulty process and product designs from going downstream.

3.2.4 Interaction table

In most Robust Design experiments, we prefer not to estimate any interaction effects among control factors. Indeed, we usually treat interactions as noise and try to identify control factors that are insensitive to noise. This is because, unless control factors are strong enough to overcome their interaction effects, they will not be of much use for Robust Design anyway. But there may be situations where certain interaction effects have to be considered. In such cases the interaction table is used to assign interactions. Figure 3:2.10 shows an $L_8(2^7)$ orthogonal array. Figure 3:2.11 shows the corresponding interaction table.

	1	2	3	4	5	6	7
1	1	1	1	1	1	1	1
2	1	1	1	2	2	2	2
3	1	2	2	1	1	2	2
4	1	2	2	2	2	1	1
5	2	1	2	1	2	1	2
6	2	1	2	2	1	2	1
7	2	2	1	1	2	2	1
8	2	2	1	2	1	1	2

Figure 3:2.10   The $L_8(2^7)$ orthogonal array.

Column	Column						
	1	2	3	4	5	6	7
1	1	3	2	5	4	7	6
2		2	1	6	7	4	5
3			3	7	6	5	4
4				4	1	2	3
5					5	3	2
6						6	1
7							7

Figure 3:2.11   The $L_8(2^7)$ interaction table.

The interaction table shows the columns in which an interaction is confounded with (or contained in) every pair of columns in the $L_8(2^7)$ orthogonal array. Thus, the interaction table can be used to determine which column of the $L_8(2^7)$ orthogonal array should be left unassigned to a factor in order to estimate a particular interaction. Two examples of interaction assignments now follow:

Example 1
Suppose factor A is assigned to column 1 and factor B is assigned to column 2. From Figure 3:2.11, we see that the interaction of columns 1 and 2 is confounded with column 3 (intersection of row 1 and column 2). Therefore, if we wish to study the interaction of factors A and B, we should leave column 3 unassigned. This will enable the analysis of the A×B interaction effect from column 3. In experimental terms, column 1 is assigned to A, column 2 is assigned to B and column 3 is left unassigned to any other factor.

Example 2
If another factor (D) is to be added to the experiment, using Figure 3:2.11, the factor may be assigned to column 4, in which case the interaction of columns 1 (factor A) and 4 (factor D) will be given in column 5 (intersection of row 1 and column 4). Hence, if it is necessary to establish the effect of the (A×D) interaction then column 5 must be left unassigned. Similarly, the interaction of columns 2 (factor B) and 4 (factor D) is given in column 6 (intersection of row 2 and column 4). Hence, if it is necessary to establish the interaction effect B×D then column 6 must be left unassigned, and so on. If however, we substitute a factor (E) into column 5 (which would contain the A×D interaction effect) then the effect of factor E will be confounded with the effect of A×D

and the result will be [effect of factor E + effect of A×D interaction]. Unless the effect of A×D interaction = 0, we cannot establish the effect of E. For this reason it is imperative to conduct a confirmation experiment.

Note that the interaction between columns a (e.g. 1) and b (e.g. 2) is the same as that between columns b (2) and a (1). That is, the interaction table is a symmetric matrix. Hence, only the upper triangle is given in the interaction table. Also, the diagonal terms in bold indicate that there is no real meaning to an interaction between a column and itself. The interaction table contains all the relevant information needed for assigning factors to columns of the orthogonal array so that all main effects and desired interactions can be estimated without confounding. The interaction tables for all standard orthogonal arrays have been prepared by Taguchi.

3.2.5 Linear graphs
Another engineering tool frequently used in Robust Design is the linear graph[2]. The linear graph represents factor and interaction assignments in diagrammatic form. The linear graphs for the $L_8(2^7)$ orthogonal array are shown in Figure 3:2.12.

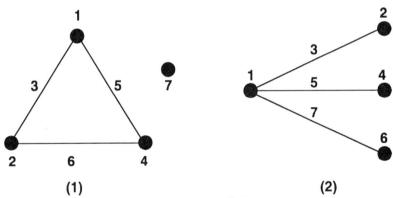

Figure 3:2.12   Linear graphs for the $L_8(2^7)$ orthogonal array.

A linear graph is a series of numbered lines and dots which have a one-to-one correspondence to the columns of the related orthogonal array. Each linear graph is associated with one orthogonal array. However, a given orthogonal array can have several linear graphs. The linear graphs facilitate the assignment of factors to specific columns of the orthogonal array. Linear graphs represent the factor and interaction

2   Taguchi and Konishi, *Orthogonal Arrays and Linear Graphs: Tools for Quality Engineering*, 1987, ASI Press.

information graphically and make it easy to assign factors and interactions to the various columns of an orthogonal array.

In a linear graph, the columns of an orthogonal array are represented by dots and lines. When two dots are connected by a line, it means that the interaction of the two columns represented by the dots is contained in the column represented by the line. Each dot and each line has a distinct column number(s) associated with it. Further, every column of the array is represented in its linear graph once and only once.

The $L_8(2^7)$ orthogonal array has two linear graphs (1) and (2), as shown in Figure 3:2.12. In (1), there are four dots corresponding to columns 1, 2, 4, and 7. It also has three lines representing columns 3, 5, and 6. These lines correspond to the interactions between columns 1 and 2, between columns 1 and 4, and between columns 2 and 4, respectively. Using the interaction table, we can verify that the interaction of columns 1 and 2 is given in 3, columns 1 and 4 is given in 5 and columns 2 and 4 is given in 6. Thus, the linear graph (1) can be verified to correspond to the interactions given in Figure 3:2.11. Linear graph (2) can also be verified from the interaction table, Figure 3:2.11. Although in principle we can always derive a linear graph from a given interaction table, the linear graphs provide readily available templates for comparing different experimental configurations, as shown in Figure 3:2.12, (1) and (2) above.

In general, a linear graph does not show the interactions between every pair of columns of the orthogonal array. An interaction table, however, shows the interaction relationship between any two columns but not the specific relations of columns. Also, a given orthogonal array can have only one interaction table but many linear graphs. All the linear graphs of a particular array can be interpreted from the interaction table.

3.2.6 Degrees of freedom of interactions

The number of degrees of freedom of an interaction of two factors is the product of the degrees of freedom of each factor. Let n_A and n_B be the number of levels of factors A and B. Then there are $n_A n_B$ total combinations of the levels of these two factors. From that, one degree of freedom is subtracted for the overall mean, $(n_A - 1)$ for the degree of freedom for A, and $(n_B - 1)$ for the degree of freedom for B. Mathematically,

$$v_{A \times B} = n_A n_B - 1 - (n_A - 1) - (n_B - 1)$$

$$= n_A n_B - n_A - n_B + 1$$

$$= (n_A - 1) \times (n_B - 1)$$

so that the number of degrees of freedom for two interacting factors is simply the product of their degrees of freedom. We follow with two examples of the calculation of degrees of freedom:

Example 1
Factor A, temperature, is studied at two levels, say 100 °C and 120 °C. Factor B, time, is studied at three levels, say 10 minutes, 5 minutes and 1 minute. How many degrees of freedom are there for the interaction of factors A and B?

$$v_A = (2 - 1) = 1$$

$$v_B = (3 - 1) = 2$$

$$\therefore v_{A \times B} = v_A \times v_B = 1 \times 2 = 2$$

Example 2
An engineer wishes to study one 2-level factor (A), five 3-level factors (B, C, D, E and F) and the interaction A×B. How many degrees of freedom are required for this experiment?
In this case, it may be convenient to tabulate the degrees of freedom as follows:

Factor / Interaction	Degrees of freedom	Total
A	$2 - 1$	1
B, C, D, E and F	$5 \times (3 - 1)$	10
A×B	$(2 - 1) \times (3 - 1)$	2
Total degrees of freedom		13

Thus, the total degrees of freedom required for the experiment is 13.

3.2.7 Assigning factors to an orthogonal array
After establishing the required degrees of freedom for experimentation, the next step is to assign the factors and interaction to an orthogonal array. The following is a guide to using an orthogonal array, interaction table or linear graph to assign control factors or interactions in an experiment:
1.   Count the total degrees of freedom needed for the experiment based on the number of factors and the factor levels. Note the numbers of 2-level factors, 3-level factors and interactions in the experiment.
2.   Select an orthogonal array that has at least the required number of degrees of freedom. Match the numbers of 2-level and 3-level columns in the orthogonal array to the numbers of 2-level and 3-level factors in the experiment. To avoid an unnecessarily complex factor assignment, it may be necessary to fit the

numbers of 2-level and 3-level factors in the experiment to the numbers of 2-level and 3-level columns in the orthogonal array.

3.   Draw the required linear graph. Remember that factors are assigned to points, and an interaction between two factors is assigned to the line segment connecting the two corresponding points. If the interaction between two factors is not to be determined, then a factor may be assigned to the corresponding line segment.

4.   Select an appropriate standard linear graph. There may be many choices. Decide on the most suitable one.

5.   Fit the required linear graph to one of the standard linear graphs of the orthogonal array selected.

6.   Assign each main effect and interaction to the appropriate column.

3.2.8 Factor assignment
We shall now work through two examples of factor assignment:
- $L_8(2^7)$ orthogonal array,
- $L_{16}(2^{15})$ orthogonal array.

1. $L_8(2^7)$ orthogonal array
Suppose an engineer wishes to study the 2-level factors A, B, C, D and the interactions A×B and B×C.

1.   The total degrees of freedom required:

Four 2-level factors $= 4 \times (2 - 1)$ $= 4$

Two interactions $= 2 \times (2 - 1) \times (2 - 1)$ $= 2$

2.   Thus, the total degrees of freedom $= 6$. Therefore, an $L_8(2^7)$ orthogonal array, which has seven degrees of freedom, should be sufficient.

3.   Draw the required linear graph. This is shown in Figure 3:2.13.

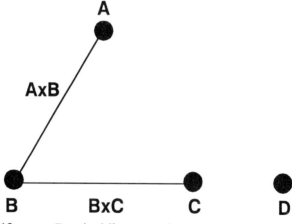

Figure 3:2.13   Required linear graph.

4.   Referring to a table of linear graphs, choose a standard linear graph that
        closely resembles the required linear graph. This is shown in Figure
        3:2.14.

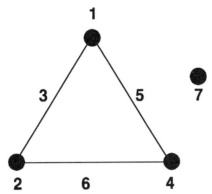

Figure 3:2.14   One of the two possible standard linear graphs.

5.   Assign A → 1, B → 2, A×B → 3, C → 4, e → 5, B×C → 6 and D → 7.
        The empty column is usually denoted as 'e' representing error.
6.   Assign factors and interactions to the orthogonal array. This is shown in
        Figure 3:2.15.

Exp	1	2	3	4	5	6	7
	A	B	A×B	C	e	B×C	D
1	1	1	1	1	1	1	1
2	1	1	1	2	2	2	2
3	1	2	2	1	1	2	2
4	1	2	2	2	2	1	1
5	2	1	2	1	2	1	2
6	2	1	2	2	1	2	1
7	2	2	1	1	2	2	1
8	2	2	1	2	1	1	2

Figure 3:2.15 $L_8(2^7)$ orthogonal array with factor assignment.

2.   **L₁₆(2¹⁵) orthogonal array**

An experiment requires eight 2-level factors: A, B, C, D, E, F, G and H, and 5 interaction effects, A×B, B×C, B×D, A×D and C×E. Assign the factors and interactions to a suitable orthogonal array.

1.   The total degrees of freedom required:

Eight 2-level factors $= 8 \times (2 - 1)$ $= 8$

Five interactions $= 5 \times (2 - 1) \times (2 - 1)$ $= 5$

2.   Thus, the total degrees of freedom = 13. Hence, an $L_{16}(2^{15})$ orthogonal array, which has 15 degrees of freedom, should be sufficient.

3.   Draw the required linear graph. This is shown in Figure 3:2.16.

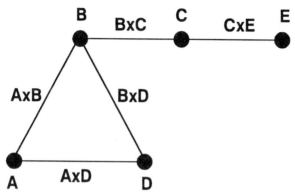

Figure 3:2.16   Required linear graph.

4.   Referring to a table of linear graphs, choose a standard linear graph that closely resembles the required linear graph. This is shown in Figure 3:2.17.

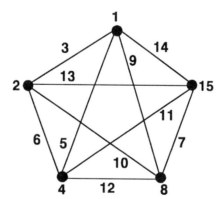

Figure 3:2.17   One of the six possible linear graphs for $L_{16}(2^{15})$.

5.   The required linear graph is then matched to the standard linear graph. Where interactions are not needed, those lines can be assigned to factors as shown in Figure 3:2.18.

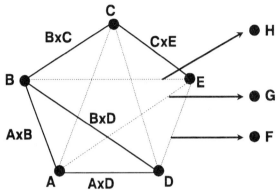

Figure 3:2.18   Matching the linear graphs.

6.   Assign factors and interactions to the orthogonal array. The factor assignment is shown in Figure 3:2.19. Assign C → 1, B → 2, B×C → 3, A → 4, e → 5, A×B → 6, F → 7, D → 8, e → 9, B×D → 10, G → 11, A×D → 12, H → 13, C×E → 14 and E → 15. Empty (or unassigned) columns are denoted as 'e' representing error. For convenience of reference, the interaction table is shown in Figure 3:2.20.

	1	2	3	4	5	6	7	8	9	10	11	12	13	14	15
	C	B	B×C	A	e	A×B	F	D	e	B×D	G	A×D	H	C×E	E
1	1	1	1	1	1	1	1	1	1	1	1	1	1	1	1
2	1	1	1	1	1	1	1	2	2	2	2	2	2	2	2
3	1	1	1	2	2	2	2	1	1	1	1	2	2	2	2
4	1	1	1	2	2	2	2	2	2	2	2	1	1	1	1
5	1	2	2	1	1	2	2	1	1	2	2	1	1	2	2
6	1	2	2	1	1	2	2	2	2	1	1	2	2	1	1
7	1	2	2	2	2	1	1	1	1	2	2	2	2	1	1
8	1	2	2	2	2	1	1	2	2	1	1	1	1	2	2
9	2	1	2	1	2	1	2	1	2	1	2	1	2	1	2
10	2	1	2	1	2	1	2	2	1	2	1	2	1	2	1
11	2	1	2	2	1	2	1	1	2	1	2	2	1	2	1
12	2	1	2	2	1	2	1	2	1	2	1	1	2	1	2
13	2	2	1	1	2	2	1	1	2	2	1	1	2	2	1
14	2	2	1	1	2	2	1	2	1	1	2	2	1	1	2
15	2	2	1	2	1	1	2	1	2	2	1	2	1	1	2
16	2	2	1	2	1	1	2	2	1	1	2	1	2	2	1

Figure 3:2.19   The $L_{16}(2^{15})$ orthogonal array.

1	2	3	4	5	6	7	8	9	10	11	12	13	14	15
1	3	2	5	4	7	6	9	8	11	10	13	12	15	14
	2	1	6	7	4	5	10	11	8	9	14	15	12	13
		3	7	6	5	4	11	10	9	8	15	14	13	12
			4	1	2	3	12	13	14	15	8	9	10	11
				5	3	2	13	12	15	14	9	8	11	10
					6	1	14	15	12	13	10	11	8	9
						7	15	14	13	12	11	10	9	8
							8	1	2	3	4	5	6	7
								9	3	2	5	4	7	6
									10	1	6	7	4	5
										11	7	6	5	4
											12	1	2	3
												13	3	2
													14	1
														15

Figure 3:2.20   The $L_{16}(2^{15})$ interaction table.

3.2.9 Optimum factor selection for interactions

Suppose an engineer performed an experiment on the factors A, B, C, D and E, and the interactions A×B and A×C for a smaller-the-better characteristic. The results of the experiment in arbitrary units are shown in Figure 3:2.21.

Exp	A	B	A×B	C	A×C	D	E	Results
1	1	1	1	1	1	1	1	2.23
2	1	1	1	2	2	2	2	3.25
3	1	2	2	1	1	2	2	2.10
4	1	2	2	2	2	1	1	1.93
5	2	1	2	1	2	1	2	3.28
6	2	1	2	2	1	2	1	2.60
7	2	2	1	1	2	2	1	3.18
8	2	2	1	2	1	1	2	4.10

Figure 3:2.21   The $L_8(2^7)$ orthogonal array.

The mean response for each factor and interaction level, i.e. A, B, A×B, C, A×C, D and E are given in the response table in Figure 3:2.22.

	A	B	A×B	C	A×C	D	E
Level 1	2.38	2.84	3.19	2.70	2.76	2.88	2.48
Level 2	3.29	2.83	2.48	2.97	2.91	2.78	3.18
Difference	0.91	0.01	0.71	0.27	0.15	0.10	0.70
Rank	1	7	2	4	5	6	3

Figure 3:2.22   Response table of factor effects.

Since an $L_8(2^7)$ orthogonal array has seven degrees of freedom, we could take about half the degrees of freedom (say three) as important effects. From the response table, the significant effects are therefore A, A×B, and E. Effects C, A×C, D and B may be regarded as insignificant. The difficulty here is how to choose a factor level for an interaction? We cannot choose a level A×B1 or A×B2 since A×B1 has two combinations (A1B1 and A2B2) and A×B2 also has two combinations (A1B2 and

A2B1). To overcome this problem we need to perform an interaction breakdown. This is done by looking at the columns for factors A and B simultaneously. The effects of factors A and B for the four combinations of factors are as follows:

The average effect of A1B1 is $(2.23 + 3.25)/2 = 2.74$

The average effect of A1B2 is $(2.10 + 1.93)/2 = 2.01$

The average effect of A2B1 is $(3.28 + 2.60)/2 = 2.94$

The average effect of A2B2 is $(3.18 + 4.10)/2 = 3.64$

This result is best tabulated as in Figure 3:2.23.

Interaction	B1	B2
A1	2.74	2.01
A2	2.94	3.64

Figure 3:2.23   Interaction breakdown.

The response table and the interaction breakdown can now be used to draw the response graph (Figure 3:2.24). Similarly, we draw the interaction breakdown effects of A×C. Comparing the interaction effect of A×B to the AB factor effect, we notice that the lines cross over. Comparing the interaction effect of A×C to the AC factor effect, we notice that the lines are almost parallel.

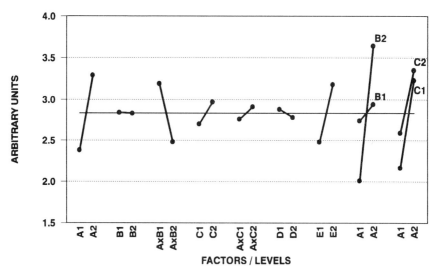

Figure 3:2.24   Response graph of factors and interactions.

Recalling that the quality characteristic is smaller-the-better, the significant factor levels recommended are A1, A1B2 (chosen from the interaction graph) and E1.

3.2.10   Interpretation of AB and A×B

The mean effect of A×B over the experimental average is the mean effect of factors A and B together over the experimental average, minus the mean effect of factor A over the experimental average, minus the mean effect of factor B over the experimental average. Mathematically:

$$(\overline{A \times B} - \bar{y}) = (\overline{AB} - \bar{y}) - (\overline{A} - \bar{y}) - (\overline{B} - \bar{y})$$

$$\therefore (\overline{AB} - \bar{y}) = (\overline{A} - \bar{y}) + (\overline{B} - \bar{y}) + (\overline{A \times B} - \bar{y})$$

If $(\overline{A \times B} - \bar{y}) = 0$, then $(\overline{AB} - \bar{y}) = (\overline{A} - \bar{y}) + (\overline{B} - \bar{y})$ and we say that the factors A and B are additive. Note that $(\overline{A \times B} - \bar{y})$ is the interaction effect whereas $(\overline{AB} - \bar{y})$ is the effect of factors A and B. This can be illustrated by analogy with a practical example. Suppose that Alison (A) alone can lift 100 kg and Brian (B) alone can lift 150 kg. Suppose further that both of them together (AB) can lift 300 kg. Clearly the interaction effect (A×B) is:

$$A \times B = AB - A - B$$

corresponding to

$$A \times B = 300 - 100 - 150 = 50 \text{ kg}$$

3.2.11   Prediction of the process average

The process average $\mu_{Predicted}$ at the optimum condition can now be calculated as the effects of the optimum conditions over the experimental mean $(\bar{y})$. Mathematically,

$$\mu_{Predicted} = \bar{y} + (\overline{A1} - \bar{y}) + \left[(\overline{A1B2} - \bar{y}) - (\overline{A1} - \bar{y}) - (\overline{B2} - \bar{y})\right] + (\overline{E1} - \bar{y})$$

$$= \overline{A1} + \overline{A1B2} - \overline{A1} - \overline{B2} + \overline{E1}$$

$$= \overline{A1B2} - \overline{B2} + \overline{E1}$$

$$= 2.01 - 2.83 + 2.48$$

$$= 1.66 \text{ unit}$$

3.2.12   Sliding levels

Besides assigning interactions to a column in an orthogonal array, there is another method of reducing interaction effects. This method uses sliding levels for interacting factors. Consider, for example, factor A (baking temperature) and factor B (baking time) for a baking process. Obviously, when the baking temperature is high, the baking time needs to be short and when the baking temperature is low, the baking time must be long. Suppose that at 150 °C (A1) the baking times we need to study are 3 hours (B1) and 4 hours (B2), then, at 120 °C (A2) the baking times may be set to 4 hours (B1) and 5 hours (B2). We therefore set the sliding levels as shown in Figure 3:2.25 with arbitrary yield values in parentheses.

		Baking time (B)	
		Level 1	Level 2
Baking temperature (A)	Level 1 150 °C	3 hours (50 %)	4 hours (60 %)
	Level 2 120 °C	4 hours (80 %)	5 hours (90 %)

Figure 3:2.25   Setting sliding levels for factors.

With this set-up, the best time for each temperature may be investigated, and combinations of experiments such as (120 °C and 3 hours) or (150 °C and 5 hours) which are definitely unfavourable are not included in the experiment. The response graph for sliding levels is shown in Figure 3:2.26.

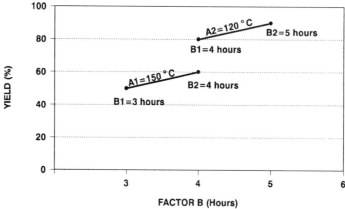

FACTOR B (Hours)

Figure 3:2.26   Interaction effects of sliding levels.

3.2.13   **Self-assessment questions**

1.   Draw the medicine interaction response graph for the following data.

Exp	A	B	A×B	Results
1	1	1	1	0.0
2	1	2	2	5.0
3	2	1	2	5.0
4	2	2	1	7.5

2.   Draw the medicine interaction response graph for the following data.

Exp	A	B	A×B	Results
1	1	1	1	0.0
2	1	2	2	5.0
3	2	1	2	5.0
4	2	2	1	5.0

3.   Draw the medicine interaction response graph for the following data.

Exp	A	B	A×B	Results
1	1	1	1	0.0
2	1	2	2	5.0
3	2	1	2	5.0
4	2	2	1	2.5

4.   Compare these response graphs with the other two interaction graphs you have studied in the text.

5.   An experimenter wishes to study the following 2-level factors and 2-factor interactions:
Factors: A, B, C, D, E, F, G and H.

Two-factor interactions: $A \times B$, $A \times C$, $A \times D$, $A \times E$, $D \times E$, $F \times G$ and $F \times H$.

Using an $L_{16}(2^{15})$ interaction table (see page 110) construct a suitable linear graph for the experiment.

6.   An engineer performed an experiment on the control factors A, B, C, D and E, including the interactions $A \times B$ and $A \times C$, affecting the flash thickness (smaller-the-better) in moulded components. The results of the experiment (in μm) are given in the $L_8(2^7)$ orthogonal array below.

Exp	A	B	A×B	C	A×C	D	E	Results			
1	1	1	1	1	1	1	1	50	48	45	46
2	1	1	1	2	2	2	2	64	64	65	60
3	1	2	2	1	1	2	2	40	46	44	45
4	1	2	2	2	2	1	1	55	52	53	55
5	2	1	2	1	2	1	2	45	42	43	47
6	2	1	2	2	1	2	1	33	32	32	35
7	2	2	1	1	2	2	1	28	24	30	31
8	2	2	1	2	1	1	2	36	33	35	31

1.   Determine the average response for each factor and interaction level and draw a response table. Hint: You need to average the results of each experiment before you proceed with the response table.
2.   Comment on the significant factors and interactions.
3.   Draw the response graph including any significant interaction.
4.   Predict the process average at the optimum condition.

3.2.14   Answers to self-assessment questions

1.   Draw the medicine interaction response graph for the following data.

Exp	A	B	A×B	Results
1	1	1	1	0.0
2	1	2	2	5.0
3	2	1	2	5.0
4	2	2	1	7.5

Answer

The response table with the interaction breakdown is shown below:

	A	B	A×B		B1	B2
Level 1	2.50	2.50	3.75	A1	0.00	5.00
Level 2	6.25	6.20	5.00	A2	5.00	7.50
Difference	3.75	3.75	1.25	-	-	-
Rank	1	1	3	-	-	-

The reponse graph is shown below:

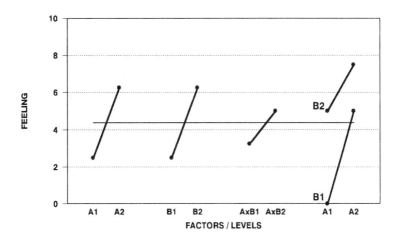

2.   Draw the medicine interaction response graph for the following data.

Exp	A	B	A×B	Results
1	1	1	1	0.0
2	1	2	2	5.0
3	2	1	2	5.0
4	2	2	1	5.0

Answer

The response table with the interaction breakdown is shown below:

	A	B	A×B		B1	B2
Level 1	2.50	2.50	2.50	A1	0.00	5.00
Level 2	5.00	5.00	5.00	A2	5.00	5.00
Difference	2.50	2.50	2.50	-	-	-
Rank	3	3	3	-	-	-

The reponse graph is shown below:

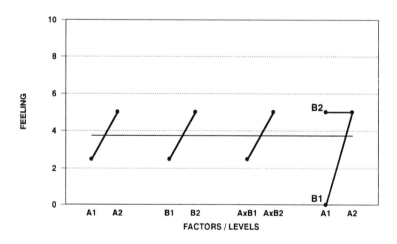

3.   Draw the medicine interaction response graph for the following data.

Exp	A	B	A×B	Results
1	1	1	1	0.0
2	1	2	2	5.0
3	2	1	2	5.0
4	2	2	1	2.5

Answer

The response table with the interaction breakdown is shown below:

	A	B	A×B		B1	B2
Level 1	2.50	2.50	1.25	A1	0.00	5.00
Level 2	3.75	3.75	5.00	A2	5.00	2.50
Difference	1.25	1.25	3.75	-	-	-
Rank	3	3	1	-	-	-

The reponse graph is shown below:

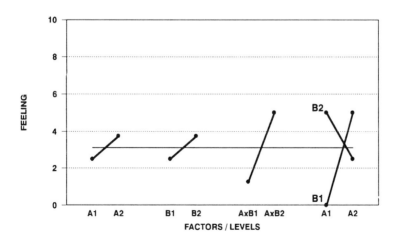

4.   Compare these response graphs with the other two interaction graphs you have studied in the text.

Answer

From the two interaction graphs in the text and the three interaction graphs above, it can be seen that when the factor effects are additive, the interaction effect (in A×B) is zero and the interaction breakdown (A1B1, A2B1 and A1B2, A2B2) is parallel. When the interaction effect is gradually increased (in A×B) the interaction breakdowns (A1B1, A2B1 and A1B2, A2B2) are no longer parallel and eventually become diagonal and additivity is lost completely. See the figures below.

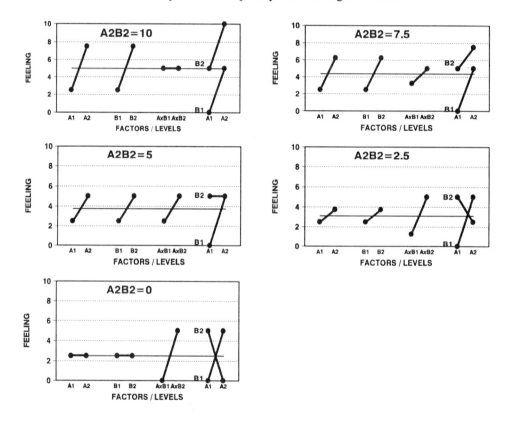

5.   An experimenter wishes to study the following 2-level factors and 2-factor interactions:

Factors: A, B, C, D, E, F, G and H.

Two-factor interactions: A×B, A×C, A×D, A×E, D×E, F×G and F×H. Using an $L_{16}(2^{15})$ interaction table (see page 110) construct a suitable linear graph for the experiment.

Answer

From the statement of the problem, the required linear graph is:

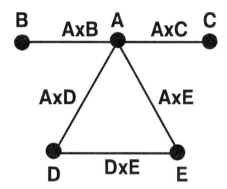 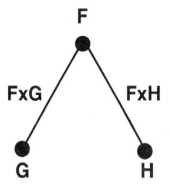

Using an $L_{16}(2^{15})$ interaction table (or a standard linear graph) the factors and interactions can now be assigned as follows:

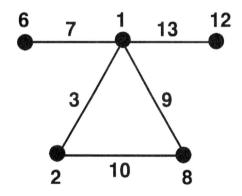 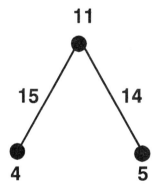

Using the interaction table:

Assign A → 1,

assign D → 2 so that the interaction effect A×D → 3,

assign B → 6 so that the interaction effect A×B → 7,

assign E → 8 so that the interaction effect A×E → 9,

assign C → 12 so that the interaction effect A×C → 13.

Since, D → 2 and E → 8 so the interaction effect D×E → 10.

Assign F → 11,

assign G → 4 so that the interaction effect F×G → 15,

assign H → 5 so that the interaction effect F×H → 14.

6.   An engineer performed an experiment on the control factors A, B, C, D and E, including the interactions A×B and A×C, affecting the flash thickness (smaller-the-better) in moulded components. The results of the experiment (in μm) are given in the $L_8(2^7)$ orthogonal array below.

Exp	A	B	A×B	C	A×C	D	E	Results			
1	1	1	1	1	1	1	1	50	48	45	46
2	1	1	1	2	2	2	2	64	64	65	60
3	1	2	2	1	1	2	2	40	46	44	45
4	1	2	2	2	2	1	1	55	52	53	55
5	2	1	2	1	2	1	2	45	42	43	47
6	2	1	2	2	1	2	1	33	32	32	35
7	2	2	1	1	2	2	1	28	24	30	31
8	2	2	1	2	1	1	2	36	33	35	31

1.   Determine the average response for each factor and interaction level and draw a response table. Hint: You need to average the results of each experiment before you proceed with the response table.

Answer

First, we need to obtain the row averages. This is done by taking the average of the four observations. For experiment 1, the average of the observations is:

$$\bar{y}_1 = \frac{50 + 48 + 45 + 46}{4}$$

$$= 47.25 \ \mu m$$

The averages of experiments 2 to 8 are similarly calculated and entered into the results:

Exp	A	B	A×B	C	A×C	D	E	Results				$\bar{y}$
1	1	1	1	1	1	1	1	50	48	45	46	47.25
2	1	1	1	2	2	2	2	64	64	65	60	63.25
3	1	2	2	1	1	2	2	40	46	44	45	43.75
4	1	2	2	2	2	1	1	55	52	53	55	53.75
5	2	1	2	1	2	1	2	45	42	43	47	44.25
6	2	1	2	2	1	2	1	33	32	32	35	33.00
7	2	2	1	1	2	2	1	28	24	30	31	28.25
8	2	2	1	2	1	1	2	36	33	35	31	33.75

Then the response table is constructed by taking the average of experiments at the appropriate factor levels. For factor A level 1:

$$\overline{A1} = \frac{47.25 + 63.25 + 43.75 + 53.75}{4}$$

$$= 52.00 \ \mu m$$

The completed response table is as follows:

	A	B	A×B	C	A×C	D	E
Level 1	52.00	46.94	43.13	40.88	39.44	44.75	40.56
Level 2	34.81	39.88	43.69	45.94	47.38	42.06	46.25
Difference	17.19	7.06	0.56	5.06	7.94	2.69	5.69
Rank	1	3	7	5	2	6	4

2.   Comment on the significant factors and interactions.

Answer

Since the $L_8(2^7)$ orthogonal array has seven degrees of freedom we can choose about half (say four) factors as significant. The significant factors and interactions are therefore A, A×C, B and E.

3.   Draw the response graph including any significant interaction.

Answer

From the response table, the A×B interaction is insignificant while the A×C interaction is significant. The interaction breakdown for A×B is as follows:

	A1	A2
B1	55.25	38.63
B2	48.75	31.00

The interaction breakdown for A×C is as follows:

	A1	A2
C1	45.50	36.25
C2	58.50	33.38

The response graph including the interaction breakdowns is shown below. Note that the A×B interaction is insignificant and the corresponding interaction breakdown is (practically) parallel. The A×C interaction is significant and the interaction breakdown is not parallel (i.e. intersecting).

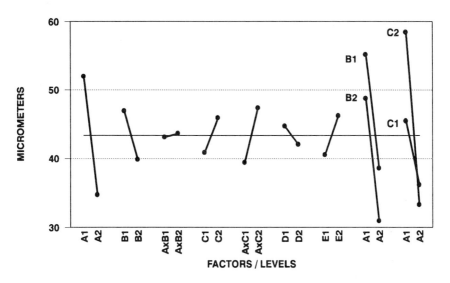

4.   Predict the process average at the optimum condition.

Answer

From the response table and the interaction breakdown, the optimum condition (smaller-the-better) is A2, A2C2, B2 and E1. The predicted process average ($\mu_{Predicted}$) at the optimum condition is therefore:

$$\mu_{Predicted} = \bar{y} + (\overline{A2} - \bar{y}) + \left[(\overline{A2C2} - \bar{y}) - (\overline{A2} - \bar{y}) - (\overline{C2} - \bar{y})\right]$$

$$+ (\overline{B2} - \bar{y}) + (\overline{E1} - \bar{y})$$

$$= \overline{A2C2} - \overline{C2} + \overline{B2} + \overline{E1} - \bar{y}$$

$$= 33.38 - 45.94 + 39.88 + 40.56 - 43.41$$

$$= 24.47 \ \mu m$$

The predicted process average is therefore 24.47 μm.

3.3 Selection of Quality Characteristics

3.3.1 Introduction

A quality characteristic is the object of interest of a product or process. Selecting the correct quality characteristic is important so that the characteristic is both additive and monotonic. In Section 3.2 we saw how interaction between factors can cause the loss of additivity. Losing additivity is synonymous with unpredictable results and hence inefficient experimentation or even wrong conclusions about the optimum levels. Some quality characteristics are intrinsically poor in additivity while others maintain good additive properties in experiments. This section explains the importance of the need for careful selection of quality characteristics.

3.3.2 Selection of quality characteristics

The primary objective of robust design is to select efficiently, robust conditions which are reproducible in manufacturing conditions (See Figure 3:3.1). The selection of quality characteristics, i.e. what to measure, with good additivity is a prerequisite for efficient, reliable and reproducible experimentation. The selection of quality characteristics requires an engineer to draw upon his/her specialized knowledge of the process or product. The selection of a quality characteristics is the most important stage in the design of an experiment. Experimental reliability, the reproducibility of conclusions on process or product quality downstream, in effect, is the evaluation of quality characteristics with respect to additivity.

Small scale Experimentation

Control Factors A, B, C...
Noise Factors: N, P, Q

CONCLUSIONS
[UPSTREAM]

Large Scale Manufacturing

User Conditions
Control Factors A, B, C...
Noise Factors: N, P, Q, R, S, T, U, V, W, X...

REPRODUCIBLE?
[DOWNSTREAM]

Figure 3:3.1 Reproducibility of experiments.

It is difficult to generalize a discussion regarding quality characteristics. However, the following will help to illustrate the selection of characteristics.

1. Experiments using orthogonal arrays designed with a minimum number of interactions or no interactions are recommended.

2. A successful confirmation run indicates that we have selected main effects which have overcome interactions and are robust against noise.

3. An unsuccessful confirmation run points to results which are not reproducible downstream and therefore indicate many interactions:

    - the effect depends on different conditions,
    - factor effects have poor additivity,
    - the effect depends on different factor levels,
    - poor reproducibility of conclusions.

    In this case, the quality characteristics need to be re-evaluated.

5. Characteristics that can be measured are not necessarily good characteristics.

6. When characteristics are selected wrongly, the conclusions could be wrong (many interactions) and may not be reproducible.

3.3.3 Guidelines for selecting quality characteristics

In designing a product we are usually interested in increasing the reliability of the product. In designing a manufacturing process we want to maximize the yield by reducing the number of defects. The final success of the process or the product depends on how well such responses (reliability, yield, etc.) meet the customer's expectations. However, such responses are not necessarily suitable quality characteristics for optimizing process or product design. The following guidelines may be useful in selecting good quality characteristics. The concepts behind these guidelines will become clear through the examples presented in the following sections. To select quality characteristics, we should consider:

- ideal functions,
- continuous variables,
- monotonic functions,
- ease of measurement,
- complete quality characteristics,
- modularization.

1.   Ideal functions

Identify the ideal function or the ideal input-output relationship for the product or the process. The quality characteristic should be directly related to the energy transfer associated with the basic mechanism of the process or the product.

2.   Continuous variables

As far as possible, choose continuous variables for quality characteristics.

3.   Monotonic functions
The quality characteristic should be monotonic, at least in the range of the experiment. That is, the effect of each control factor should be in a consistent direction, even when the settings of other control factors are changed. It is often difficult to judge the monotonicity of a quality characteristic before conducting experiments. Hence, matrix experiments followed by confirmation experiments may be the only way to determine whether the quality characteristic is monotonic.

4.   Ease of measurement
Use a quality characteristics that is easy to measure. The availability of appropriate measurement techniques is often an important consideration in the selection of a quality characteristic.

5.   Complete quality characteristics
Ensure that quality characteristics are complete. That is, they should cover all dimensions of the ideal function or the input-output relationship.

6.   Modularization
Complex products should be divided into convenient modules. Each module should be optimized separately and the modules should then be integrated together. While optimizing a particular module, the variation in other modules should be treated as noise. This is important for smooth system integration.

Finding quality characteristics that satisfy all of these guidelines is sometimes difficult or simply impossible. However, a Robust Design experiment will be inefficient to the extent these guidelines are not satisfied. The following are some considerations of quality characteristics.
- feeling of patient,
- percentage yield,
- difference,
- operating window,
- non-negative nominal-the-best.

3.3.4 Feeling of patient
The following example shows how a characteristic such as the feeling of a patient is not a good quality characteristic. Suppose an experiment in which,

y =   feeling of patient,
A1   =   did not take medicine A,
A2   =   took medicine A,
B1   =   did not take medicine B, and
B2   =   took medicine B.

A quality characteristic such as y = feeling of patient tends to have poor additivity, as illustrated earlier in this chapter. If taking the two medicines together has an *additive effect* then the *quality characteristic is predictable* for any other combination of factor settings. If taking the two medicines together *does not have an additive effect* then the *quality characteristic is unpredictable* (for any combination of factor settings) and hence is not useful in an engineering (or medical) sense.

The interaction in the Medicine example in Section 3.2 can be explained as follows. Assume that the active ingredient in both Medicines A and B is a substance called *Actogen* and that the concentration of Actogen in both medicines is 5 mg per capsule. Hence, if a patient takes either medicine, the patient feels well. If both medicines are taken, the combined Actogen concentration exceeds the threshold for feeling well and the patient feels ill. Thus, studying the quality characteristic *feeling of a patient* with respect to Medicines A and B is not monotonic. That is, feeling well does not increase correspondingly.

A better quality characteristic is to study the *concentration of Actogen* since this can predict that taking both medicines will cause the patient to feel ill due to the high concentration of Actogen. Thus, *feeling of patient* is unscientific data with low effectiveness, while *Actogen concentration* constitutes scientific data with high effectiveness. We can therefore say that Actogen concentration is a better quality characteristic. If it is not possible to find characteristics that reflect the effects of the individual factors regardless of the influence of other factors then the efficiency of research will drop drastically. The experimental result of selecting a poor characteristic is poor reproducibility downstream.

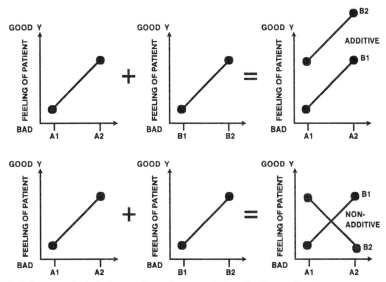

Figure 3:3.2 Interaction in quality characteristic: feeling of patient.

3.3.5 Percentage yield

Consider another example of a poor quality characteristic for a crushing process as shown in Figure 3:3.3. The mesh of a strainer provides a measure of the coarseness of crushed material, based on the number of wires per inch. A strainer of 15 mesh allows larger particles to fall through than would a strainer of 50 mesh. Crushed products are shaken through a 15 mesh strainer to remove large particles and then shaken through a 50 mesh strainer to remove small particles. The particles retained in the 50 mesh strainer form the useful or acceptable portion. Many quality control activities measure the yield of similar processes. However, measuring the yield of the useful portion (of such and similar processes) is not a good quality characteristic. This may be explained as follows:

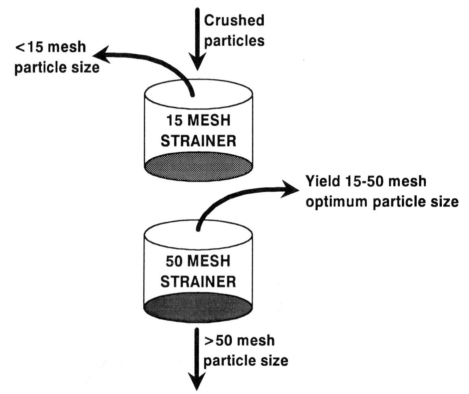

Figure 3:3.3 Interaction in quality characteristic: percentage yield.

Suppose an experiment based on the factors A, B and C. If the result of the experiment measured not only the yield of the useful particles (15 – 50 mesh) but also the oversized (<15 mesh) and the undersized (>50 mesh) particles as presented in Figure 3:3.4 (1) and (2). Based on that yield result alone (1), an engineer may choose factors A2 and B2 as the optimum factor levels since these levels have the higher yield levels. From the results of particle size (2), clearly, this will result in low yield since the amount of fine particles will increase. Additionally, factor C does not appear to have any significant effect on the yield. However, the choice of factor level C2 would also result in particles which are too fine. The choice of A2, B2 and C2 would therefore clearly result in very poor yield.

Factors	(1)	(2)		
	Yield	Particle size		
	15 – 50 mesh	<15 mesh	15 – 50 mesh	>50 mesh
A1	40 %	60 %	40 %	0 %
A2	70 %	0 %	70 %	30 %
B1	30 %	70 %	30 %	0 %
B2	90 %	0 %	90 %	10 %
C1	40 %	60 %	40 %	0 %
C2	40 %	0 %	40 %	60 %

Figure 3:3.4 Crushing process.

This effect is easily understood if we regard the quality characteristic y as the *particle size* instead of yield. The quality characteristic would then be nominal-the-best with a target of 32.5 mesh. For a nominal-the-best characteristic, the non-additivity of yield becomes clear when we consider the following three distributions of yield for Processes P, Q and R as shown in Figure 3:3.5. In Process P, the yield is poor because the mean particle size is too large. In Process Q, the yield is poor because the mean particle size is too small. In Process R, the yield is poor because there is too much variability. If the useful portions are about the same, then from a yield point of view all the processes are the same. However, from an engineering point of view they are three distinct problems. In general, therefore, y = yield and y = number of defects tend to be poor indicators of quality.

In reality, quality characteristics such as yield and number of defects do not express the true task an engineer has to accomplish. Yield is just a by-product of the crushing process. In this example, the reduction in variability of particle size and an adjustment to the nominal value is what an engineer would like to achieve. Particle size uniformity is a better indicator for a quality characteristic than percentage yield. Thus, it is better to select y = particle size as a nominal-the-best quality characteristic than to evaluate y = yield as a quality characteristic.

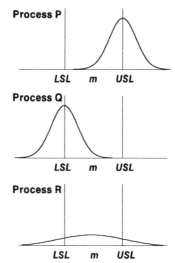

Figure 3:3.5 Three processes with 50 % yield.

3.3.6 Difference

The alignment of a car body and car door has been approached as a function of $y = z - x$, where z is the body dimension and x is the door dimension. See Figure 3:3.6.

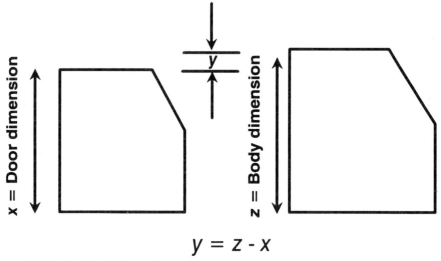

Figure 3:3.6 Car body-door alignment.

The function of $y = z - x$, is a poor quality characteristic. Variability arises from the variability of x and z, the stamping processes of door and body components. In reality there is a stacking up of the tolerances of x and z. y approaches a specified value only as $x = z$, with x and z both being either large or small. If y is taken as the quality characteristic to be measured, the measurements of y may be full of interactions. The real engineering problem is the variability in the stamping process. Stacking of tolerances is just a result of treating x and z as dependent engineering problems. The conclusion drawn from this experiment has a poor chance of being reproducible downstream. A better quality characteristics is to regard $x =$ door dimension and $z =$ body dimension as independent quality characteristics to be measured or optimized. Both x and y can be treated as non-negative nominal-the-best values. This would greatly improve the reproducibility of results for the car body-door alignment.

3.3.7 Operating window

In some cases, faults are measured as the *number of problems per thousand operations* Examples include the number of wrong items per thousand items dispensed from a vending machine, or the number of paper jams out of a thousand sheets of paper fed through a photocopier.

The characteristic $y =$ number of paper jams per thousand sheets is a poor quality characteristic. In this case a larger *operating window* would provide a better quality characteristic. The idea of the operating window is explained with reference to a simplified paper feeding mechanism as shown in Figure 3:3.7. The force applied on the paper tray depends on the force f maintained by the spring. Suppose:

$f = x$ is the force (misfeed force) at which one sheet of paper starts to feed (paper feeding threshold) and

$f = z$ is the force (multifeed force) at which more than one sheet of paper starts to feed (paper multifeed threshold).

If the objective is set to:

Minimize x as a smaller-the-better characteristic and

Maximize z as a larger-the-better characteristic,

then the objective can be made to maximize $z - x$.

Hence, an experiment can be conducted to find the control factor levels A, B, C, etc., that are robust to noise factors P, Q, etc., so that x is minimized and z is maximized. The concept of increasing this operating window, i.e. the range that allows one sheet of paper to be fed through without jamming, is the same as that of robust design (see Figure 3:3.8). It is important to realize that the paper feeding example is not a statistical problem but entirely an engineering one. Generally, increasing the operating window is an excellent quality characteristic.

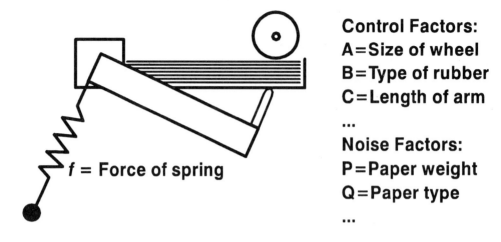

Control Factors:
A=Size of wheel
B=Type of rubber
C=Length of arm

...

Noise Factors:
P=Paper weight
Q=Paper type

...

Figure 3:3.7 Simplified paper feeding mechanism.

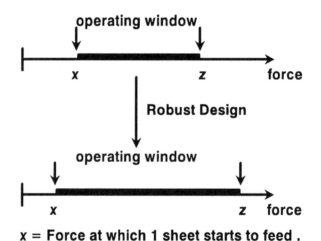

x = Force at which 1 sheet starts to feed .
z = Force at which >1 sheet starts to feed.

Figure 3:3.8 Increasing the operating window.

We must distinguish the operating window from the stacked tolerance of the car body-door alignment example. In that example, variability arose due to variability in car-body and car-door components. Ideally, the car-body and the car-door components should each attain the mean value of the specification (nominal-the-best). In that way the mean difference between the two distributions will ensure that the alignment is neither too tight nor too slack and so there will no stacking of tolerance. Additionally, there is no objective (or need) to increase or reduce this difference (alignment tolerance).

In the paper feeding mechanism, however, the objective is to increase the difference between the misfeed force at which one sheet feeds and the multifeed force at which more than one sheet feeds. That is, we want to increase the operating window. This can be done by studying control factors such as the size of the wheel (A), the type of rubber (B), the length of arm (C), etc. and including noise factors such as paper weight (P), paper type (Q), etc. Under the optimum control factor condition, then, the new paper misfeed threshold $x_{Optimum}$ should be smaller than the original paper misfeed threshold x and likewise, the new paper multifeed threshold $z_{Optimum}$ should be larger than the original multifeed threshold z.

3.3.8 Non-negative nominal-the-best

Heat exchangers are used to transfer heat from one fluid to another. Heat exchangers are commonly used in car radiators and refrigerators. In a refrigerator, heat exchange occurs between the food in the *refrigerator compartments and the refrigerant fluid* as well as between the *refrigerant fluid and the radiator* outside a refrigerator. Of course the refrigerant fluid is a heat carrier which is continuously pumped through the compartment and the radiator. In effect the heat is transferred from the food to the ambient.

Figure 3:3.9 shows a schematic of a heat exchanger used to cool the fluid inside an inner tube. Suppose the inlet temperature of the fluid is T_1. As the fluid moves on, it loses heat to the fluid entering the exchanger and exits at a temperature T_2. The inlet temperature of the refrigerant (or cooling fluid) is T_3 and its exit temperature is T_4. Suppose further that the target temperature of T_2 is T_0 and the customer requirement is specified to be $T_0 \pm 10 \, °C$. In other words,

$$|T_2 - T_0| < 10 \, °C$$

Is this a good quality characteristic in this example? Suppose we choose T_0 as the reference temperature and define:

$$y = |T_2 - T_0|$$

as smaller-the-better quality characteristic with target zero. The problem with this quality characteristic is that both the positive and negative deviations from T_0 will be indistinguishable and are regarded similarly. Hence, interactions become important and reproducibility is poor.

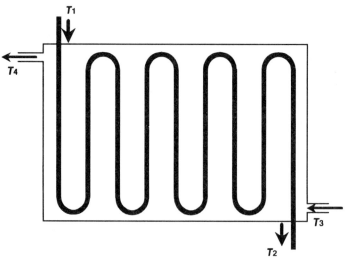

Figure 3:3.9 Schematic diagram of a heat exchanger.

A better method is to use T_3 as a reference temperature since that is the lowest temperature that can be reached by T_2 and would represent a design effort that aimed to make the heat exchanger efficient. Thus, we may choose T_3 as the reference temperature and define:

$$y' = T_2 - T_3$$

as a nominal-the-best quality characteristic with target value:

$$y' = T_0 - T_3$$

This quality characteristic does not have the same problems as that of the previous quality characteristic with regards to interactions. Additionally, if a different customer requires a change in T_0, the experimental information obtained from an initial experiment could still be useful and we would merely have to find a suitable way of adjusting to the new target value. If however, y had been used, we would have to re-run another experiment for the new target value. Clearly, a waste of engineering effort.

3.3.9 Insights regarding quality characteristics
Four important insights regarding quality characteristics are as follows:
- energy related,
- non-negative nominal-the-best,

- compounded noise factors,
- assignment of main factors to orthogonal array experiments.

1.   Energy related
Energy related (direct or indirect) quality characteristics are good measures of quality. Variability occurs as a function of time and energy fluctuations. If the energy input is zero, we cannot produce a product, or that product does not function. In fact, if energy is zero, there is no variability.

2.   Non-negative nominal-the-best
Non-negative nominal-the-best characteristics tend to have good reproducibility.

3.   Compounded noise factors
It is recommended that noise factors (especially compounded noise factors) are included; this will ensure better reproducibility downstream where numerous noise factors exist. The inclusion of one or two important noise factors simulates these unknown noises and thus allows a process or product to be designed to be robust against them. If it is difficult to introduce noise factors because of cost or impracticality, a simple repetition and sample to sample variation can be used to simulate noise. The greater the ability of a process or product to remain robust against noise, the greater the assurance that the results will be reproducible in manufacturing and in the customer's hands.

4.   Assignment of main factors to orthogonal array experiments
Where possible, experiments using orthogonal arrays are better designed without interactions in order to validate the selection of a good quality characteristic. If the result of a confirmation run is close enough to the predicted value, we can expect good reproducibility downstream. A good quality characteristic is thus a measure of engineering technology and engineering progress.

3.3.10   The case of more than one quality characteristic
Often more than one quality characteristic of a product has to be considered. The level of importance of each characteristic may vary, as well as the type of response, i.e. smaller-the-better, larger-the-better and nominal-the-best. In such cases, each quality characteristic must be analyzed separately from the available experimental data. A review of a summary table of factors and effects must be conducted. An overall selection of factors that minimizes the sacrifice of quality characteristics may require a trade-off of factor levels. Hence it may be necessary to review the selection of factor levels.

3.3.11   Conflicting quality characteristic

Occasionally, it is possible that an improvement of one quality characteristic may result in the deterioration of another. To reduce such effects it is important to find a factor which affects the one quality characteristic but not the other. This strategy becomes important when optimizing conflicting characteristics such as fuel efficiency and power. By including many factors in an experiment, the likelihood of identifying factors that can be used selectively to improve both quality characteristics simultaneously may be increased.

3.3.12   Course of action for poor results

If the result of a confirmation experiment from a design of experiment does not match the predicted mean then it is unlikely that the results will be reproducible. The possible cause(s) for this lack of additivity should be investigated as follows:

- poor selection of quality characteristics,
- poor selection of control factors,
- factor levels are too narrow or too broad,
- noise factor effects,
- lack of balance among control factors.

1.   Poor selection of quality characteristics

Strong interactions among control factors nearly always result in poor additivity. The experiment may need to be performed with a good quality characteristic as suggested in this chapter.

2.   Poor selection of control factors

It is likely that one or more of the factors selected for the experiment are not in effect control factors, i.e. they do not affect the results. In this case, it may be necessary to research for engineering knowledge of the particular quality characteristic.

3.   Factor levels are too narrow or too broad

The effects of factor level selection on the results of an experiment have been discussed before. Again, engineering knowledge may be needed to improve the range of factor levels, or it may be necessary to use 3-level factors.

4.   Noise factor effects

It is possible that an unexpected noise factor was not included in the experiment. Further research should be done to include likely noise factors.

5.   Lack of balance among control factors

Sometimes it is possible to make a mistake in the preparation of experiments based on the factor level combination. If this is suspected it may necessary to repeat the experiment. It is also recommended that an orthogonal array of factor levels is translated into engineering terms, especially if the experiment is conducted by a less technical person.

Valuable information can usually be obtained even from an experiment which is not reproducible. A final resort, assuming that parameter design has been exhausted, is to consider tolerance design. Sometimes, however, present day engineering technology may be inadequate to correct the problem. In that case, a system design will be necessary.

3.3.13   Self-assessment questions

1.   Discuss the importance of selecting good quality characteristics.

2.   What are the guidelines for selecting quality characteristics?

3.   Sagging is a common defect in spray painting. It is caused by the formation of large paint drops that flow downward due to gravity. Is the distance through which the paint drops sag a good quality characteristic?

4.   There are many chemical processes that begin with a chemical A, which after reaction becomes chemical B and, if the reaction is allowed to continue, turns into chemical C. If B is the desired product, is yield a good quality characteristic?

5.   Paper jamming is an important customer observable problem in a photocopier operation. Is the number of jams per thousand sheets of paper a good quality characteristic?

3.3.14   Answers to self-assessment questions

1.   Discuss the importance of selecting good quality characteristics.

Answer
The primary importance of selecting a good quality characteristic is to ensure that a laboratory scale experiment is reproducible in a large scale production environment. Under laboratory conditions, experimental test conditions do not vary very much. In large scale production environments and, in particular, customer usage conditions, the conditions of use vary a great deal. Thus, it is important to select a quality characteristic that ensures reproducibility of results. Hence, it is necessary to find quality characteristics that are additive.

2.   What are the guidelines for selecting quality characteristics?

Answer
Identify the ideal function or the ideal input-output relationship for the product or the process. The quality characteristic should be directly related to the energy transfer associated with the basic mechanism of the product or the process. As far as possible, choose continuous variables as quality characteristics. The quality characteristics should be monotonic. That is, the effect of each control factor on robustness should be in a consistent direction, even when the settings of other control factors are changed. Usually, it is difficult to judge the monotonicity of a quality characteristic before conducting experiments. Try to use quality characteristics that are easy to measure. The availability of appropriate measurement techniques is often an important consideration in the selection of quality characteristics. Ensure that quality characteristics are complete; that is, they should cover all dimensions of the ideal function or the input-output relationship.

3.   Sagging is a common defect in spray painting. It is caused by the formation of large paint drops that flow downward due to gravity. Is the distance through which the paint drops sag a good quality characteristic?

Answer
No, because this distance is primarily controlled by gravity, and is not related to the basic energy transfer in spray painting. However, the size of the drops created by the spray nozzle is directly related to energy transfer, and is thus a better quality characteristic. By taking the size of the drops as the quality characteristic, we can block out the effect of gravity, which is an extraneous phenomenon in the spray painting process.

4.   There are many chemical processes that begin with a chemical A, which after reaction becomes chemical B and, if the reaction is allowed to continue, turns into chemical C. If B is the desired product, is yield a good quality characteristic?

Answer

If B is the desired product of the chemical process, then considering the yield of B as a quality characteristic is a poor choice because yield is not a monotonic characteristic. A better quality characteristic for this experiment is the concentration of each of the three chemicals. The concentration of A and the concentration of A plus B possess the needed monotonic property.

5.   Paper jamming is an important customer observable problem in a photocopier operation. Is the number of jams per thousand sheets of paper a good quality characteristic?

Answer

The number of jams per thousand sheets of paper fed through a photocopier is not a good quality characteristic since the number of sheets that would need to be fed through before paper jam occurs would be very large. The two main problems that arise in the paper feeding mechanism are: no sheet fed or multiple sheets fed. Therefore, a quality characteristic that measures the force needed to pick up one sheet rather than two or more sheets of paper would be a better quality characteristic.

CHAPTER 4

OBJECTIVE FUNCTIONS IN ROBUST DESIGN

AIMS:
The aim of this chapter is to provide methods of calculating signal-to-noise ratios for a number of quality characteristic types and, using these signal-to-noise ratios, to perform data analysis for process optimization.

OBJECTIVES:
When you have completed this chapter you should be able to:
- explain the need for signal-to-noise ratios,
- calculate signal-to-noise ratios for smaller-the-better, larger-the-better and nominal-the-best characteristics,
- perform operating window analysis,
- perform Omega transformations on percent defective data.

OVERVIEW:
This chapter introduces signal-to-noise ratios (SN ratios) for static problems. Static problems are distinguished from dynamic problems. SN ratios for static problems are developed for smaller-the-better, nominal-the-best, larger-the-better, signed-target and percent defective quality characteristics. The use of SN ratios and noise factors in experimentation is illustrated with a detailed example of a smaller-the-better characteristic, showing the method of calculating signal-to-noise ratios and analyzing the data for optimum conditions. The method of predicting process improvement through optimization is also given. A method for increasing the operating limits for a process is illustrated using the operating window analysis. Methods for improving additivity in percent defective data are illustrated using the Omega (Ω) transformation.

4.1 Signal-to-noise ratios for Static Problems

4.1.1 Introduction

Taguchi has extended the audio concept of the signal-to-noise ratio (SN ratio) to experiments involving many factors. Such experiments are frequently called multifactor experiments. The formulae for signal-to-noise ratio are designed so that an experimenter can always select the largest factor level setting to optimize the quality characteristic of an experiment. Therefore, the method of calculating the signal-to-noise ratio depends on whether the quality characteristic has a smaller-the-better, larger-the-better or nominal-the-best response.

4.1.2 SN ratio for static problems

Design of experiments can be classified into two main types:
- static functions,
- dynamic functions.

1. Static functions

In many experiments, the quality characteristic has a specific value to be optimized. Attaining a target value of 115 volts for a voltage output circuit, minimizing the surface roughness of a machined plate or maximizing the wire tear-apart force of soldered wires are all examples of static functions. In all these cases, there are fixed target values (115 volts nominal, zero and infinity, respectively). The study of such systems requires static functions.

2. Dynamic functions

In some experiments, the quality characteristic has a variable target value. The braking mechanism of a car, for example, has no predetermined target value. Sometimes we need to brake slowly and sometimes sharply. In this case, the braking action must follow the intent of the driver depending upon the need of the moment. The study of such a system requires dynamic functions.

In Chapter 2, we considered the Parameter Diagram, which is shown again for convenience in Figure 4:1.1. The response (y) is a function of noise factors (X), signal factors (M), control factors (Z) and scaling factors (R). In particular, a signal factor changes the magnitude of the response variable. In static functions, the signal factor is effectively a constant and we usually only evaluate control and noise factors. Thus, attaining a target voltage of 115 volts for colour television voltage, minimizing surface roughness or maximizing tear-apart force all have fixed targets and are static problems. In dynamic functions, the signal factor is variable and we need to evaluate the control and noise factors alongside the signal factor. In the car braking system, the harder the driver presses on the brake pedal, the sharper is the braking action. Thus, the pressure on the brake pedal is a signal factor.

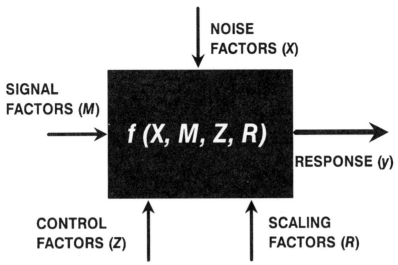

Figure 4:1.1 Factors affecting a response.

4.1.3 Importance of the SN ratio

The objective of multifactor experiment in terms of robust design is to minimize the sensitivity of a quality characteristic to noise factors. Such a decision may be based on the mean squared deviation:

$$MSD = \sigma^2 + (\bar{y} - m)^2$$

Such a measurement without adjusting the mean value $(\bar{y})$ to the target value (m) is influenced not only by the sensitivity to noise (σ^2) but also by the deviation from the target value $(\bar{y} - m)$. Consequently, we risk the possibility of not choosing the factor level that minimizes sensitivity to noise.

We now develop a mathematical model of a robust design experiment with the various parameters affecting the response shown in Figure 4:1.1 as:

$$y = f(X, M, Z, R) \qquad (4:1.1)$$

where y (response variable) is a function of X (noise), M (signal), Z (control) and R (scaling). In reality, the function may be regarded as consisting of two parts:

- the predictable part,
- the unpredictable part.

1.   The predictable part

The predictable part is the desirable part g $(M; Z, R)$ usually referred to as the *signal*.

2.   The unpredictable part

The unpredictable part is the undesirable part e $(X, M; Z, R)$ which is usually referred to as the *noise*.

Thus, Equation 4:1.1 can be re-written as:

$$y = g \ (M; Z, R) + e \ (X, M; Z, R) \tag{4:1.2}$$

When a linear relationship is desired between y and M, g $(M; Z, R)$ must be a linear function with all non-linear effects included in e $(X, M; Z, R)$. Of course, the effect of all noise factors will be included in e $(X, M; Z, R)$. The objective of robust design is to maximize the predictable part and minimize the unpredictable part.

However, Equation 4:1.2 is not very useful directly. Through experimentation however, we can find the variance of the predictable part (V_g) and the unpredictable part (V_e). We can then use Taguchi's suggestion to evaluate quality by the definition:

$$\eta \ = \ \frac{signal}{noise}$$

$$= \ \frac{variance \ of \ the \ predictable \ part}{variance \ of \ the \ unpredictable \ part}$$

$$= \ \frac{V_g}{V_e} \tag{4:1.3}$$

By analogy with communications theory, we may take η (eta) in the decibel (dB) scale:

$$\eta \ = \ 10 \ \log_{10} \left[\frac{variance \ of \ the \ predictable \ part}{variance \ of \ the \ unpredictable \ part} \right]$$

$$= \ 10 \ \log_{10} \left[\frac{V_g}{V_e} \right] \tag{4:1.4}$$

Expressed in this way, the predictable part is the signal and the unpredictable part is the noise; hence, the equation is called the signal-to-noise ratio.

The objective of robust design is to maximize the predictable part and minimize the unpredictable part. This is best done by choosing suitable levels of Z and R so that the SN ratio is clearly a function of both Z and R and:

$$\eta(Z,\ R) = 10 \ \log_{10} \left[\frac{V_g(Z,\ R)}{V_e(Z,\ R)} \right]$$

However, R is a factor that is used to change the overall mean effect of the response and hence the predictable part. Therefore, it is not necessary to perform experiments with various values of R so that:

$$\eta(Z) \ = \ \max \ \eta \ (Z,\ R \ constant)$$

and we need to maximize the SN ratio in the domain of Z.

The maximum value of η may then be attained by searching through its values, usually by a set of noise, signal, control and scaling factors assigned to one or more orthogonal arrays. The optimization strategy is therefore to measure the SN ratio of the response for selected values of the control (and scaling) factors by emulating the noise factors in the desired range of the signal factor. Upon evaluating $\eta(Z)$ for all the control factors, the effect of each of the control factors (Z_i) on η is established, so that the corresponding optimum level of Z_i may be determined to form the set of optimum control factor levels Z_o. The next step is to perform a confirmation experiment at Z_o.

When the signal factor is constant, the problem reduces to one of a static function and the signal-to-noise ratio can be defined as:

$$\eta \ = \ -10 \ \log_{10} \ [MSD] \qquad\qquad (4:1.5)$$

where *MSD* is the mean squared deviation from target.

The SN ratio also has a definite advantage over the mean squared deviation. When the target value is changed, the optimum conditions obtained by maximizing the SN ratio will still be valid except for adjustment of the mean. However the same cannot be said of using the mean squared deviation from target as the objective function. We would have to perform the optimization again.

For nominal-the-best characteristics, the problem of minimizing the variance while keeping the mean on target is a problem of constrained optimization. By using the SN ratio, the problem can be converted into an unconstrained optimization problem that is much easier to solve. The property of unconstrained optimization is the basis for our ability to separate the actions for minimizing sensitivity to noise factors and the adjustment of the mean on target. The SN ratio is thus a very useful[1] way of evaluating the quality of a process or product. This ratio measures the level of performance against the level of noise factors on performance. It is an evaluation of the stability of the performance of an output characteristic.

Better performance, as measured by a high SN ratio, implies a smaller loss than that measured by the corresponding quality loss function. Like the quality loss function, the SN ratio is an objective function of quality that takes both the effect of the mean and the variation into account. Finding the correct objective function to maximize in an engineering design problem is very important. Failure to do so can lead to inefficiencies in experimentation and even wrong conclusions about optimum levels. The task of finding what adjustments are meaningful in a particular problem and determining the correct objective function (or SN ratio) is not always easy. We now develop more specifically the SN ratios for:

- smaller-the-better,
- nominal-the-best,
- larger-the-better,
- signed-target,
- fraction defective,
- operating window.

4.1.4 SN ratio – smaller-the-better

Here, the quality characteristic is continuous and non-negative, that is, it can take any value from 0 to ∞. Its most desired value is zero. These problems are characterized by the absence of a scaling factor or any other adjustment factor. Surface roughness is an example of this type of problem.

Another example of a smaller-the-better type problem is the pollution from a power plant. We can reduce the total pollutants emitted by reducing the power output of the plant. However, reducing pollution by reducing power consumption does not signify any quality improvement for the power plant. Hence, it is inappropriate to think of the power output as an adjustment factor. In fact, we should consider the pollution per megawatt-hour of power output as the quality characteristic to be improved instead of the pollution itself.

Additional examples of smaller-the-better type problems are electromagnetic radiation from telecommunications equipment, leakage current in integrated circuits and corrosion of metals. Because there is no adjustment factor in these problems, we should simply minimize the quality loss without adjustment. That is, we should minimize:

---

1   The signal-to-noise ratio is not without criticism. The interested reader is referred to Dorian Shainin, *Better than Taguchi Orthogonal Tables.* Quality and Reliability Engineering International, Volume 4, pages 143 – 149.

$$Loss = k \, [MSD]$$

$$= k \left[\frac{1}{n} \sum_{i=1}^{n} y_i^2 \right] \tag{4:1.6}$$

Minimizing quality loss is equivalent to maximizing the signal-to-noise ratio, η, defined by Equation (4:1.7). Note that η does not depend on the cost coefficient k. Also, since the signal factor is a constant, and the target value is equal to zero, η merely measures the effect of noise.

$$\eta = -10 \, \log_{10} \, [MSD]$$

$$= -10 \, \log_{10} \left[\frac{1}{n} \sum_{i=1}^{n} y_1^2 \right]$$

$$= -10 \, \log_{10} \, [\sigma^2 + \bar{y}^2] \tag{4:1.7}$$

4.1.5 SN ratio – nominal-the-best

In nominal-the-best type problems, the quality characteristic is continuous and non-negative: that is, it can take any value from 0 to ∞. Its target value is non-zero and finite. In these problems, when the mean becomes zero, the variance also becomes zero. Additionally, we can use an adjustment factor to move the mean to target. This type of problem occurs frequently in engineering designs. The objective function to be maximized for such a problem is:

$$\eta = 10 \, \log_{10} \left[\frac{\mu^2}{\sigma^2} \right] \tag{4:1.8}$$

where

$$\mu = \frac{1}{n} \sum_{i=1}^{n} y_i \tag{4:1.9}$$

and

$$\sigma^2 = \frac{1}{n} \sum_{i=1}^{n} (y_i - \mu)^2 \tag{4:1.10}$$

Adjustment factors can often be identified readily through engineering expertise or experimentation. The optimization of the nominal-the-best problems can be accomplished in the *two-step optimization process:*

1.   Maximize η or minimize sensitivity to noise. During this step we select the control factor levels that minimize noise.
2.   Adjust the mean on target. During this step we use an adjustment factor to bring the mean on target without changing η.

Figure 4:1.2 shows the two-step optimization process.

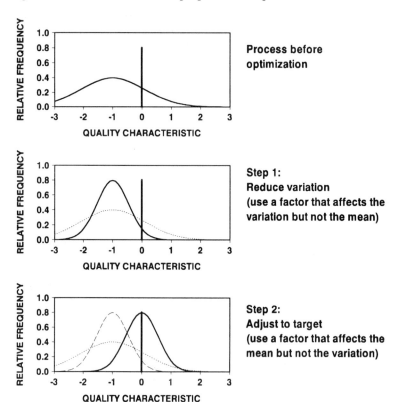

Process before optimization

Step 1:
Reduce variation
(use a factor that affects the variation but not the mean)

Step 2:
Adjust to target
(use a factor that affects the mean but not the variation)

4.1.6 SN ratio – larger-the-better

Here, the quality characteristic is continuous and non-negative, that is, it can take any value from 0 to ∞. Its target value is non-zero and, ideally, as large as possible. These problems are also characterized by the absence of a scaling factor or any other adjustment factor. Examples of such problems are the mechanical strength of a wire per unit cross-section area and the fuel efficiency of an automobile. These problems can be transformed into smaller-the-better type problems by considering the reciprocal (inverse) of the quality characteristic. That is, if y is a larger-the-better characteristic, then taking the reciprocal ($1/y$) transforms the quality characteristic into a smaller-the-better characteristic. The objective function to be maximized in this case is given by:

$$\eta = -10 \log_{10} [MSD]$$

$$= -10 \log_{10} \left[\frac{1}{n} \sum_{i=1}^{n} \frac{1}{y_i^2} \right] \tag{4:1.11}$$

and the corresponding loss-to-society is :

$$Loss = k [MSD]$$

$$= k \left[\frac{1}{n} \sum_{i=1}^{n} \frac{1}{y_i^2} \right] \tag{4:1.12}$$

4.1.7 SN ratio – signed-target

In this class of problems, the quality characteristic can take positive as well as negative values. Often, the target value for the quality characteristic is zero. If not, the target value can be made zero by selecting an appropriate reference value for the quality characteristic. Here, we can find an adjustment factor that can move the mean without changing the standard deviation. The objective function to be maximized in this case is given by:

$$\eta = -10 \log_{10} [MSD]$$

$$= -10 \log_{10} \left[\frac{1}{n} \sum_{i=1}^{n} (y_i - \mu)^2 \right]$$

$$= -10 \log_{10} \sigma^2 \tag{4:1.13}$$

Note that signed-target problems are different from smaller-the-better type problems although in both cases the target value is zero. In signed-target problems, the quality characteristic can take positive as well as negative values, whereas in the smaller-the-better type problems the quality characteristic cannot take negative values.

The range of possible values for the quality characteristic also distinguishes signed-target problems from nominal-the-best. In signed-target problems the range of values is $(-\infty$ to $+\infty)$ whereas in nominal-the-best problems, the range of values is $(0$ to $\infty)$.

The signed-target problem is also different from the nominal-the-best quality characteristic in another way. In the signed-target problem, when the mean is zero, the standard deviation is not zero. In nominal-the-best problems, when the mean is zero, the standard deviation is also zero.

An example of a signed-target problem is the direct current offset voltage of a differential operational amplifier. The offset voltage could be positive or negative. If the offset voltage is consistently off zero, then we can easily compensate for the offset voltage in the receiver circuit of the differential operational amplifier without affecting the standard deviation.

4.1.8 SN ratio – fraction defective

Another problem occurs when the quality characteristic is a proportion, such as the fraction defective, denoted p, which can take values between 0 and 1. Frequently, p is expressed as a percentage where it can take values between 0 % and 100 %. With fraction defective, the best value for p is zero. Sometimes, p can denote the percentage yield, the best value for which is 1 (or 100 %). Fraction defective problems do not have an adjustment factor. When the average fraction defective is p, we have to manufacture $1/(1 - p)$ pieces to produce one good piece. Thus, for every good piece produced, there is a loss that is equivalent to the cost of processing $(1/(1 - p) - 1) = p/(1 - p)$ pieces. Thus, the quality loss is given by:

$$Loss = k \frac{p}{(1 - p)} \qquad (4:1.14)$$

where k is the cost of processing one piece. The objective function to be maximized is:

$$\eta = -10 \log_{10} \left[\frac{1}{p} - 1 \right] \qquad (4:1.15)$$

and is frequently called the Omega (Ω) transformation. The relationship between proportion (p) and signal-to-noise ratio (η) is shown in Figure 4:1.3. Note that the range of possible values of proportion p is 0 to 1, but the range of possible values of η is $-\infty$ to $+\infty$. Therefore, the additivity of factor effects is better for η than for p.

4.1.9 SN ratio – operating window

In the case of the operating window, the objective characteristic depends on two quality characteristics. Paper jams in a photocopier can arise because no paper has been fed into the system (misfeed) or more than one sheet of paper has been fed into the system (multifeed). Supposing that among other control factors, misfeed arises because of low spring force (f_s) and multifeed arises because of high spring force (f_L). In order to increase the range within which only one sheet of paper is fed into the system, it is necessary to minimize the misfeed force and to maximize the multifeed force. Therefore, misfeed is regarded as a smaller-the-better characteristic and multifeed is regarded as a larger-the-better characteristic. Hence, it is necessary to measure two sets

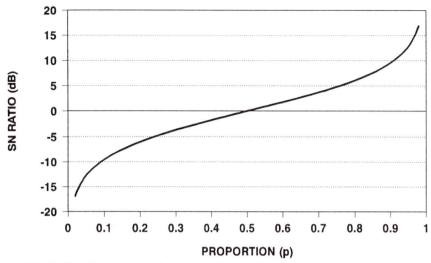

Figure 4:1.3 The Omega transformation.

of data (misfeed and multifeed) under the same control factor and noise factor conditions. Using these data, the signal-to-noise ratios for misfeed $\eta_{Misfeed}$ and multifeed $\eta_{Multifeed}$ can be calculated. The objective characteristic is then defined as:

$$\eta = \eta_{Misfeed} + \eta_{Multifeed} \qquad (4{:}1.16)$$

and must be maximized to improve quality. Since $\eta_{Misfeed}$ is a smaller-the-better characteristic and $\eta_{Multifeed}$ is a larger-the-better characteristic:

$$\eta_{Misfeed} = \text{signal-to-noise ratio for misfeed data}$$

$$= -10 \ \log_{10} \left[\frac{1}{n} \left(y_{S1}^2 + y_{S2}^2 + \ldots + y_{Sn}^2 \right) \right] \qquad (4{:}1.17)$$

and

$$\eta_{Multifeed} = \text{signal-to-noise ratio for multifeed data}$$

$$= -10 \ \log_{10} \left[\frac{1}{n} \left(\frac{1}{y_{L1}^2} + \frac{1}{y_{L2}^2} + \ldots + \frac{1}{y_{Ln}^2} \right) \right] \qquad (4{:}1.18)$$

and:

$$\eta = \eta_{Misfeed} + \eta_{Multifeed}$$

$$= -10 \log_{10} \left[\frac{1}{n} \left(y_{S1}^2 + y_{S2}^2 + ... + y_{Sn}^2 \right) \right] -10 \log_{10} \left[\frac{1}{n} \left(\frac{1}{y_{L1}^2} + \frac{1}{y_{L2}^2} + ... + \frac{1}{y_{Ln}^2} \right) \right]$$

After optimization, the optimum spring force may be determined by minimizing the loss-to-society due to the two failure modes. Since it is likely that misfeed and multifeed conditions have different quality cost coefficients, the loss-to-society may be calculated as:

$$Loss = k_1 \left[MSD_{Misfeed} \right] + k_2 \left[MSD_{Multifeed} \right]$$

$$= k_1 \left[\frac{1}{n} \sum_{i=1}^{n} y_{Si}^2 \right] + k_2 \left[\frac{1}{n} \sum_{i=1}^{n} \frac{1}{y_{Li}^2} \right] \qquad (4:1.19)$$

where k_1 is the cost coefficient for misfeed and k_2 is the cost coefficient for multifeed.

4.1.10   Relationship between SN ratio and quality loss
When comparing the loss-to-society at the optimum condition to the existing condition we need only compare the loss-per-piece at the optimum condition $L_{Optimum}$ with the loss-per-piece at the existing condition $L_{Existing}$. This is because the loss-to-society is the aggregate loss of the loss-per-piece and can be calculated from the product of the loss-per-piece and the total number of pieces. The difference in loss-per-piece between the optimum condition and the existing condition is the gain in loss-per-piece, denoted $\mathfrak{I}$.

$$\mathfrak{I} = L_{Existing} - L_{Optimum}$$

$$= k \left[MSD \right]_{Existing} - k \left[MSD \right]_{Optimum}$$

$$= k \left(\left[MSD \right]_{Existing} - \left[MSD \right]_{Optimum} \right) \qquad (4:1.20)$$

The gain in loss-per-piece can also be expressed in terms of the signal-to-noise ratio. Using the SN ratio,

$$\eta = -10 \log_{10} \left[MSD \right] \qquad (4:1.21)$$

If η_O and M_O are the SN ratio and the mean squared deviation, respectively, at the optimum condition then,

$$\eta_O = -10 \log_{10} M_O$$

$$\therefore M_O = 10^{\frac{-\eta_O}{10}} \qquad (4:1.22)$$

Similarly, if η_E and M_E are the SN ratio and the mean squared deviation, respectively, at the existing condition, then,

$$\eta_E = -10 \log_{10} M_E$$

$$\therefore M_E = 10^{\frac{-\eta_E}{10}} \qquad (4:1.23)$$

If the ratio of the loss reduction is $\Re$, then,

$$\Re = \frac{M_O}{M_E}$$

$$= 10^{\left(\frac{-\eta_O}{10} - \frac{-\eta_E}{10}\right)}$$

$$= 10^{-\left(\frac{\eta_O - \eta_E}{10}\right)}$$

$$= 10^{\frac{-X}{10}} \qquad (4:1.24)$$

where $X = (\eta_{Optimum} - \eta_{Existing})$ is the gain in signal-to-noise ratio. If we choose to expressed $\Re$ in another way:

$$\Re = 0.5^{\frac{X}{\kappa}} \qquad (4:1.25)$$

and

$$\Re = 0.5^{\frac{X}{\kappa}}$$

$$10^{\frac{-X}{10}} = 0.5^{\frac{X}{\kappa}}$$

$$\therefore 10^{\frac{-1}{10}} = 0.5^{\frac{1}{\kappa}} \qquad (4:1.26)$$

Taking logarithms on both sides:

$$\log_{10} 10^{\frac{-1}{10}} = \log_{10} 0.5^{\frac{1}{\kappa}}$$

$$\frac{-1}{10} = \frac{1}{\kappa} \log_{10} 0.5$$

$$\kappa = -10 \log_{10} 0.5$$

$$= 3.01$$

$$\approx 3 \tag{4:1.27}$$

Therefore,

$$\mathfrak{R} = \text{Loss reduction}$$

$$= 10^{\frac{-X}{10}}$$

$$= 0.5^{\frac{X}{3}}$$

$$= 0.5^{\left(\frac{\eta_{Optimum} - \eta_{Existing}}{3}\right)} \tag{4:1.28}$$

Note that the loss reduction $\mathfrak{R}$ is simply a factor by which the existing mean squared deviation $[MSD_{Existing}]$, has been reduced by. By definition:

$$\mathfrak{R} = \frac{[MSD]_{Optimum}}{[MSD]_{Existing}}$$

$$\therefore \ [MSD]_{Optimum} = \mathfrak{R} \times [MSD]_{Existing} \tag{4:1.29}$$

and we may express the gain in loss-per-piece $\mathfrak{G}$ as:

$$\mathfrak{S} = k \left[MSD_{Existing} - MSD_{Optimum}\right]$$

$$= k \left[MSD_{Existing} - \mathfrak{R} \times MSD_{Existing}\right]$$

$$= k \, MSD_{Existing} \times \left[1 - \mathfrak{R}\right]$$

$$= k \, MSD_{Existing} \times \left[1 - 0.5^{\left(\frac{\eta_{Optimum} - \eta_{Existing}}{3}\right)}\right] \tag{4:1.30}$$

This equation can now be used to evaluate the loss-per-piece and hence the loss-to-society purely from a knowledge of the *MSD* at the current process, the gain in signal-to-noise ratio following optimization and the total number of products.

4.1.11   **Self-assessment questions**

1.   What are the advantages of signal-to-noise ratio analysis?

2.   Given the data below for existing and optimum conditions, calculate the signal-to-noise ratios for smaller-the-better quality characteristics.

Existing	0.25	0.28	0.35	0.26	0.22
Optimum	0.13	0.18	0.15	0.14	0.12

3.   What is the gain in signal-to-noise ratio for the above data?

4.   Relate the gain in loss-to-society to the gain in signal-to-noise ratio for the above data.

4.1.12   Answers to self-assessment questions

1.   What are the advantages of signal-to-noise ratio analysis?

Answer

The signal-to-noise ratio looks at both the average and the variability. The signal-to-noise ratio is a data transformation that gives a measure of performance against noise. The signal-to-noise ratio is related to cost.

2.   Given the data below for existing and optimum conditions, calculate the signal-to-noise ratios for smaller-the-better quality characteristics.

Existing	0.25	0.28	0.35	0.26	0.22
Optimum	0.13	0.18	0.15	0.14	0.12

Answer

The mean value for the existing condition is $\bar{y} = 0.2720$ and the sample standard deviation, $\sigma = 0.0487$. The sample standard deviation can be calculated using Equation (4:1.10) or a statistical calculator.

$$\eta_{Existing} = -10 \log_{10} (\bar{y}^2 + \sigma^2)$$

$$= -10 \log_{10} (0.2720^2 + 0.0487^2)$$

$$= -10 \log_{10} 0.0764$$

$$= 11.17 \text{ dB}$$

The mean value at the optimum condition is $\bar{y} = 0.1440$ and the sample standard deviation, $\sigma = 0.0230$.

$$\eta_{Optimum} = -10 \log_{10} (\bar{y}^2 + \sigma^2)$$

$$= -10 \log_{10} (0.1440^2 + 0.0230^2)$$

$$= -10 \log_{10} (0.0213)$$

$$= 16.72 \text{ dB}$$

3.   What is the gain in signal-to-noise ratio for the above data?

Answer

The gain in signal-to-noise ratio for the above data is:

$$Gain = \eta_{Optimum} - \eta_{Existing}$$

$$= 16.72 - 11.17 \text{ dB}$$

$$= 5.55 \text{ dB}$$

4.   Relate the gain in loss-to-society to the gain in signal-to-noise ratio for the above data.

Answer

The gain in loss-to-society can be related to the gain in signal-to-noise ratio through the gain in loss-per-piece ($\mathfrak{I}$),

$$\mathfrak{I} = k \left[MSD_{Existing} - MSD_{Optimum} \right]$$

$$= k \left[MSD_{Existing} - \mathfrak{R} \times MSD_{Existing} \right]$$

$$= k \, MSD_{Existing} \times [1 - \mathfrak{R}]$$

$$= k \, MSD_{Existing} \times \left[1 - 0.5^{\left(\frac{\eta_{Optimum} - \eta_{Existing}}{3} \right)} \right]$$

$$= k \, MSD_{Existing} \times \left[1 - 0.5^{\left(\frac{5.55}{3} \right)} \right]$$

$$= k \, MSD_{Existing} \times 0.723$$

Since $k \, MSD_{Existing}$ is the existing loss-to-society, the gain in loss-to-society is 0.723 or 72.3 %.

4.2 Application of the SN ratio – Smaller-the-better

4.2.1 Introduction
In this section, an example of the SN ratio calculation and the method of analysis is given for the smaller-the-better characteristic. The theory for this analysis has been developed in the previous section.

4.2.2 Optimization of smaller-the-better characteristic
In performing robust designs, it is preferable to study only main factors. That is, we use only control factors and do not include interactions. In particular, it is important to include noise factors. In the following example, we study only seven control factors and three noise factors. The analysis focuses on SN ratio calculations.

4.2.3 Example of smaller-the-better parameter design
Consider an experiment to reduce the flash thickness in moulded components. The quality characteristic is smaller-the-better and is measured in μm. Suppose the experiment is conducted on seven 2-level control factors A, B, C, D, E, F and G, where the existing levels are all level 1 and the proposed levels are all level 2. Consider also three 2-level noise factors H, I and J. Since noise factors are not intended to be controlled, there are no existing or proposed levels although we set them at two levels for the purpose of the experiment. We shall use an $L_8(2^7)$ orthogonal array for the control factors (inner array) and an $L_4(2^3)$ orthogonal array for the noise factors (outer array), as shown in Figure 4:2.1. Such an experimental design arrangement is called a direct product design. Notice that there are eight main experiments in the $L_8(2^7)$ orthogonal array. For each of these eight experiments, there are four combinations of noise factors settings as shown by the $L_4(2^3)$ orthogonal array, giving a total of 32 different experiments.

The direct product design lays out the way the experiments are conducted. For example, the $L_8(2^7)$ orthogonal array experiment 1 has control factors A in level 1, B in level 1, C in level 1, D in level 1, E in level 1, F in level 1 and G in level 1. The first combination of noise factor settings is H in level 1, I in level 1 and J in level 1. The result of the experiment conducted with this combination of control and noise factor settings is entered in the result column R1. The second combination retains the same control factor setting but the noise factor setting is H in level 1, I in level 2 and J in level 2. The result of this combination of control and noise factor settings is entered in the result column R2, and so on, for R3 and R4. We then proceed to the $L_8(2^7)$ orthogonal array experiment 2. The results of the entire experiment are given in Figure 4:2.2.

Exp	A	B	C	D	E	F	G	J	1	2	2	1
								I	1	2	1	2
								H	1	1	2	2
Exp	A	B	C	D	E	F	G	R1	R2	R3	R4
1	1	1	1	1	1	1	1				
2	1	1	1	2	2	2	2				
3	1	2	2	1	1	2	2				
4	1	2	2	2	2	1	1				
5	2	1	2	1	2	1	2				
6	2	1	2	2	1	2	1				
7	2	2	1	1	2	2	1				
8	2	2	1	2	1	1	2				

Figure 4:2.1 A direct product design with an $L_8(2^7)$ and an $L_4(2^3)$ orthogonal array.

The four results R1, R2, R3 and R4 of each experiment is then used to calculate the SN ratio. For smaller-the-better characteristic, the SN ratio η, is:

$$\eta = -10 \log_{10} \left[\sigma^2 + \bar{y}^2 \right]$$

Thus, for the $L_8(2^7)$ orthogonal array experiment 1, the SN ratio is denoted η_1 and calculated as follows. The data (R1, R2, R3 and R4) for this experiment are: 2.2, 2.1, 2.3 and 2.3. The sample standard deviation for these data is $\sigma = 0.0957$ and the mean $\bar{y} = 2.225$. Since

$$\eta_1 = -10 \log_{10} \left[\sigma^2 + \bar{y}^2 \right]$$

$$= -10 \log_{10} \left[0.0957^2 + 2.225^2 \right]$$

$$= -10 \log_{10} 4.96$$

$$= -6.95 \text{ dB}$$

For convenience, this result has also been entered into the column η (SN ratio) in Figure 4:2.2. The remaining SN ratio values, η_2 to η_8 are calculated similarly and entered in Figure 4:2.2.

							J	1	2	2	1	
							I	1	2	1	2	η
							H	1	1	2	2	
Exp	A	B	C	D	E	F	G	R1	R2	R3	R4	
1	1	1	1	1	1	1	1	2.2	2.1	2.3	2.3	−6.95
2	1	1	1	2	2	2	2	0.3	2.5	2.7	0.3	−5.88
3	1	2	2	1	1	2	2	0.5	3.1	0.4	2.8	−6.98
4	1	2	2	2	2	1	1	2.0	1.9	1.8	2.0	−5.70
5	2	1	2	1	2	1	2	3.0	3.1	3.0	3.0	−9.62
6	2	1	2	2	1	2	1	2.1	4.2	1.0	3.1	−9.36
7	2	2	1	1	2	2	1	4.0	1.9	4.5	2.2	−10.64
8	2	2	1	2	1	1	2	2.0	1.9	1.9	1.8	−5.58

Figure 4:2.2 Direct product design with results.

From then on, the response table is completed as shown in Figure 4:2.3. From the response table, the significant factors are A, D and F. As a rule of thumb, only three factors have been chosen since in an $L_8(2^7)$ orthogonal array experiment we may only take about half the degrees of freedom for important factors.

	A	B	C	D	E	F	G
Level 1	−6.38	−7.95	−7.26	−8.55	−7.22	−6.96	−8.16
Level 2	−8.80	−7.23	−7.91	−6.63	−7.96	−8.22	−7.01
Difference	2.42	0.73	0.65	1.92	0.74	1.25	1.15
Rank	1	6	7	2	5	3	4

Figure 4:2.3 Response table of experimental data.

Although the quality characteristic is smaller-the-better, recall that the SN ratio is defined in such a way as to *always* transform the quality characteristic into a larger-the-better characteristic. Thus, from Figure 4:2.3 the *optimum process* is therefore, A1, D2 and F1. The predicted SN ratio at this condition is:

$$\eta_{Optimum} = \bar{\eta} + (\overline{A1} - \bar{\eta}) + (\overline{D2} - \bar{\eta}) + (\overline{F1} - \bar{\eta})$$

$$= \overline{A1} + \overline{D2} + \overline{F1} - 2 \times \bar{\eta}$$

$$= -6.38 - 6.63 - 6.96 - 2 \times (-7.59)$$

$$= -4.79 \text{ dB}$$

The corresponding levels in the *existing process* are A1, D1 and G1. The predicted SN ratio at this condition is:

$$\eta_{Existing} = \bar{\eta} + (\overline{A1} - \bar{\eta}) + (\overline{D1} - \bar{\eta}) + (\overline{F1} - \bar{\eta})$$

$$= \overline{A1} + \overline{D1} + \overline{F1} - 2 \times \bar{\eta}$$

$$= -6.38 - 8.55 - 6.96 - 2 \times (-7.59)$$

$$= -6.71 \text{ dB}$$

The gain in loss-per-piece is therefore:

$$\Im = k \, MSD_{Existing} \times \left[1 - 0.5^{\left(\frac{\eta_{Optimum} - \eta_{Existing}}{3}\right)}\right]$$

$$= k \, MSD_{Existing} \times \left[1 - 0.5^{\left(\frac{-4.79 + 6.71}{3}\right)}\right]$$

$$= k \, MSD_{Existing} \times \left[1 - 0.5^{0.64}\right]$$

$$= k \, MSD_{Existing} \times \left[1 - 0.64\right]$$

$$= k \, MSD_{Existing} \times 0.36$$

This corresponds to a saving of 36 % of the original loss-to-society. Notice that the SN ratio value for the *existing process* was taken as $\eta_{Optimum} = -6.71$ dB based on the factors A1, D1 and F1 only. Alternatively, we could used $\eta_1 = -6.95$ dB from experiment 1 of the $L_8(2^7)$ orthogonal array experiment. Doing so would include the effects of the remaining unimportant factors not included in the optimum setting calculation of $\eta_{Existing} = -4.79$ dB and hence result in a less conservative estimate. The response graph of factor effects is shown in Figure 4:2.4.

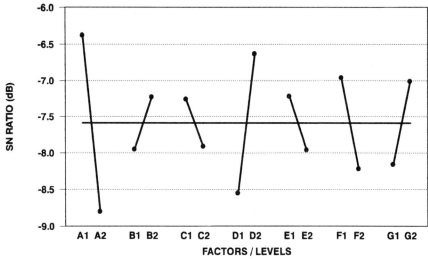

Figure 4:2.4 Response graph of factor effects.

4.2.4 Self-assessment questions

1.   Assuming a smaller-the-better characteristic, calculate the SN ratios for the following data:

							J	1	2	2	1	
							I	1	2	1	2	η
							H	1	1	2	2	
	A	B	C	D	E	F	G	R1	R2	R3	R4	
1	1	1	1	1	1	1	1	5.29	4.84	5.76	5.76	
2	1	1	1	2	2	2	2	4.76	6.76	7.84	6.35	
3	1	2	2	1	1	2	2	7.36	8.22	5.25	8.41	
4	1	2	2	2	2	1	1	4.41	4.00	3.61	4.41	
5	2	1	2	1	2	1	2	9.61	10.24	9.61	9.61	
6	2	1	2	2	1	2	1	3.45	4.56	1.21	3.45	
7	2	2	1	1	2	2	1	6.67	8.00	12.48	5.29	
8	2	2	1	2	1	1	2	4.41	7.34	5.00	3.61	

2.   Calculate the response table for the above data.

	A	B	C	D	E	F	G
Level 1							
Level 2							
Difference							
Rank							

3.   Draw the response graph of factor effects.

4.   Calculate the SN ratio at the optimum condition.

5.   Calculate the SN ratio at the existing condition.

6.   Calculate the reduction in loss-to-society.

4.2.5 Answers to self-assessment questions

1. Assuming a smaller-the-better characteristic, calculate the SN ratios for the following data:

								J	1	2	2	1	
								I	1	2	1	2	η
								H	1	1	2	2	
	A	B	C	D	E	F	G	R1	R2	R3	R4	
1	1	1	1	1	1	1	1	5.29	4.84	5.76	5.76	
2	1	1	1	2	2	2	2	4.76	6.76	7.84	6.35	
3	1	2	2	1	1	2	2	7.36	8.22	5.25	8.41	
4	1	2	2	2	2	1	1	4.41	4.00	3.61	4.41	
5	2	1	2	1	2	1	2	9.61	10.24	9.61	9.61	
6	2	1	2	2	1	2	1	3.45	4.56	1.21	3.45	
7	2	2	1	1	2	2	1	6.67	8.00	12.48	5.29	
8	2	2	1	2	1	1	2	4.41	7.34	5.00	3.61	

Answer

								J	1	2	2	1	
								I	1	2	1	2	η
								H	1	1	2	2	
	A	B	C	D	E	F	G	R1	R2	R3	R4	
1	1	1	1	1	1	1	1	5.29	4.84	5.76	5.76	-14.70
2	1	1	1	2	2	2	2	4.76	6.76	7.84	6.35	-16.33
3	1	2	2	1	1	2	2	7.36	8.22	5.25	8.41	-17.45
4	1	2	2	2	2	1	1	4.41	4.00	3.61	4.41	-12.31
5	2	1	2	1	2	1	2	9.61	10.24	9.61	9.61	-19.80
6	2	1	2	2	1	2	1	3.45	4.56	1.21	3.45	-10.81
7	2	2	1	1	2	2	1	6.67	8.00	12.48	5.29	-18.78
8	2	2	1	2	1	1	2	4.41	7.34	5.00	3.61	-14.55

2.   Calculate the response table for the data above.

	A	B	C	D	E	F	G
Level 1							
Level 2							
Difference							
Rank							

Answer

	A	B	C	D	E	F	G
Level 1	−15.20	−15.41	−16.09	−17.68	−14.37	−15.34	−14.14
Level 2	−15.98	−15.77	−15.09	−13.49	−16.80	−15.84	−17.03
Difference	0.78	0.36	1.00	4.19	2.43	0.50	2.89
Rank	5	7	4	1	3	6	2

3.   Draw the response graph of factor effects.

Answer

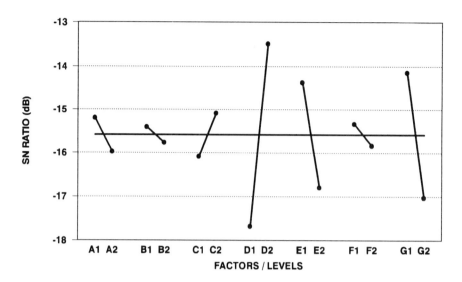

4.   Calculate the SN ratio at the optimum condition.

Answer

$$\eta_{Optimum} = \bar{\eta} + (\overline{D2} - \bar{\eta}) + (\overline{G1} - \bar{\eta}) + (\overline{E1} - \bar{\eta})$$

$$= \overline{D2} + \overline{G1} + \overline{E1} - 2 \times \bar{\eta}$$

$$= -13.49 - 14.14 - 14.37 - 2 \times (-15.59)$$

$$= -10.84 \text{ dB}$$

5.   Calculate the SN ratio at the existing condition.

Answer

$$\eta_{Existing} = \bar{\eta} + (\overline{DI} - \bar{\eta}) + (\overline{GI} - \bar{\eta}) + (\overline{EI} - \bar{\eta})$$

$$= \overline{DI} + \overline{GI} + \overline{EI} - 2 \times \bar{\eta}$$

$$= -17.68 - 14.14 - 14.37 - 2 \times (-15.59)$$

$$= -15.02 \text{ dB}$$

Alternatively, we may use η_1 (-14.70) as the SN ratio at the existing conditon.

6.   Calculate the reduction in loss-to-society.

Answer
The reduction in loss-to-society is:

$$\mathfrak{L} = k \, MSD_{Existing} \times \left[1 - 0.5^{\left(\frac{\eta_{Optimum} - \eta_{Existing}}{3}\right)}\right]$$

$$= k \, MSD_{Existing} \times \left[1 - 0.5^{\left(\frac{-10.84 + 15.02}{3}\right)}\right]$$

$$= k \, MSD_{Existing} \times \left[1 - 0.5^{1.40}\right]$$

$$= k \, MSD_{Existing} \times \left[1 - 0.38\right]$$

$$= k \, MSD_{Existing} \times 0.62$$

corresponding to a saving of 62 % of the original loss-to-society.

4.3 Fraction Defective Analysis

4.3.1 Introduction
Fraction defective type of data can be analyzed through fraction defective analysis. However, additivity in percentage data is poor when the data is close to zero (0) or one (1). A better method of analysis is to use the Omega transformation (Ω) for improving the additivity of such characteristics.

4.3.2 Fraction defective analysis
To illustrate a fraction defective analysis, we shall consider a wave soldering process. After wave soldering, some holes in a printed circuit board get blocked by solder bridging. Solder bridging is a defect since components cannot be inserted into these bridged holes. The defects are not continuously measurable and are treated as a classified attribute. They were qualitatively rated and grouped into three categories.

y = number of holes with bridging (none, some, severe) where
none   means that no hole in a printed circuit board has been bridged,
some   means that 5 % or less of the holes in a printed circuit board have been bridged,
severe means that more than 5 % of the holes in a printed circuit board have been bridged.

The factors shown in Figure 4:3.1 were investigated using an $L_8(2^7)$ orthogonal array.

	Factor	Level 1	Level 2
A	Flux type	Existing	New
B	Flux density	Low	High
C	Solder temperature	Low	High
D	Solder wave height	Low	High
E	Preheat setting	3	6
F	Flux air-knife angle	45°	90°
G	Hot air blast	Yes	No

Figure 4:3.1 Factors and level settings.

Twenty repetitions of eight experimental runs were conducted. The results are given in Figure 4:3.2. Classified attributes are less sensitive than continuous variables; therefore more data points are needed. Although only 20 repetitions were used in each experiment, a comparison of A1 and A2 for example has 160 data points (20 × 8).

Exp	A	B	C	D	E	F	G	None	Some	Severe	Total
1	1	1	1	1	1	1	1	15	4	1	20
2	1	1	1	2	2	2	2	5	13	2	20
3	1	2	2	1	1	2	2	8	12	0	20
4	1	2	2	2	2	1	1	2	11	7	20
5	2	1	2	1	2	1	2	17	2	1	20
6	2	1	2	2	1	2	1	4	15	1	20
7	2	2	1	1	2	2	1	7	13	0	20
8	2	2	1	2	1	1	2	3	11	6	20

Figure 4:3.2 Orthogonal array and results.

4.3.3 Completing the response table
The response table for fraction defective data is constructed by determining the *totals* for each factor level in each category as shown in Figure 4:3.3. For each factor, the total of the data in level 1 and level 2 should be the same (160 in this case).

		A	B	C	D	E	F	G
None	Level 1	30	41	30	47	30	37	28
	Level 2	31	20	31	14	31	24	33
Some	Level 1	40	34	41	31	42	28	43
	Level 2	41	47	40	50	39	53	38
Severe	Level 1	10	5	9	2	8	15	9
	Level 2	8	13	9	16	10	3	9
Difference None		1	21	1	33	1	13	5
Difference Some		1	13	1	19	3	25	5
Difference Severe		2	8	0	14	2	12	0
Rank		6	3	7	1	5	2	4

Figure 4:3.3 Response table of factor effects (totals).

The difference between level 1 and level 2 for each category must then be calculated. For example, for category None, level 1 = 30 and level 2 = 31. Therefore, the difference between these levels is 31 − 30 = 1. This value is entered into the row Difference None. Similarly, the difference between level 1 (41) and level 2 (40) for category Some is 1 and is entered into the row Difference Some. This analysis must be done for all the factors and categories of defects.

For fraction defective data, it is necessary to draw the response graph as a 100 % chart. The data have to be collected by factor level. Thus, for factor A level 1, None = 30, Some = 40 and Severe = 10. The total is of course 80. Expressed as percentages, this corresponds to None = 37.5 %, Some = 50 % and Severe = 12.5 %. The calculations would then need to be repeated for all the factors and levels. This is then drawn into the response graph in a cumulative form as shown in Figure 4:3.4.

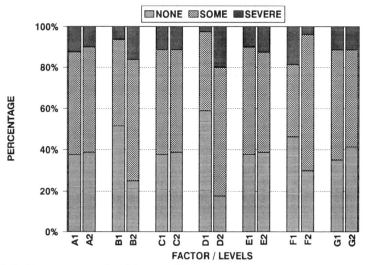

Figure 4:3.4 Response graph of factor effects.

4.3.4 Establishing significant factors

To find the important factor effects, we need to examine each factor and each category. This is best done from the response table. Looking at the differences in the Difference None, Some and Severe rows, the largest difference (33) occurs for factor D. So, factor D is ranked 1. The next largest difference (25) occurs for factor F, and this factor is therefore ranked 2. Similarly, all the factors are ranked. Since the orthogonal array used in this experiment is an $L_8(2^7)$ orthogonal array with seven degrees of freedom, we take about half the degrees of freedom (3 in this case) for important factors. These are factors D, F and B.

4.3.5 Establishing optimum conditions

The factor level selection of optimum condition depends on what category we want to maximize (or minimize). In the present example we could choose to maximize the None category. Alternatively, we could minimize the Severe category. Occasionally, it may be appropriate to maximize the Some category.

Suppose we want to minimize the Severe category. To do that, we take the important factor levels with the smaller fraction defective values for the Severe category. Thus, for factor D, we choose level 1. Similarly, for factor F we choose level 2 and for factor B we choose level 1. The average fraction defective $\bar{y}$ for Severe defects is:

$$\bar{y} = \frac{total\ number\ in\ category}{total\ number\ of\ observations}$$

$$= \frac{18}{160}$$

$$= 0.1125$$

The process average at the predicted optimum condition $P_{Predicted}$ is:

$$P_{Predicted} = \bar{y} + (\overline{D1} - \bar{y}) + (\overline{F2} - \bar{y}) + (\overline{B1} - \bar{y})$$

$$= \overline{D1} + \overline{F2} + \overline{B1} - 2 \times \bar{y}$$

$$= \frac{2}{80} + \frac{3}{80} + \frac{5}{80} - 2 \times \frac{18}{160}$$

$$= -0.10$$

$$= -10\ \%$$

From experience, it is infeasible for fraction defective to be negative. However, the results above predict a negative value. This is because the quality characteristic namely, fraction defective, is not additive. For characteristics such as percent defective (or yield), arithmetic additivity is poor in the neighbourhood of 0 or 1. To overcome this problem, it is recommended to use the *Omega transformation* to convert non-additive fraction defective data into a form that possesses additivity. The Omega transformation formula is:

$$\Omega = -10\ \log_{10} \left(\frac{1}{p} - 1 \right)\ dB$$

where p is the fraction defective. The Omega transformation is the conversion from a range of 0 to 1 to a range of negative infinity to positive infinity. Mathematically:

$$\text{Range of } p \; [0 \text{ to } 1] \rightarrow \text{Range of } \Omega \; [-\infty \text{ to } +\infty]$$

When the fraction defective for a factor level is known, for example, D1 Severe is (2/80), i.e. 0.025, we use the Omega transformation to establish the dB value. Substituting $p = 0.025$ into:

$$\Omega_{D1} = -10 \log_{10} \left(\frac{1}{p} - 1 \right)$$

$$= -10 \log_{10} \left(\frac{1}{0.025} - 1 \right)$$

$$= -10 \log_{10} (40 - 1)$$

$$= -15.91 \text{ dB}$$

Similarly, the Omega transformation for the factor levels F2, B1 and the overall mean ($\bar{\Omega}$) for the Severe category is calculated. These are -14.09, -11.76 and -8.97 dB, respectively. The predicted Omega value at the predicted optimum condition is:

$$\Omega_{Predicted} = \bar{\Omega} + (\Omega_{D1} - \bar{\Omega}) + (\Omega_{F2} - \bar{\Omega}) + (\Omega_{B1} - \bar{\Omega})$$

$$= \Omega_{D1} + \Omega_{F2} + \Omega_{B1} - 2 \times \bar{\Omega}$$

$$= -15.91 - 14.09 - 11.76 - 2 \times (-8.97)$$

$$= -23.82 \text{ dB}$$

The predicted Omega value is then transformed back into the fraction defective. We should be able to transform the formula to obtain p in fraction defective using the following formula:

$$p = \frac{1}{1 + 10^{\frac{\Omega}{-10}}}$$

Substituting $\Omega = -23.82$ dB,

$$p = \frac{1}{1 + 10^{-\frac{\Omega}{10}}}$$

$$= \frac{1}{1 + 10^{\frac{23.82}{10}}}$$

$$= 0.0041$$

$$= 0.41 \ \%$$

Using the Omega transformation, the value of percent defective at the optimum condition for category Severe is 0.41 %. This is a more sensible estimate compared to the value of -10 % obtained using percent defective data. Of course, a confirmation experiment should always be carried out at the optimum condition and compared with the predicted value to verify the reproducibility of the results.

4.3.6 Self-assessment questions

1.   What is the main problem encountered when percent defective is used as a measure of quality?

2.   How can this problem be overcome?

3.   A scientist found that drugs A, B and C can be used to treat wheezing in asthmatic patients. If no drug is given, the patient's condition is poor. If only drug A is given the patient feels good. If only drug B is given, the patient feels very good. If only drug C is given the patient feels good (similar to A). Can the three drugs be considered as three control factors?

4.   Defects known as streaks occur during the polishing operation on optical lenses. An experiment was designed to minimize these streaks. Seven factors were studied at two levels each. The lenses were graded as Good, Fair and Bad, where only the Bad lenses were unacceptable to the customer. Hence, the objective is to maximize Good and Fair lenses. From the data below:

Exp	A	B	C	D	E	F	G	Good	Fair	Bad	Total
1	1	1	1	1	1	1	1	1	9	5	15
2	1	1	1	2	2	2	2	1	10	4	15
3	1	2	2	1	1	2	2	4	10	1	15
4	1	2	2	2	2	1	1	8	7	0	15
5	2	1	2	1	2	1	2	11	3	1	15
6	2	1	2	2	1	2	1	1	4	10	15
7	2	2	1	1	2	2	1	3	10	2	15
8	2	2	1	2	1	1	2	14	1	0	15

1.   Construct a response table and determine the significant factors.
2.   Construct a response graph for the factor effects.
3.   What are the optimum factor levels for reducing the number of Bad streaks?
4.   Estimate the fraction defective at the optimum factor setting using percentage data. Why is this estimation not useful?

5. Estimate the fraction defective at the optimum factor setting using the Omega transformation. Why is this a better estimation than that found using percentage data?

6. What is the next step? On what do you base your decision?

4.3.7 Answers to self-assessment questions

1.   What is the main problem encountered when percent defective is used as a measure of quality?

Answer

Percent defective data lie in the range of 0 to 1. Additivity is poor in the neighbourhood of 0 or 1. Therefore, a quality characteristic based on percent defective data does not have good additivity. Hence, the experimental result may not be reproducible.

2.   How can this problem be overcome?

Answer

This problem can be overcome by choosing a quality characteristic with good additivity through the selection of a function which has a monotonic property. Alternatively, the Omega transformation may be used.

3.   A scientist found that drugs A, B and C can be used to treat wheezing in asthmatic patients. If no drug is given, the patient's condition is poor. If only drug A is given the patient feels good. If only drug B is given, the patient feels very good. If only drug C is given the patient feels good (similar to A). Can the three drugs be considered as three control factors?

Answer

No. Suppose we take a close look at these drugs and find that they all contain Actophine as the active ingredient. Suppose, drug A has 50 % full dose, drug B has 100 % full dose and drug C has 150 % full dose. During experimentation, it is possible that one or more experiments require the patient to take all three drugs. This could lead to an overdose of Actophine and could worsen the patient's condition. Clearly, Actophine concentration would be a better quality characteristic.

4.   Defects known as streaks occur during the polishing operation on optical lenses. An experiment was designed to minimize these streaks. Seven factors were studied at two levels each. The lenses were graded as Good, Fair and Bad, where only the Bad lenses were unacceptable to the customer. Hence, the objective is to maximize Good and Fair lenses. From the data below:

Exp	A	B	C	D	E	F	G	Good	Fair	Bad	Total
1	1	1	1	1	1	1	1	1	9	5	15
2	1	1	1	2	2	2	2	1	10	4	15
3	1	2	2	1	1	2	2	4	10	1	15
4	1	2	2	2	2	1	1	8	7	0	15
5	2	1	2	1	2	1	2	11	3	1	15
6	2	1	2	2	1	2	1	1	4	10	15
7	2	2	1	1	2	2	1	3	10	2	15
8	2	2	1	2	1	1	2	14	1	0	15

1.   Construct a response table and determine the significant factors.

Answer

		A	B	C	D	E	F	G
Good	Level 1	14	14	19	19	20	34	13
	Level 2	29	29	24	24	23	9	30
Fair	Level 1	36	26	30	32	24	20	30
	Level 2	18	28	24	22	30	34	24
Bad	Level 1	10	20	11	9	16	6	17
	Level 2	13	3	12	14	7	17	6
Difference Good		15	15	5	5	3	25	17
Difference Fair		18	2	6	10	6	14	6
Difference Bad		3	17	1	5	9	11	11
Rank		2	3	7	5	6	1	3

The most important factors are F, A, B and G.

2.   Construct a response graph for the factor effects.

Answer

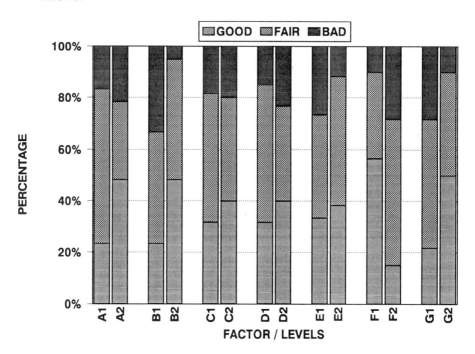

3.   What are the optimum factor levels for reducing the number of Bad streaks?

Answer

From the response table, the optimum factor setting is F1, A1, B2 and G2.

4.   Estimate the fraction defective at the optimum factor setting using percentage data. Why is this estimation not useful?

Answer

The average fraction acceptable $\bar{y}$ for Good and Fair products is the total of lenses in the Good and Fair classification:

$$\bar{y} = \frac{\text{total number in categories}}{\text{total number of observations}}$$

$$= \frac{97}{120}$$

$$= 0.808$$

The predicted process average at the optimum condition $P_{Predicted}$ is:

$$P_{Predicted} = \bar{y} + (\overline{F1} - \bar{y}) + (\overline{A1} - \bar{y}) + (\overline{B2} - \bar{y}) + (\overline{G2} - \bar{y})$$

$$= \overline{F1} + \overline{A1} + \overline{B2} + \overline{G2} - 3 \times \bar{y}$$

$$= \frac{54}{60} + \frac{50}{60} + \frac{57}{60} + \frac{54}{60} - 3 \times \frac{97}{120}$$

$$= 1.158$$

$$= 115.8 \%$$

Note that this estimation of 115.8 % is not meaningful. This problem arises because percentage streaks is not additive. The purpose of this question is to compare the results with the Omega transformation.

5.   Estimate the fraction defective at the optimum factor setting using the Omega transformation. Why is this a better estimation than that found using percentage data?

Answer

The fraction defective for F1 is 54/60. Substituting into:

$$\Omega_{F1} = -10 \, \log_{10} \left(\frac{1}{p} - 1 \right)$$

$$= -10 \, \log_{10} \left(\frac{60}{54} - 1 \right)$$

$$= -10 \, \log_{10} 0.11$$

$$= 9.54 \text{ dB}$$

Similarly, the Omega transformations for factor levels A1, B2 and G2 are calculated. These values are 6.99, 12.79, and 9.54 dB, respectively. The mean (Ω) for Good and Fair lenses is 97/120. The corresponding Omega transformation is 6.25 dB. The predicted Omega value at the optimum condition is:

$$\Omega_{Predicted} = \bar{\Omega} + (\Omega_{FI} - \bar{\Omega}) + (\Omega_{AI} - \bar{\Omega}) + (\Omega_{B2} - \bar{\Omega}) + (\Omega_{G2} - \bar{\Omega})$$

$$= \Omega_{FI} + \Omega_{AI} + \Omega_{B2} + \Omega_{G2} - 3 \times \bar{\Omega}$$

$$= 9.54 + 6.99 + 12.79 + 9.54 - 3 \times 6.25$$

$$= 20.11 \text{ dB}$$

The predicted Omega value is then transformed back into fraction defective p, using the following formula:

$$p = \frac{1}{1 + 10^{\frac{\Omega}{-10}}}$$

Substituting $\Omega = 20.11$ dB,

$$p = \frac{1}{1 + 10^{\frac{\Omega}{-10}}}$$

$$= \frac{1}{1 + 10^{\frac{20.11}{-10}}}$$

$$= 0.990$$

$$= 99.0 \%$$

The predicted process average calculated using the fraction defective is 115.8 %, whereas the predicted process average calculated using the Omega transformation is 99.0 %. The Omega transformation has better additivity in the neighbourhood of 0 or 1. This is because the Omega transformation transforms a range of 0 to 1 to a range of $-\infty$ to $+\infty$. Hence additivity is improved in the neighbourhood of 0 or 1.

6.   What is the next step? On what do you base your decision?

Answer

The next step is to conduct a confirmation experiment. If the result of the confirmation experiment is close to the predicted process average then I would proceed to implement the optimum factor setting on a larger scale. I will base my decisions on the fact that if the confirmation experiment results is close to the predicted process average then the optimum factor settings may be considered reproducible.

4.4 Application of the SN ratio – Operating Window

4.4.1 Introduction
In this section, an example of the SN ratio calculation and the method of analysis is given for an operating window characteristic. The theory for this analysis has been developed in Section 4.1.9.

4.4.2 Optimization of the operating window
In performing Robust Design, it is preferable to study only main factors. That is, we use only control factors and do not include interactions for reasons already explained. In particular, it is important to include noise factors. In the following example, we shall study only seven main factors and three noise factors. Again, the analysis focuses on SN ratio calculations.

4.4.3 Example of the operating window
Consider an experiment to improve paper feeding in a photocopier. As discussed earlier, the misfeed threshold is a smaller-the-better quality characteristic and the multifeed threshold is a larger-the-better characteristic. Suppose an experiment is conducted with seven 2-level control factors A, B, C, D, E, F and G where the existing levels are all level 1 and the proposed levels are all level 2. Consider also three 2-level noise factors H, I and J. Since noise factors are not intended to be controlled, there are no existing or proposed levels although we set them at two levels for the purpose of the experiment.

     We will use an $L_8(2^7)$ orthogonal array for the control factors (inner array) and an $L_4(2^3)$ orthogonal array for the noise factors (outer array), as shown in Figure 4:4.1. Recall that such an experimental design is called the direct product design.

     The response measured is the spring tension, which is a continuous variable. With control factors set as in experiment 1, and noise factor combinations as in column R1; see Figure 4:4.1, the spring force is adjusted slowly until a sheet of paper is fed. This is the threshold point of misfeed. This result is recorded as $y_{S:1,1}$. The spring force is then adjusted further until more than one sheet of paper is fed. This is the point of multifeed. This result is recorded as $y_{L:1,1}$. Next, the noise factor conditions are changed to those in column R2 and the threshold points of misfeed $y_{S:1,2}$ and multifeed $y_{L:1,2}$ are recorded.

     Notice that there are eight main experiments in the $L_8(2^7)$ orthogonal array. For each of these eight experiments, there are four combinations of noise factor settings as shown by the $L_4(2^3)$ orthogonal array. Further, since two quality characteristics, misfeed (y_S) and multifeed (y_L) are studied, there are a total of 64 ($8 \times 4 \times 2$) different experiments.

								J	1	2	2	1			η_{Mis}
								I	1	2	1	2	$\bar{y}$	η	$+$
								H	1	1	2	2			η_{Mul}
	A	B	C	D	E	F	G	R1	R2	R3	R4			
1	1	1	1	1	1	1	1	$y_{S:1,1}$	$y_{S:1,2}$	$y_{S:1,3}$	$y_{S:1,4}$	$\bar{y}_S$	η_{Mis}	η_1
								$y_{L:1,1}$	$y_{L:1,2}$	$y_{L:1,3}$	$y_{L:1,4}$	$\bar{y}_L$	η_{Mul}	
2	1	1	1	2	2	2	2							
3	1	2	2	1	1	2	2							
4	1	2	2	2	2	1	1							
5	2	1	2	1	2	1	2							
6	2	1	2	2	1	2	1							
7	2	2	1	1	2	2	1							
8	2	2	1	2	1	1	2							

Figure 4:4.1 A direct product design with an $L_8(2^7)$ and an $L_4(2^3)$ orthogonal array. For brevity, we use *Mis* for misfeed and *Mul* for multifeed.

Using the first row of each experiment, namely the four results R1, R2, R3 and R4 of the misfeed data, the SN ratio for smaller-the-better characteristic $\eta_{1:Misfeed}$ is calculated using the equation:

$$\eta_{1:Misfeed} = -10 \log_{10} \frac{1}{n} \sum y_{Si}^2$$

$$= -10 \log_{10} \frac{1}{n} \left[y_{S:1,1}^2 + y_{S:1,2}^2 + ... + y_{S:1,n}^2 \right]$$

Similarly, using the second row of each experiment, namely the four results R1, R2, R3 and R4 of the multifeed data, the SN ratio for larger-the-better characteristic $\eta_{1:Multifeed}$ is calculated using the equation:

$$\eta_{1:Multifeed} = -10 \log_{10} \frac{1}{n} \sum \frac{1}{y_{Li}^2}$$

$$= -10 \log_{10} \frac{1}{n} \left[\frac{1}{y_{L:1,1}^2} + \frac{1}{y_{L:1,2}^2} + ... + \frac{1}{y_{L:1,n}^2} \right]$$

In operating window analysis, it is necessary to maximize the range of the operating window. Therefore, we calculate:

$$\eta_1 = \eta_{Misfeed} + \eta_{Multifeed}$$

$$= \eta_{SI} + \eta_{LI}$$

Similarly, the SN ratios (η_2 to η_8) for the remaining seven experiments need to be calculated.

4.4.4 Experimental data analysis
Figure 4:4.2 shows the result of an operating window characteristic experiment. For the $L_8(2^7)$ orthogonal array Experiment 1, we calculate the:
- mean for misfeed data,
- mean for multifeed data,
- SN ratio for misfeed data,
- SN ratio for multifeed data,
- sum of SN ratios.

1.   Mean for misfeed data
The mean $\bar{y}_{SI}$ is calculated as:

$$\bar{y}_{SI} = \frac{1}{n}\sum y_{Si}$$

$$= \frac{1}{n}\left[y_{S:1,1} + y_{S:1,2} + ... + y_{S:1,n}\right]$$

$$= \frac{20 + 25 + 32 + 39}{4}$$

$$= 29.00$$

2.   Mean for multifeed data

The mean $\bar{y}_{LI}$ is calculated as:

$$\bar{y}_{LI} = \frac{1}{n}\sum y_{Li}$$

$$= \frac{1}{n}\left[y_{L:1,1} + y_{L:1,2} + ... + y_{L:1,n}\right]$$

$$= \frac{55 + 64 + 72 + 73}{4}$$

$$= 66.00$$

3.   SN ratio for misfeed data

The SN ratio denoted η_{SI} is calculated as follows:

$$\eta_{SI} = -10 \log_{10} \frac{1}{n}\left[y_{S:1,1}^2 + y_{S:1,2}^2 + ... + y_{S:1,n}^2\right]$$

$$= -10 \log_{10} \frac{1}{4}\left[20^2 + 25^2 + 32^2 + 39^2\right]$$

$$= -10 \log_{10} 892.5$$

$$= -29.51 \text{ dB}$$

4.   SN ratio for multifeed data

The SN ratio denoted η_{LI} is calculated as follows:

$$\eta_{LI} = -10 \log_{10} \frac{1}{n} \left[\frac{1}{y_{L:1,1}^2} + \frac{1}{y_{L:1,2}^2} + ... + \frac{1}{y_{L:1,n}^2} \right]$$

$$= -10 \log_{10} \frac{1}{4} \left[\frac{1}{55^2} + \frac{1}{64^2} + \frac{1}{72^2} + \frac{1}{73^2} \right]$$

$$= -10 \log_{10} 0.0002388$$

$$= 36.22 \text{ dB}$$

5.   Sum of SN ratios
The sum of the SN ratios is:

$$\eta_1 = \eta_{Misfeed} + \eta_{Multifeed}$$

$$= \eta_{SI} + \eta_{LI}$$

$$= -29.51 + 36.22$$

$$= 6.71 \text{ dB}$$

These values are entered in Figure 4:4.2. These calculations need to be repeated for the remaining data. The complete set of results are also shown in Figure 4:4.2. From these results we draw the response table (Figure 4:4.3) for the sum of the SN ratios (last column of Figure 4:4.2).

From Figure 4:4.3, the optimum factor levels based on SN ratio (larger-the-better) are E1, D1, F2 and A1. Only four factors have been chosen since in an $L_8(2^7)$ orthogonal array experiment we may only take about half the degrees of freedom for significant factors.

							J	1	2	2	1			η_{Mis}
							I	1	2	1	2	$\bar{y}$	η	$+$
							H	1	1	2	2			η_{Mul}
	A	B	C	D	E	F	G	R1	R2	R3	R4			
1	1	1	1	1	1	1	1	20	25	32	39	29.0	−29.5	6.7
								55	64	72	73	66.0	36.2	
2	1	1	1	2	2	2	2	25	28	30	40	30.8	−29.9	2.7
								38	41	45	50	43.5	32.6	
3	1	2	2	1	1	2	2	15	18	22	23	19.5	−25.9	12.3
								80	81	82	83	81.5	38.2	
4	1	2	2	2	2	1	1	30	33	33	34	32.5	−30.2	−0.4
								25	33	33	38	32.3	29.9	
5	2	1	2	1	2	1	2	30	31	30	31	30.5	−29.7	3.7
								45	46	47	48	46.5	33.3	
6	2	1	2	2	1	2	1	30	34	35	36	33.8	−30.6	5.1
								55	61	63	67	61.5	35.7	
7	2	2	1	1	2	2	1	26	28	29	30	28.3	−29.0	1.6
								25	35	43	45	37.0	30.6	
8	2	2	1	2	1	1	2	29	35	35	38	34.3	−30.7	2.2
								40	43	47	49	44.8	32.9	

Figure 4:4.2 Misfeed and multifeed data for operating window experiment.

	A	B	C	D	E	F	G
Level 1	5.34	4.56	3.31	6.07	6.59	3.05	3.27
Level 2	3.15	3.93	5.17	2.42	1.90	5.44	5.22
Difference	2.19	0.62	1.86	3.65	4.68	2.40	1.96
Rank	4	7	6	2	1	3	5

Figure 4:4.3 Response table for combined SN ratio data.

We now characterize the:
- misfeed process,
- multifeed process.

1.   Misfeed process

In order to characterize the misfeed process, we need to calculate the predicted SN ratio and the predicted mean for misfeed data. The SN ratio response table in shown in Figure 4:4.4.

	A	B	C	D	E	F	G
Level 1	−28.89	−29.92	−29.79	−28.54	−29.19	−30.04	−29.84
Level 2	−30.01	−28.98	−29.11	−30.37	−29.72	−28.86	−29.06
Difference	1.12	0.94	0.68	1.83	0.53	1.18	0.78
Rank	3	4	6	1	7	2	5

Figure 4:4.4 Response table for misfeed data (SN ratio).

The mean response table for misfeed data is shown in Figure 4:4.5.

	A	B	C	D	E	F	G
Level 1	27.94	31.00	30.56	26.81	29.13	31.56	30.88
Level 2	31.69	28.63	29.06	32.81	30.50	28.06	28.75
Difference	3.75	2.38	1.05	6.00	1.38	3.50	2.13
Rank	2	4	6	1	7	3	5

Figure 4:4.5 Response table for misfeed data (Mean).

Using only the levels for E1, D1, F2 and A1, **as obtained from the combined SN ratio analysis (Figure 4:4.3)**, the predicted SN ratio from the response table for the SN ratio of the misfeed data (Figure 4:4.4) is:

$$\eta_{Predicted} = (\overline{E1} - \overline{\eta}) + (\overline{D1} - \overline{\eta}) + (\overline{F2} - \overline{\eta}) + (\overline{A1} - \overline{\eta})$$

$$= \overline{E1} + \overline{D1} + \overline{F2} + \overline{A1} - 3 \times \overline{\eta}$$

$$= -29.19 - 28.54 - 28.86 - 28.89 + 3 \times 29.45$$

$$= -27.12 \text{ dB}$$

Using only the levels for E1, D1, F2 and A1, **as obtained from the combined SN ratio analysis (Figure 4:4.3)**, the predicted mean from the response table for the mean of the misfeed data (Figure 4:4.5) is:

$$y_{Predicted} = (\overline{E1} - \overline{y}) + (\overline{D1} - \overline{y}) + (\overline{F2} - \overline{y}) + (\overline{A1} - \overline{y})$$

$$= \overline{E1} + \overline{D1} + \overline{F2} + \overline{A1} - 3 \times \overline{y}$$

$$= 29.13 + 26.81 + 28.06 + 27.94 - 3 \times 29.81$$

$$= 22.50$$

Knowing the predicted SN ratio and the mean, we can calculate the variance at this condition:

$$\eta = -10 \log_{10} (\sigma^2 + \overline{y}^2)$$

$$\therefore \sigma^2 + \overline{y}^2 = 10^{\left(\frac{\eta}{-10}\right)}$$

$$\sigma^2 = 10^{\left(\frac{\eta}{-10}\right)} - \overline{y}^2$$

$$\sigma = \sqrt{10^{\left(\frac{\eta}{-10}\right)} - \overline{y}^2}$$

Substituting, $\eta = -27.12$ and $\overline{y} = 22.50$;

$$\sigma = \sqrt{10^{\left(\frac{\eta}{-10}\right)} - \bar{y}^2}$$

$$\sigma = \sqrt{10^{\left(\frac{-27.12}{-10}\right)} - 22.50^2}$$

$$\sigma = \sqrt{515.08 - 506.25}$$

$$\sigma = \sqrt{8.83}$$

$$\sigma = 2.97$$

2. Multifeed process

In order to characterize the multifeed process, we need to calculate the predicted SN ratio and the predicted mean for multifeed data. The SN ratio response table in shown in Figure 4:4.6.

	A	B	C	D	E	F	G
Level 1	34.23	34.48	33.11	34.61	35.77	33.09	33.11
Level 2	33.16	32.91	34.28	32.79	31.62	34.30	34.28
Difference	1.08	1.56	1.18	1.82	4.15	1.21	1.17
Rank	7	3	5	2	1	4	6

Figure 4:4.6 Response table for multifeed data (SN ratio).

The mean response table for multifeed data is shown in Figure 4:4.7.

	A	B	C	D	E	F	G
Level 1	55.81	54.38	47.81	57.75	63.44	47.38	49.19
Level 2	47.44	48.88	55.44	45.50	39.81	55.88	54.06
Difference	8.38	5.50	7.63	12.25	23.63	8.50	4.88
Rank	4	6	5	2	1	3	7

Figure 4:4.7 Response table for multifeed data (Mean).

Using only the levels for E1, D1, F2 and A1, **as obtained from the combined SN ratio analysis (Figure 4:4.3)**, the predicted SN ratio from the response table for the SN ratio of the multifeed data (Figure 4:4.6) is:

$$\eta_{Predicted} = (\overline{E1} - \overline{\eta}) + (\overline{D1} - \overline{\eta}) + (\overline{F2} - \overline{\eta}) + (\overline{A1} - \overline{\eta})$$

$$= \overline{E1} + \overline{D1} + \overline{F2} + \overline{A1} - 3 \times \overline{\eta}$$

$$= 35.77 + 34.61 + 34.30 + 34.23 - 3 \times 33.70$$

$$= 37.83 \text{ dB}$$

Using only the levels for E1, D1, F2 and A1, **as obtained from the combined SN ratio analysis (Figure 4:4.3)**, the predicted mean from the response table for the mean of the multifeed data (Figure 4:4.7) is:

$$y_{Predicted} = (\overline{E1} - \overline{y}) + (\overline{D1} - \overline{y}) + (\overline{F2} - \overline{y}) + (\overline{A1} - \overline{y})$$

$$= \overline{E1} + \overline{D1} + \overline{F2} + \overline{A1} - 3 \times \overline{y}$$

$$= 63.44 + 57.75 + 55.88 + 55.81 - 3 \times 51.63$$

$$= 78.00$$

Knowing the predicted SN ratio and the mean, we can calculate the standard deviation at this condition:

$$\eta = -10 \log_{10}\left[\frac{1}{\overline{y}^2}\left[1 + \frac{3\sigma^2}{\overline{y}^2}\right]\right]$$

$$\frac{\eta}{-10} = \log_{10}\left[\frac{1}{\overline{y}^2}\left[1 + \frac{3\sigma^2}{\overline{y}^2}\right]\right]$$

$$\therefore \frac{1}{\overline{y}^2}\left[1 + \frac{3\sigma^2}{\overline{y}^2}\right] = 10^{\left(\frac{\eta}{-10}\right)}$$

$$\left[1 + \frac{3\sigma^2}{\overline{y}^2}\right] = \overline{y}^2 \times 10^{\left(\frac{\eta}{-10}\right)}$$

$$\left[1 + \frac{3\sigma^2}{\bar{y}^2}\right] = \bar{y}^2 \times 10^{\left(\frac{\eta}{-10}\right)}$$

$$\frac{3\sigma^2}{\bar{y}^2} = \bar{y}^2 \times 10^{\left(\frac{\eta}{-10}\right)} - 1$$

$$\sigma^2 = \frac{\bar{y}^2 \left(\bar{y}^2 \times 10^{\left(\frac{\eta}{-10}\right)} - 1\right)}{3}$$

$$\sigma = \sqrt{\frac{\bar{y}^2 \left(\bar{y}^2 \times 10^{\left(\frac{\eta}{-10}\right)} - 1\right)}{3}}$$

Substituting, $\eta = 37.83$ and $\bar{y} = 78.00$;

$$\sigma = \sqrt{\frac{\bar{y}^2 \left(\bar{y}^2 \times 10^{\left(\frac{\eta}{-10}\right)} - 1\right)}{3}}$$

$$= \sqrt{\frac{78.00^2 \left(78.00^2 \times 10^{\left(\frac{37.83}{-10}\right)} - 1\right)}{3}}$$

$$= 2.73$$

4.4.5 Optimum process

The best spring force $f_{Optimum}$ is taken as the geometric mean of the misfeed force and multifeed force. Mathematically:

$$f_{Optimum} = \sqrt{f_{Misfeed} \times f_{Multifeed}}$$

$$= \sqrt{22.50 \times 78.00}$$

$$= 41.89$$

The gain in going from the existing process (experiment 1) to the optimum process is:

$$\eta_{Gain} = (\eta_{Misfeed} + \eta_{Multifeed})_{Optimum} - (\eta_{Misfeed} + \eta_{Multifeed})_{Existing}$$

$$= (-27.12 + 37.83) - (-29.51 + 36.22)$$

$$= 4.00 \text{ dB}$$

This corresponds to a reduction $\Re$, by a factor of:

$$\Re = \frac{1}{0.5^{\left(\frac{\eta_{Gain}}{3}\right)}}$$

$$= \frac{1}{0.5^{\left(\frac{4.00}{3}\right)}}$$

$$= 2.52$$

Assume that the current process can be characterised with the data from Experiment 1. The mean misfeed force is 29.00 with a standard deviation of 8.29. The mean multifeed force is 66.00 with a standard deviation of 8.37. These, together with the predicted optimum process, can be characterized as shown in Figure 4:4.8.

	Before optimization		After optimization	
	Misfeed	Multifeed	Misfeed	Multifeed
mean	29.00	66.00	22.50	78.00
standard deviation	8.29	8.37	2.97	2.73

Figure 4:4.8 Comparison of process characteristics before and after optimization.

Using these results, we may compare the process improvement using the normal distribution plots:

$$\Phi = \frac{1}{\sqrt{(2\pi\sigma^2)}} e^{-\frac{1}{2}\left(\frac{y-\bar{y}}{\sigma}\right)^2} \tag{4:4.21}$$

where Φ is the relative frequency, σ is the standard deviation and $\bar{y}$ is the mean. The normal plots for existing conditions of misfeed and multifeed are compared to those of the optimum conditions of misfeed and multifeed in Figure 4:4.9.

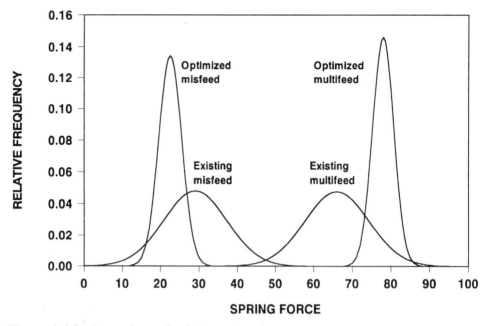

Figure 4:4.9 Comparison of existing and optimum processes.

4.4.6 Discussion of operating windows

In this example, the optimum condition was determined from the sum of SN ratios for misfeed and multifeed data. The SN ratio for the optimum condition could also have been determined directly from the combined SN ratio response data. However, we chose to calculate the SN ratio of the misfeed data from the response table for the misfeed SN ratio and, similarly, the SN ratio of the multifeed data from the response table for the multifeed SN ratio. Doing so allows us to predict the performance of both misfeed and multifeed processes and make a direct comparison with the confirmation results for each process. However, one disadvantage of this is that some factors may have a strong effect on one of the two quality characteristics without having a significant impact on the operating window. This tendency could lead to an under estimation of the individual response. A confirmation experiment may therefore appear to indicate that a major factor effect is missing, whereas in fact all the key factors may have been properly identified.

4.4.7 Self-assessment questions

1.   Complete the operating window analysis for the following data.

								J	1	2	2	1			η_{Mis} + η_{Mul}
								I	1	2	1	2	$\bar{y}$	η	
								H	1	1	2	2			
	A	B	C	D	E	F	G	R1	R2	R3	R4			
1	1	1	1	1	1	1	1	12	13	17	20			
								31	33	38	38			
2	1	1	1	2	2	2	2	13	15	15	21			
								20	21	24	26			
3	1	2	2	1	1	2	2	8	10	12	12			
								41	41	43	43			
4	1	2	2	2	2	1	1	15	18	18	19			
								15	18	19	21			
5	2	1	2	1	2	1	2	16	16	15	17			
								25	26	26	26			
6	2	1	2	2	1	2	1	17	17	20	18			
								29	32	34	35			
7	2	2	1	1	2	2	1	14	15	17	15			
								15	21	23	26			
8	2	2	1	2	1	1	2	16	20	18	21			
								25	23	27	22			

4.4.8 Answers to self-assessment questions

1.   Complete the operating window analysis for the following data.

								J	1	2	2	1			η_{Mis} + η_{Mul}
								I	1	2	1	2	$\bar{y}$	η	
								H	1	1	2	2			
	A	B	C	D	E	F	G	R1	R2	R3	R4			
1	1	1	1	1	1	1	1	12	13	17	20			
								31	33	38	38			
2	1	1	1	2	2	2	2	13	15	15	21			
								20	21	24	26			
3	1	2	2	1	1	2	2	8	10	12	12			
								41	41	43	43			
4	1	2	2	2	2	1	1	15	18	18	19			
								15	18	19	21			
5	2	1	2	1	2	1	2	16	16	15	17			
								25	26	26	26			
6	2	1	2	2	1	2	1	17	17	20	18			
								29	32	34	35			
7	2	2	1	1	2	2	1	14	15	17	15			
								15	21	23	26			
8	2	2	1	2	1	1	2	16	20	18	21			
								25	23	27	22			

Answer

								J	1	2	2	1			η_{Mis} + η_{Mul}
								I	1	2	1	2	$\bar{y}$	η	
								H	1	1	2	2			
	A	B	C	D	E	F	G		R1	R2	R3	R4			
1	1	1	1	1	1	1	1		12	13	17	20	15.5	−23.9	6.8
									31	33	38	38	35.0	30.8	
2	1	1	1	2	2	2	2		13	15	15	21	16.0	−24.2	2.8
									20	21	24	26	22.8	27.0	
3	1	2	2	1	1	2	2		8	10	12	12	10.5	−20.5	11.9
									41	41	43	43	42.0	32.4	
4	1	2	2	2	2	1	1		15	18	18	19	17.5	−24.9	0.1
									15	18	19	21	18.3	25.0	
5	2	1	2	1	2	1	2		16	16	15	17	16.0	−24.1	4.1
									25	26	26	26	25.8	28.2	
6	2	1	2	2	1	2	1		17	17	20	18	18.0	−25.1	5.0
									29	32	34	35	32.5	30.1	
7	2	2	1	1	2	2	1		14	15	17	15	15.3	−23.7	2.3
									15	21	23	26	21.3	25.9	
8	2	2	1	2	1	1	2		16	20	18	21	18.8	−25.5	2.1
									25	23	27	22	24.3	27.6	

The response table for the combined SN ratios is shown in Figure 4:4.13.

	A	B	C	D	E	F	G
Level 1	5.40	4.68	3.49	6.28	6.47	3.29	3.57
Level 2	3.39	4.12	5.31	2.51	2.33	5.51	5.23
Difference	2.01	0.56	1.82	3.77	4.14	2.22	1.66
Rank	4	7	5	2	1	3	6

Figure 4:4.13　　　　Response table for combined SN ratio data.

The optimum factor levels based on the SN ratio (larger-the-better) are:
E1, D1, F2 and A1. Only four factors have been chosen since in an $L_8(2^7)$ orthogonal array experiment we may only take about half the degrees of freedom for significant factors. We now characterize the:
● 　　misfeed process,
● 　　multifeed process.

　　1.　　**Misfeed process**
　　In order to characterise the misfeed process, we need to calculate the predicted SN ratio and the predicted mean for misfeed data. The SN ratio response table is shown in Figure 4:4.14.

	A	B	C	D	E	F	G
Level 1	−23.41	−24.36	−24.35	−23.07	−23.79	−24.62	−24.42
Level 2	−24.60	−23.65	−23.66	−24.94	−24.23	−23.39	−23.59
Difference	1.19	0.71	0.69	1.86	0.44	1.23	0.83
Rank	3	5	6	1	7	2	4

Figure 4:4.14　　　　Response table for misfeed data (SN ratio).

Using only the levels for E1, D1, F2 and A1 the predicted SN ratio from the response table for SN ratio of the misfeed data is:

$$\eta_{Predicted} = (\overline{E1} - \overline{\eta}) + (\overline{D1} - \overline{\eta}) + (\overline{F2} - \overline{\eta}) + (\overline{A1} - \overline{\eta})$$

$$= \overline{E1} + \overline{D1} + \overline{F2} + \overline{A1} - 3 \times \overline{\eta}$$

$$= -23.79 - 23.07 - 23.39 - 23.41 + 3 \times 24.01$$

$$= -21.65 \text{ dB}$$

Similarly, the mean response table for misfeed data is shown in Figure 4:4.15.

	A	B	C	D	E	F	G
Level 1	14.88	16.38	16.38	14.31	15.69	16.94	16.56
Level 2	17.00	15.50	15.50	17.56	16.19	14.94	15.31
Difference	2.13	0.88	0.88	3.25	0.50	2.00	1.25
Rank	2	5	5	1	7	3	4

Figure 4:4.15   Response table of misfeed data (Mean).

The mean misfeed force is:

$$y_{Predicted} = (\overline{E1} - \overline{y}) + (\overline{D1} - \overline{y}) + (\overline{F2} - \overline{y}) + (\overline{A1} - \overline{y})$$

$$= \overline{E1} + \overline{D1} + \overline{F2} + \overline{A1} - 3 \times \overline{y}$$

$$= 15.69 + 14.31 + 14.94 + 14.88 - 3 \times 15.94$$

$$= 12.00$$

Substituting, $\eta = -21.65$ and $\overline{y} = 12.00$;

$$\sigma = \sqrt{10^{\left(\frac{\eta}{-10}\right)} - \bar{y}^2}$$

$$\sigma = \sqrt{10^{\left(\frac{-21.65}{-10}\right)} - 12.00^2}$$

$$\sigma = \sqrt{146.22 - 144.00}$$

$$\sigma = \sqrt{2.22}$$

$$\sigma = 1.45$$

2.   Multifeed process

In order to characterize the multifeed process, we need to calculate the predicted SN ratio and the predicted mean for multifeed data. The SN ratio response table is shown in Figure 4:4.16.

	A	B	C	D	E	F	G
Level 1	28.82	29.04	27.84	29.36	30.25	27.91	27.99
Level 2	28.00	27.77	28.97	27.45	26.56	28.90	28.82
Difference	0.82	1.27	1.12	1.91	3.70	0.99	0.83
Rank	7	3	4	2	1	5	6

Figure 4:4.16   Response table for multifeed data (SN ratio).

Using only the levels for E1, D1, F2 and A1 the predicted SN ratio from the response table for SN ratio of the multifeed data is:

$$\eta_{Predicted} = (\overline{E1} - \bar{\eta}) + (\overline{D1} - \bar{\eta}) + (\overline{F2} - \bar{\eta}) + (\overline{A1} - \bar{\eta})$$

$$= \overline{E1} + \overline{D1} + \overline{F2} + \overline{A1} - 3 \times \bar{\eta}$$

$$= 30.25 + 29.36 + 28.90 + 28.82 - 3 \times 28.41$$

$$= 32.12 \text{ dB}$$

Similarly, the mean response table for multifeed data is shown in Figure 4:4.17.

	A	B	C	D	E	F	G
Level 1	29.50	29.00	25.81	31.00	33.44	25.81	26.75
Level 2	25.94	26.44	29.63	24.44	22.00	29.63	28.69
Difference	3.56	2.56	3.81	6.56	11.44	3.81	1.94
Rank	5	6	3	2	1	3	7

Figure 4:4.17   Response table for multifeed data (Mean).

The mean multifeed force is:

$$y_{Predicted} = (\overline{A1} - \overline{y}) + (\overline{D1} - \overline{y}) + (\overline{F2} - \overline{y}) + (\overline{A1} - \overline{y})$$

$$= \overline{E1} + \overline{D1} + \overline{F2} + \overline{A1} - 3 \times \overline{y}$$

$$= 33.44 + 31.00 + 29.63 + 29.50 - 3 \times 27.72$$

$$= 40.41$$

Substituting, $\eta = 32.12$ and $\overline{y} = 40.41$;

$$\sigma = \sqrt{\frac{\overline{y}^2 \left(\overline{y}^2 \times 10^{\left(\frac{\eta}{-10}\right)} - 1\right)}{3}}$$

$$= \sqrt{\frac{40.41^2 \left(40.41^2 \times 10^{\left(\frac{32.12}{-10}\right)} - 1\right)}{3}}$$

$$= 1.31$$

The best spring force $f_{Optimum}$ is taken as the geometric mean of the misfeed force and multifeed force. Mathematically:

$$f_{Optimum} = \sqrt{f_{Misfeed} \times f_{Multifeed}}$$

$$= \sqrt{12.00 \times 40.41}$$

$$= 22.02$$

The gain in going from the existing process (experiment 1) to the optimum process is:

$$\eta_{Gain} = (\eta_{Misfeed} + \eta_{Multifeed})_{Optimum} - (\eta_{Misfeed} + \eta_{Multifeed})_{Existing}$$

$$= (-21.65 + 32.12) - (-23.99 + 30.78)$$

$$= 3.68 \text{ dB}$$

This corresponds to a reduction $\mathfrak{R}$, by a factor of:

$$\mathfrak{R} = \frac{1}{0.5^{\left(\frac{\eta_{Gain}}{3}\right)}}$$

$$= \frac{1}{0.5^{\left(\frac{3.68}{3}\right)}}$$

$$= 2.34$$

The current process can be characterized with the data from experiment 1. The mean misfeed force is 15.00 with a standard deviation of 3.70. The mean multifeed force is 35.00 with a standard deviation of 3.56. These together with the predicted optimum process can be characterized as in Figure 4:4.18.

	Before optimization		After optimization	
	Misfeed	Multifeed	Misfeed	Multifeed
mean	15.50	35.00	12.00	40.41
standard deviation	3.70	3.56	1.45	1.31

Figure 4:4.18   Comparison of process characteristics before and after optimization.

Using these results, we may compare the process improvement using the normal distribution plots:

$$\Phi = \frac{1}{\sqrt{(2 \pi \sigma^2)}} e^{-\frac{1}{2}\left(\frac{y - \bar{y}}{\sigma}\right)^2}$$

where Φ is the relative frequency, σ is the standard deviation and $\bar{y}$ is the mean. The normal plots for existing conditions of misfeed and multifeed are compared to those of the optimum conditions of misfeed and multifeed as follows:

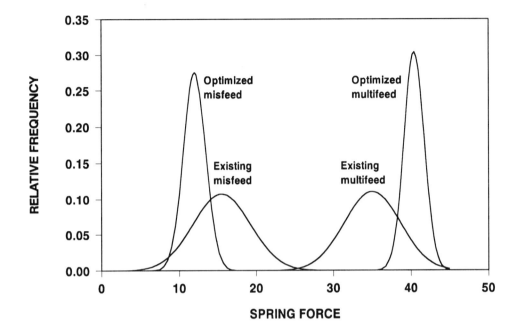

CHAPTER 5

BASIC ANALYSIS OF VARIANCE

AIMS:
The aim of this chapter is to provide a method of calculating the statistical significance of factors in experimentation.

OBJECTIVES:
When you have completed studying this chapter you should be able to:
- perform analysis of variance,
- perform pure sum of squares calculations,
- calculate contribution ratios,
- calculate confidence intervals.

OVERVIEW:
This chapter introduces you to the analysis of variance. Fundamental ideas of statistics are developed from first principles. A simple idea of analysis of variance is introduced through the no-way (or no-factor) analysis of variance and built up to a one-way (or one-factor) analysis of variance and eventually to a multi-way (or multi-factor) analysis of variance using orthogonal arrays. The concept of conserving the total sum of squares and total degrees of freedom is emphasized. Pure sums of squares are calculated for factors to establish the contribution ratios. Calculations of confidence intervals for factor levels, the predicted mean and the confirmation experiment are also included.

5.1   Analysis of Variance

5.1.1 Introduction

Analysis of variance (Anova) was first introduced by Sir Ronald Fisher, the British Statistician. Analysis of variance is a method of partitioning variability into identifiable sources of variation and the associated degrees of freedom in an experiment.

5.1.2 The number of elements

Suppose an engineer's result of some measurement is a set of five numbers, as follows: {2, 3, 5, 6, 4}.

We denote the number of items in the set as n. Since there are five numbers in this case, therefore $n = 5$.

To denote the first, third or some number from the set, we represent this by using y_i. The symbol y_i is read as y subscript i. This means that:
when $i = 1$, $y = 2$,
when $i = 2$, $y = 3$, etc. and
when $i = 5$, $y = 4$ and so on. That is,

i	1	2	3	4	5
y_i	2	3	5	6	4

Obviously, i could only be any integer between 1 and 5 inclusive. In general, if there is a set of n numbers, we could say that i can be any integer from $i = 1$ to n. The symbol y_i denotes any of the n values as above.

5.1.3 The sum of a set of numbers

If we wanted to sum all the numbers of a set, we could write;

$$sum = y_1 + y_2 + y_3 + y_4 + y_5 \qquad (5:1.1)$$

Mathematically, we could write this as

$$\sum_{i=1}^{5} y_i = y_1 + y_2 + y_3 + y_4 + y_5 \qquad (5:1.2)$$

The symbol[1] Σ means *the sum of all elements from i=1 to i=5*. In general, we may write this as:

_____

1   We merely introduce these symbols for completeness of discussion.

$$\sum_{i=1}^{n} y_i = y_1 + y_2 + ... + y_n \qquad (5{:}1.3)$$

For simplicity, when no confusion can arise, we denote the above by simply Σy, which may be read as *the sum of y from i = 1 to n*, where n is the number of elements in the set.

Example
Given a set of five numbers {2, 3, 5, 6, 4}, use the Σ notation to write the sum of all the numbers in the set.

Answer.

$$\Sigma y = 2 + 3 + 5 + 6 + 4$$

$$= 20$$

5.1.4 The arithmetic mean
The arithmetic mean is denoted $\bar{y}$ (y bar) and is also called the average:

$$\bar{y} = \frac{\Sigma y}{n} \qquad (5{:}1.4)$$

Example
Given a set of five numbers {2, 3, 5, 6, 4}, use the Σ notation to write the mean of all the numbers in the set.

Answer
The arithmetic mean of the set of five numbers {2, 3, 5, 6, 4} is:

$$\bar{y} = \frac{\Sigma y}{n}$$

$$= \frac{2 + 3 + 5 + 6 + 4}{5}$$

$$= \frac{20}{5}$$

$$= 4$$

5.1.5 The square of the sum of a set of numbers

As the name suggests, this is the *square of the sum* $(\Sigma y)^2$, of a set of numbers.

$$\left[\sum y\right]^2 = (y_1 + y_2 + y_3 + ... + y_n)^2 \qquad (5:1.5)$$

Example

Given a set of five numbers {2, 3, 5, 6, 4}, use the Σ notation to write the square of the sum of the numbers.

Answer

$$\left[\sum y\right]^2 = (2 + 3 + 5 + 6 + 4)^2$$

$$= 20^2$$

$$= 400$$

5.1.6 The total sum of squares of a set of numbers

The *total sum of squares*, *ST*, of a set of numbers is:

$$ST = \sum y^2$$

$$= y_1^2 + y_2^2 + y_3^2 + ... + y_n^2 \qquad (5:1.6)$$

Example

Given a set of five numbers {2, 3, 5, 6, 4}, use the Σ notation to write the sum of squares of the set of numbers.

Answer

$$ST = \sum y_i^2$$

$$= y_1^2 + y_2^2 + y_3^2 + ... + y_n^2$$

$$= 2^2 + 3^2 + 5^2 + 6^2 + 4^2$$

$$= 90$$

Notice that the **square of the sum** is different from the **sum of squares**. You should never confuse the two.

5.1.7 The sum of squares of the mean

The *sum of squares of the mean*, *Sm*, can be written as follows:

$$Sm = \sum \bar{y}^2$$
$$= \bar{y}^2 + \bar{y}^2 + \bar{y}^2 + ... + \bar{y}^2$$
$$= n \times \bar{y}^2 \qquad (5:1.7)$$

This can also be expressed as:

$$Sm = n \times \bar{y}^2$$
$$= n \times \left[\frac{\sum y}{n}\right]^2$$
$$= \frac{\left[\sum y\right]^2}{n} \qquad (5:1.8)$$

Example.
Given a set of five numbers {2, 3, 5, 6, 4}, use the Σ notation to write the sum of squares of the mean of the set of numbers.

Answer.
$$Sm = \sum \bar{y}^2$$
$$= n \times \bar{y}^2$$
$$= 5 \times 4^2$$
$$= 80$$

The sum of squares of the mean is frequently called the *correction factor* and denoted as *CF*.

5.1.8 The error sum of squares

The *error sum of squares*, *Se*, is the sum of the squared deviations of y_i from the mean and is calculated as follows:

$$Se = \sum (y - \bar{y})^2$$

$$= \sum (y^2 - 2y\bar{y} + \bar{y}^2)$$

$$= \sum y^2 - 2\bar{y} \sum y + \sum \bar{y}^2$$

But, $\sum y = n\bar{y}$ and $\sum \bar{y}^2 = n\bar{y}^2$, therefore,

$$Se = \sum y^2 - 2\bar{y} \, n\bar{y} + n\bar{y}^2$$

$$= \sum y^2 - 2n\bar{y}^2 + n\bar{y}^2$$

$$= \sum y^2 - n\bar{y}^2 \qquad\qquad (5:1.9)$$

The error sum of squares is also called the *residual sum of squares* or simply the residual. This quantity can also be expressed in another way:

$$Se = \sum y^2 - n\bar{y}^2$$

$$= \sum y^2 - n\left[\frac{\sum y}{n}\right]^2$$

$$= \sum y^2 - \frac{[\sum y]^2}{n}$$

$$= \frac{n\sum y^2 - [\sum y]^2}{n} \qquad\qquad (5:1.10)$$

Both forms are given together as:

$$Se = \sum (y - \bar{y})^2$$

$$= \sum y^2 - n\bar{y}^2 \qquad\qquad A$$

$$= \frac{n\sum y^2 - [\sum y]^2}{n} \qquad\qquad B$$

With the use of calculators, either form is just as easy. However, form B was preferred when calculators were less common. This is because the sum of squares due to error could be calculated as *n times the sum of squares, minus the square of the sum, whole thing over n*. The author suggests the use of form (A) for simplicity of use as the *sum of squares minus n times the square of the mean*. Notice that **error** here does not mean **mistake** but rather the differences due to **chance**.

Example
Given a set of five numbers $\{2, 3, 5, 6, 4\}$, use the Σ notation to write the error sum of squares of the set of numbers.

Answer

$$Se = \sum (y - \bar{y})^2$$
$$= \sum y^2 - n\,\bar{y}^2$$
$$= 90 - 5 \times 4^2$$
$$= 10$$

5.1.9 A measure of dispersion: the variation
The degree to which numerical data tend to spread about a mean value is called the *variance*, and is denoted σ^2. Mathematically, the variance is defined as:

$$\sigma^2 = \frac{Se}{v} \qquad\qquad (5{:}1.11)$$

where Se is the sum of squres due to error and v is the degrees of freedom associated with Se. There are two forms of variance:
● 	population,
● 	sample.

1.   Population variance

The population variance[2] is also called the *biased estimator* (σ_n), with $\nu = n$ degrees of freedom. Therefore,

$$\sigma_n^2 = \frac{Se}{n}$$

$$= \frac{\sum y^2 - n\,\bar{y}^2}{n} \qquad\qquad (5:1.12)$$

Example:

Given a set of five numbers {2, 3, 5, 6, 4} and using the Σ notation, write the population variance of the set of numbers.

Answer

$$\sum y^2 = 2^2 + 3^2 + 5^2 + 6^2 + 4^2 = 90$$

$$\bar{y} = \frac{2 + 3 + 5 + 6 + 4}{5} = 4$$

$$n = 5$$

$$\sigma_n^2 = \frac{\sum y^2 - n\,\bar{y}^2}{n}$$

$$= \frac{90 - 5 \times 4^2}{5}$$

$$= 2.0$$

2.   Sample variance

The sample variance is also called the *unbiased estimator* (σ_{n-1}), with $\nu = n - 1$ degrees of freedom.

---

2   Although we define the population variance to be the error sum of squares divided by n, unfortunately, this is not a sensible term since it is not a *population variance*. There is no way we will ever know the population variance. Thus the term population variance creates a biased estimate which on average is too small. Regrettably, many calculators also define this term to be the population variance. If this is true, there is no need to estimate population variance.

$$\sigma^2_{n-1} = \frac{Se}{n-1}$$

$$= \frac{\sum y^2 - n\,\bar{y}^2}{n-1} \qquad\qquad (5{:}1.13)$$

Example
Given a set of five numbers {2, 3, 5, 6, 4}, use the Σ notation to write the sample variance of the set of numbers.

Answer

$$\sum y^2 = 2^2 + 3^2 + 5^2 + 6^2 + 4^2 = 90$$

$$\bar{y} = \frac{2 + 3 + 5 + 6 + 4}{5} = 4$$

$$n - 1 = 4$$

$$\sigma^2_{n-1} = \frac{\sum y^2 - n\,\bar{y}^2}{n-1}$$

$$= \frac{90 - 5 \times 4^2}{4}$$

$$= 2.5$$

5.1.10   The standard deviation
The standard deviation is the square root of the variance and may be the population standard deviation σ_n or the sample standard deviation σ_{n-1}.

5.1.11   The various sums of squares
Consider the set of numbers {2, 3, 5, 6, 4}. The total sum of squares ST, was calculated as follows:

$$ST = total\ sum\ of\ squares$$

$$ST = \sum y^2$$

$$= 2^2 + 3^2 + 5^2 + 6^2 + 4^2$$

$$= 90$$

The sum of squares due to the mean *Sm* was calculated as:

$$Sm = \text{\textit{sum of squares of the mean}}$$

$$= n\,\bar{y}^{\,2}$$

$$= 5 \times 4^2$$

$$= 80$$

The sum of squares due to error *Se* was calculated as:

$$Se = \text{\textit{error sum of squares}}$$

$$= \sum (y - \bar{y})^2$$

$$= \frac{n\sum y^2 - \left[\sum y\right]^2}{n}$$

$$= \frac{5 \times 90 - 20^2}{5}$$

$$= 10$$

We shall now look at the relationship between these sums of squares.

5.1.12   The relationship between *St, Sm* and *Se*
The relationship between *ST*, *Sm* and *Se* is:

$$ST = Sm + Se \tag{5:1.14}$$

and we can verify that in the present example,

$$ST = Sm + Se$$

$$90 = 80 + 10$$

This total sum of squares is an important concept since in any calculation of sums of squares the total sum of squares must always be accounted for. Usually, the quantity *Se* is the most tedious to calculate. However, owing to the relationship above, *Se* can be more easily determined by the relationship:

$$Se = ST - Sm \qquad (5:1.15)$$

5.1.13 Degrees of freedom

The degrees of freedom, v, of a set of numbers is the number of independent fair comparisons that can be made. Like the sums of squares, the total degrees of freedom vT can be broken down into the degrees of freedom for the mean, vm, and the degrees of freedom for error, ve. Mathematically:

$$vT = vm + ve \qquad (5:1.16)$$

The total degrees of freedom is an important concept since in any calculation of the total sum of squares the total degrees of freedom must always be accounted for.

5.1.14 Analysis of variance

Let us summarize the present example in Figure 5:1.1 also referred to as an analysis of variance.

Source	Sum of squares	Degrees of freedom	Variance
Mean	80	1	-
error	10	4	2.5
Total	90	5	-

Figure 5:1.1 No-way analysis of variance.

This analysis of variance can be classified as a no-way analysis of variance since no particular factor was included in the source. Notice also that the variance of the source error is 2.5 calculated with 4 degrees of freedom or in fact $(n - 1)$ degrees of freedom in general. For this reason, the calculation of variance with $(n - 1)$ degrees of freedom is often referred to as the *unbiased estimate* since the remaining 1 degree of freedom has been assigned to the mean. The unbiased estimate of variance corresponds to the **sample variance** discussed earlier. If the sum of squares due to error is divided by n degrees of freedom, the resulting variance is a *biased estimate* corresponding to the **population variance** discussed earlier.

5.1.15   The decomposition diagram

At this point we introduce the decomposition diagram. The decomposition diagram, Figure 5:1.2, shows the breakdown of the sums of squares and the degrees of freedom. The decomposition diagram clearly shows how the no-way analysis of variance is interpreted. The decomposition diagram accounts for both the total sum of squares and the total degrees of freedom. In any analysis of variance, the total sum of squares and the total degrees of freedom must be accounted for. Together, they simply mean that of the total sum of squares of 90, 80 are due to the mean effect and (only) 10 are due to error; and that, of the five observations made, one is due to the mean effect and four are due to error.

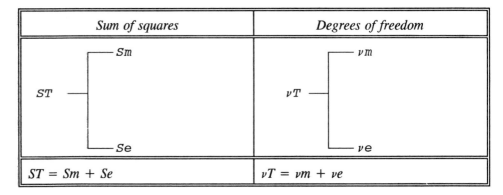

Figure 5:1.2 Decomposition of sum of squares and degrees of freedom.

5.1.16   Sample and population variance

In this section, we have used the set of data {2, 3, 5, 6, 4} to illustrate data points. In experimentation this data set can be regarded as a sample to estimate the properties of a population. Consider, for example, a machine that makes glass beads. During the operation of this machine, there is always an intrinsic variation in, say, the weight of the glass beads. Of course, if this machine is working, over the shift, over the day, over the week, month or year, there will be a *population* of glass beads. We could then weigh all the glass beads produced over the year (or decade, etc.) to calculate their average weight. But such information is not much use for day-to-day operations. We could therefore take a *sample* of glass beads to *estimate* the properties of the population. We write the population average as μ. Of course we will never know the exact value of μ because it is impractical to weigh each and every glass bead. For practical reasons we take the sample results to estimate the corresponding value of the population. In this case, we may take a sample of data, say {2, 3, 5, 6, 4}, and infer that the average weight is $\bar{y}$. To imply that μ was measured by $\bar{y}$ we write:

$$\hat{\mu} = \bar{y}$$

Two conventions are used here:
- the bar,
- the hat.

1.   The bar
The bar (¯) over the y implies that the statistic (the average in this case) was calculated from a sample.

2.   The hat
The hat (ˆ) implies that the statistic (the average in this case) is an estimate.

In the same way, if we can weigh each and every glass bead produced, then we can calculate the *population standard deviation* σ_n. Again however, that is impractical and we can never know σ_n. For this reason, we avoid the population standard deviation altogether. We can however, take a sample of data and calculate the *sample standard deviation*, denoted σ_{n-1} to estimate σ_n. Since σ_n is estimated, using the convention above we write:

$$\hat{\sigma}_n = \sigma_{n-1}$$

For ease of transcription, we shall denote:

$$\sigma = \sigma_{n-1}$$

unless we specifically mean otherwise. Henceforth, the reader should take σ to be the **sample standard deviation**.

There is also another convention where Greek letters are used to denote population parameters and Roman letters are used to denote sample estimates. Based on this convention, we can represent σ_{n-1} as S. To ensure no misunderstanding at all, we write:

$$\sigma = \sigma_{n-1} = S$$

Often an electronic scientific calculator has both σ_n and σ_{n-1} (or S) buttons or functions. In many ways this has been a hindrance to understanding basic statistics. In any case, if the population standard deviation can be calculated from a sample, there is no need to make any estimation of it. It is because we cannot know the population parameter in the first place that we need the sample. The reader should therefore be aware of this notation and use the σ_{n-1} (or S) button or function appropriately.

5.1.17 Self-assessment questions

1. The strength of an adhesive is determined by the kilogram-force (kgf) needed to break apart specimens joined by the adhesive. The breaking force recorded for ten readings of 'Supergum' adhesive is:

10.3	5.8	19.7	18.8	9.4	14.6	8.6	11.8	13.7	12.6

Conduct a no-way analysis of variance to establish the variance of the breaking force for Supergum.

2. The sales agent from 'Ultragum' (a competitor of Supergum) claims that Ultragum is a better adhesive. The breaking force recorded for ten readings of the Ultragum adhesive is:

8.3	9.5	10.2	9.8	12.5	7.9	8.9	11.3	12.5	10.5

Conduct a no-way analysis of variance to establish the variance of the breaking force for Ultragum.

3. Compare your results of the mean values and the variance for both Supergum and Ultragum. What is your conclusion?

5.1.18 Answers to self-assessment questions

1. The strength of an adhesive is determined by the kilogram-force (kgf) needed to break apart specimens joined by the adhesive. The breaking force recorded for ten readings of the 'Supergum' adhesive is:

10.3	5.8	19.7	18.8	9.4	14.6	8.6	11.8	13.7	12.6

Conduct a no-way analysis of variance to establish the variance of the breaking force for Supergum.

Answer

Source	Sum of squares	Degrees of freedom	Variance
Mean	1570.01	1	-
error	172.42	9	19.16
Total	1742.43	10	-

2. The sales agent from 'Ultragum' (a competitor of Supergum) claims that Ultragum is a better adhesive. The breaking force recorded for ten readings of the Ultragum adhesive is:

8.3	9.5	10.2	9.8	12.5	7.9	8.9	11.3	12.5	10.5

Conduct a no-way analysis of variance to establish the variance of the breaking force for Ultragum.

Answer

Source	Sum of squares	Degrees of freedom	Variance
Mean	1028.20	1	-
error	23.08	9	2.56
Total	1051.28	10	-

3.   Compare your results of the mean values and the variance for both Supergum and Ultragum. What is your conclusion?

Answer

The mean breaking force for Supergum adhesive is:

$$\bar{y}_{Supergum} = \frac{10.3 + 5.8 + 19.7 + 18.8 + 9.4 + 14.6 + 8.6 + 11.8 + 13.7 + 12.6}{10}$$

$$= 12.53$$

The mean breaking force for Ultragum adhesive is:

$$\bar{y}_{Ultragum} = \frac{8.3 + 9.5 + 10.2 + 9.8 + 12.5 + 7.9 + 8.9 + 11.3 + 12.5 + 10.5}{10}$$

$$= 10.14$$

Thus, Supergum has a mean of 12.53 and a variance of 19.16 while Ultragum has a mean of 10.14 and a variance of 2.56.

By comparing the mean breaking force, Supergum has a larger mean breaking force compared to Ultragum. By comparing the variance around the respective mean values, Ultragum has a much smaller variance compared to Supergum. Thus, although Ultragum has a better variance, its mean value is smaller than that of Supergum.

5.2   One-way Analysis of Variance

5.2.1 Introduction
The no-way analysis of variance only established the variance due to the mean and error. When a specific factor is compared (at two levels at least), we use the one-way analysis of variance to establish the variance due to the mean, the factor and the error variance.

5.2.2 An engineer's day-off
Suppose an engineer is playing darts. In this game, the objective is to throw five darts into the centre point. Observations are made by taking the distance of the dart from the centre point to the nearest centimetre.

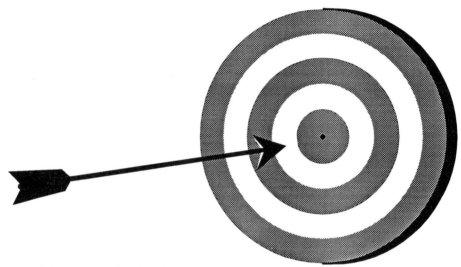

Figure 5:2.1   Engineer's day-off dart game.

Suppose that the temperature of the room in which the game of darts is played influences a player's ability. The room TEMPERATURE is a factor. In order to study the effect of temperature, we need to study at least two levels, say 15 and 20 °C. Since we intend to compare the player's results at 15 and 20 °C, we need to conduct at least two experiments. In conducting two experiments, there is only one fair comparison we can make. Hence, there is 1 degree of freedom. Obviously, it is not possible to make any more comparisons than measurements.

Suppose the effect of the factor, such as TEMPERATURE, on play results is to be compared at three levels, say 15, 20, and 25 °C, then we need to conduct at least three experiments. Having conducted three experiments, there are only two fair comparisons we can make. That is, there are two degrees of freedom. In general, the number of degrees of freedom for a factor is one less than the number of levels.

Note that *an experiment* need not only be *one measurement*. Indeed, an experiment usually consists of many observations. Thus, if we made ten observations at each of the temperature levels, say 15 and 20 °C, we would still have (only) one degree of freedom for TEMPERATURE. But ten observations at two levels is 20 observations. Twenty observations should allow 20 degrees of freedom, one of which is due to the factor TEMPERATURE and one of which is due to the overall mean. To understand what happens to the rest of the degrees of freedom, we need to conduct the one-way analysis of variance.

5.2.3 One-way analysis of variance
Let us return to the engineer's game of darts. Suppose we wish to study the effect of TEMPERATURE on the engineer's ability to play the game. To compare the effects of temperature on the engineer's ability, we shall study the game at two temperature settings. For simplicity, we denote the levels A1 and A2 corresponding to 15 and 20 °C, as shown in Figure 5:2.2.

Factor	Code	Level 1	Level 2
TEMPERATURE	A	15 °C	20 °C

Figure 5:2.2 Factor level setting.

We allow the engineer ten throws at each temperature level, giving a total of 20 throws. The results of the engineer's game of darts at levels A1 (15 °C) and A2 (20 °C) are given in Figure 5:2.3.

A1	6	4	5	3	5	5	4	3	4	6
A2	4	2	4	3	4	3	2	1	2	2

Figure 5:2.3 Results of the engineer's game of darts.

Since 20 observations have been made, we shall account for 20 degrees of freedom. We shall also determine the various sums of squares:
- ST – the total sum of squares, which comprises,
- Sm – the sum of squares due to the mean,
- SA – the sum of squares due to factor A,
- Se – the sum of squares due to error.

1.   *ST* – **the total sum of squares**
The total sum of squares is given by:

$$ST = \sum y^2$$

$$= 6^2 + 4^3 +...+ 2^2 + 2^2$$

$$= 296.00$$

2.   *Sm* – **the sum of squares due to the mean**
The sum of squares due to the mean is given by:

$$Sm = n \bar{y}^2$$

$$= 10 \times 3.60^2$$

$$= 259.20$$

3.   *SA* – **the sum of squares due to factor A**
The sum of squares due to factor A is given by:

$$SA = \frac{[Total\ of\ A1]^2}{n1} + \frac{[Total\ of\ A2]^2}{n2} - \frac{[Total\ of\ A]^2}{n}$$

$$= \frac{45^2}{10} + \frac{27^2}{10} - \frac{(45 + 27)^2}{(10 + 10)}$$

$$= 16.20$$

4.   *Se* – **the sum of squares due to error**
The sum of squares due to error is given by:

$$ST = Sm + SA + Se$$

$$Se = ST - Sm - SA$$

$$= 296.00 - 259.20 - 16.20$$

$$= 20.60$$

The breakdown of the sum of squares is shown in the decomposition diagram, Figure 5:2.4.

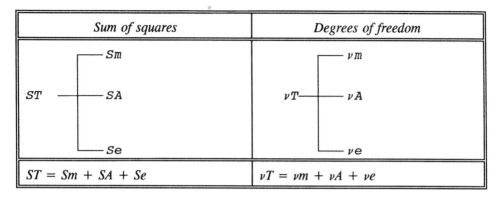

Sum of squares	Degrees of freedom
ST — Sm / SA / Se	νT — νm / νA / νe
$ST = Sm + SA + Se$	$\nu T = \nu m + \nu A + \nu e$

Figure 5:2.4 Conservation of total sum of squares and degrees of freedom.

These results are presented conveniently in Figure 5:2.5, where Sq is the sum of squares, ν is the degrees of freedom, and Mq is the mean sum of squares. Notice that the mean sum of squares (Mq) in Equation (5:2.1) is really the sum of squares (Sq) *per unit degree of freedom* and is therefore a variance quantity. The F-ratio is explained below.

$$Mq = \frac{Sq}{\nu} \qquad (5:2.1)$$

Source	Sq	ν	Mq	F-ratio
Mean	Sm	1	Vm	Fm
Factor A	SA	1	VA	FA
error	Se	$n-2$	Ve	1
Total	ST	n	-	-

Figure 5:2.5 Layout of analysis of variance.

In fact, it is because we are interested in comparing the variance due to each source, that we need to know the sum of squares and degrees of freedom of the source. The last column in Figure 5:2.5 shows the ratio of a source variance (e.g. Mq of factor A) to the variance of the error (Mq of error). This is simply a comparison of the number of times the source variance is larger than the error variance. This ratio is called the F-ratio. The larger the F-ratio, the more significant is the source. A more detailed explanation of the importance of the F-ratio is given in the next section.

5.2.4 One-way analysis of variance: excluding the mean

The one-way analysis of variance including the mean is shown in Figure 5:2.6.

Source	Sq	ν	Mq	F-ratio
Mean	259.20	1	-	-
Factor A	16.20	1	16.20	14.16
error	20.60	18	1.14	1.00
Total (ST)	246.00	20	-	-

Figure 5:2.6 One-way analysis of variance: including the mean.

Another very common way of displaying the one-way analysis of variance excludes the source Mean as shown in Figure 5:2.7. In fact, this is the classical method of displaying analysis of variance information. Why are there two forms of the analysis of variance? From an engineering point of view, this seems unnecessary.

Source	Sq	ν	Mq	F-ratio
Factor A	16.20	1	16.20	14.16
error	20.60	18	1.14	1.00
total (St)	36.80	19	-	-

Figure 5:2.7 One-way analysis of variance: excluding the mean.

Notice that in the classical method, the source, Sq, ν, Mq and the F-ratio due to the mean have all been omitted and the total sum of squares (ST) has been reduced by Sm to St. We should immediately recognize that **ST has n degrees of freedom** and **St has $n - 1$ degrees of freedom**. Normally, the effect of the mean is to uniformly raise or lower the overall effects. For this reason it is usually omitted in classical statistics. Only in the case where the quality characteristic is a smaller-the-better characteristic would the variance due to the mean be relevant. In that case, the variance due to the mean will provide a measure of how effective a factor might be in reducing the average response to zero. For this reason, the author chooses to show the mean effect in all further analysis of variance.

Additionally, especially for non-statisticians, there can be ambiguities about which *total sum of squares* is being referred to, ST or St? In order to overcome these problems permanently, the author recommends that the analysis of variance be represented as shown in Figure 5:2.8. The conservation of sums of squares and degrees of freedom shown in Figure 5:2.4 can now be modified, as shown in Figure 5:2.9.

Source	*Sq*	ν	*Mq*	*F-ratio*
Factor A	16.20	1	16.20	14.16
error	20.60	18	1.14	1.00
St	36.80	19	1.94	-
Mean	259.20	1	-	-
ST	296.00	20	-	-

Figure 5:2.8 One-way analysis of variance.

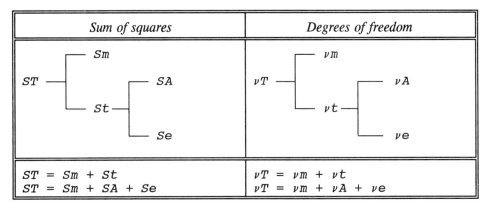

Figure 5:2.9 Conservation of total sum of squares and degrees of freedom.

5.2.5 Central limit theorem

The variance of factor A shown in Figure 5:2.8 is an estimate of the error variance based on the variation of the averages of A1 and A2 around the overall average. The error variance shown in Figure 5:2.8 is an estimate of the error variance based on the variation of the actual individual observations within a controlled group such as data points within factor A level 1 or factor A level 2. In order to understand more about the relative importance of a source in an analysis of variance, it is necessary to introduce the *central limit theorem*. The central limit theorem has three tenets:

1.   Sample averages tend to be normally distributed regardless of the distribution of the individuals.

2.   The average of the distribution of sample averages will approach the average of the distribution of the individuals.

3.   The variance of sample averages will be equal to the variance of the individuals divided by the sample size used to obtain the sample averages.

All of the above are predicated on the fact that one population is being sampled. Therefore, the sample has a constant average and a constant variance. The sample variance can be estimated in two ways:
- from sample averages,
- from individual variance.

We shall estimate the sample variance for the engineer's game of darts in both these ways and then build a relationship to the F-ratio. The F-ratio is formally defined as:

$$F\text{-}ratio = \frac{Estimate\ of\ population\ variance\ based\ on\ sample\ means}{Estimate\ of\ population\ variance\ based\ on\ individual\ means}$$

5.2.6 Estimating sample variance – from sample averages

The formula that describes the third and most important tenet in the central limit theorem is:

$$\sigma_{\bar{y}}^2 = \frac{\sigma_y^2}{n} \tag{5:2.2}$$

where σ_y is the variance of the individuals about the general mean, $\sigma_{\bar{y}}$ is the variance of sample averages about the general mean and n is the sample size. In other words, this formula states that the variance of a sample average will be equal to the variance of the individuals divided by the sample size used to obtain that sample average. We may rewrite this equation as follows:

$$\sigma_y^2 = n \times \sigma_{\bar{y}}^2 \tag{5:2.3}$$

Of course, statistically, we can never know the population standard deviation, σ_n. As an estimate of σ_n, however, we can use the sample standard deviation, σ_{n-1} but use the notation S to avoid over subscripting. Thus,

$$S_y^2 = n \times S_{\bar{y}}^2$$

We can now use this formula to make an estimate of individual variance by taking the variance of the sample averages and multiplying by the sample size (equal sample sizes for all samples are required). This property of the central limit theorem can be applied to the engineer's dart game since both levels A1 and A2 have sample sizes of $n = 10$. Regarding A1 and A2 as two samples with means of 4.30 and 2.10, respectively,

$$\sum y^2 = 4.50^2 + 2.70^2$$

$$= 27.54$$

$$\bar{y} = \frac{\sum y}{n}$$

$$= \frac{4.50 + 2.70}{2}$$

$$= 3.60$$

$$n = 2$$

$$S^2_{n_A-1} = \frac{\sum y^2 - n\,\bar{y}^2}{n - 1}$$

$$= \frac{27.54 - 2 \times 3.60^2}{2 - 1}$$

$$= 1.62$$

Since we know that the individual variance $S^2_y = n \times S^2_{\bar{y}}$, we can calculate the individual variance, S^2_{y1}, where the subscript 1 is used to denote the first of the two methods for calculating the variance of the distribution of individuals. Hence,

$$S^2_{y1} = n \times S^2_{n_A-1}$$

$$= 10 \times 1.62$$

$$= 16.2$$

Notice that this value is equal to the variance of factor A in the analysis of variance shown in Figure 5:2.8. Indeed, the variance of factor A is really an *estimate of individual variance based on the variance of sample averages*.

5.2.7 Estimating sample variance – from individual variance

We shall now estimate the sample variance based on the individual variance. Using the formula for individual variance we calculate the variance for group A1. That is, we calculate the variance of all the observations in group A1 about the mean value of A1:

$$S_{A1}^2 = \frac{\sum y^2 - n\,\bar{y}^2}{n - 1}$$

$$= \frac{213.00 - 202.50}{10 - 1}$$

$$= 1.17$$

Similarly, we calculate the variance for group A2 as follows:

$$S_{A2}^2 = \frac{\sum y^2 - n\,\bar{y}^2}{n - 1}$$

$$= \frac{83.00 - 72.90}{10 - 1}$$

$$= 1.12$$

Each of these is an estimate of individual variance with 9 degrees of freedom. Since two estimates are available, it would be better to estimate the individual variance S_{y2}^2 by averaging this information. The subscript 2 denotes that this is the second of the two methods of estimating individual variance.

$$S_{y2}^2 = \frac{1.17 + 1.12}{2}$$

$$= 1.14$$

Notice that this second method corresponds to the variance of the source error in Figure 5:2.8. Now there are *two independent estimates* of individual variance:

S_{y1}^2, which is determined from the variation of averages, and

S_{y2}^2, which is determined from the variation of individuals.

However, these estimates appear to be considerably different. The ratio S_{y1}^2 / S_{y2}^2 is called the F-ratio. For the present example,

$$F\text{-}ratio = \frac{S_{y1}^2}{S_{y2}^2}$$

$$= \frac{16.20}{1.14}$$

$$= 14.16$$

This is also the F-ratio value tabulated in Figure 5:2.8.

5.2.8 F-test for variance comparison

Statistically, there is a method which provides a decision at some confidence level as to whether these estimates are *significantly* different. This method is called the F-test, after Sir Ronald Fisher, who invented the analysis of variance. The F-test is simply a ratio of sample variances.

$$F\text{-}ratio = \frac{\textit{Estimate of population variance based on sample means}}{\textit{Estimate of population variance based on individual means}}$$

The denominator is a good estimate of the population variance *based on individual means*. The numerator is another estimate of the population variance *based on sample means*. If the variance about the sample means is not significantly different from the individual variance, then the numerator and denominator must be about equal and the F-ratio must be approximately one. Conversely, if the F-ratio becomes larger than one, we would be inclined to believe that the population variances are unequal. When the F-ratio becomes large enough at some *confidence level*, then the two sample variances can be accepted as being unequal. To determine whether an F-ratio of two sample variances is statistically large enough, three pieces of information have to be considered:
● the alpha risk,
● numerator degrees of freedom,
● denominator degrees of freedom.

1.   **The alpha risk**
      The alpha risk, α, is related to the confidence level (*CL*) by the equation:
      CL = 1 − alpha risk
      = 1 − α.
      The alpha risk is explained in more detail later in this section.

2.   **Numerator degrees of freedom**
      The numerator degrees of freedom, $v1$, is the number of degrees of freedom associated with the sample variance and is equal to the degrees of freedom of the source variance.

3.   **Denominator degrees of freedom**
      The denominator degrees of freedom $v2$, is the the number degrees of freedom associated with the sample variance and is equal to the degrees of freedom of the error variance.

Each combination of risk, numerator degrees of freedom and denominator degrees of freedom has a particular F-ratio[3] associated with it. The syntax for representing this information is $F_{\alpha,v1,v2}$.

The confidence level is also frequently expressed as a percentage. So, if the required confidence level is 95 %, we express the confidence level as 0.95. The corresponding risk is $1 - 0.95 = 0.05$. Turning to the engineer's game of darts and referring to Figure 5:2.8, the degrees of freedom for the numerator (source factor A) is $v1 = 1$ and the degrees of freedom for the denominator (source error) is $v2 = 18$. The necessary F-ratio to look for in the statistical tables is $F_{0.05,1,18}$. Referring to the appropriate table, $F_{0.05,1,18} = 4.41$. In the present example the ratio of the two estimates of individual variance is 14.16. Comparing the F-ratio of a source with the tabulated F-ratio is called an *F-test*. The F-test in this case shows:

$F_{Calculated}$ » $F_{0.05,1,18}$ since,

14.16   »   4.41

Statistically, this means that with at least 95 % confidence the two estimates of variance are believed to be unequal. What this means from a practical viewpoint is that the estimate of the *individual variance based on variation of averages* is inappropriately too high when compared to the estimate of the *individual variance based on the variation of individuals*. Averages vary much more than would be expected from the individual variation that is present. Therefore, rather than believe that only one population has been sampled, it is reasonable to believe that two (or more) different populations are being sampled.

The way F-ratio is applied in the analysis of variance depends on the alpha-risk. This risk is explained with reference to criminal justice and manufacturing sampling in Figure 5:2.10. In a criminal trial, suppose a jury found a (genuinely) innocent person *not guilty*, then a correct verdict has been reached and no error is committed. Similarly, if a (genuinely) guilty person is found *guilty*, then again a correct verdict has been reached and no error is committed. However, if a (genuinely) innocent person is *proven guilty* then a wrong verdict has been reached and an error is committed. This error is

---

3   F-tables listing the required F-ratios to achieve a stated confidence level can be found in most standard textbooks or statistical tables such as Murdoch and Barnes, *Statistical Tables for Science, Engineering, Management and Business Studies*, MacMillan Education Ltd.

called the **alpha-error**. If however, a (genuinely) guilty person is *proven not guilty* then again a wrong verdict has been reached and an error is committed. This error is called the **beta-error**. This distinction is necessary because society places different social values on the two types of error. For instance, *reasonable people* would rather commit an alpha-error than a beta-error.

		Criminal Justice				*Batch Sampling*		
		Verdict					Test result	
		Not-guilty	Guilty				Accept	Reject
T	Innocent	No error	error (α)		T	Good	No error	error (α)
r					r			
u	Guilty	error (β)	No error		u	Bad	error (β)	No error
e					e			

Figure 5:2.10   Alpha and beta errors.

In manufacturing, batches of material are often sampled for acceptance. Of course sampling is always subject to sampling error. When a good batch is accepted or a bad batch is rejected, no error is made. Occasionally, however, a good batch can be rejected (alpha-error) or a bad batch can be accepted (beta-error) due to sampling errors. In acceptance sampling (unlike criminal justice), a quality inspector would rather reject a good batch (commit a beta-error) than accept a bad batch (commit an alpha-error). This is because a manufacturer would want to make sure that a faulty product does not reach the customer.

In an analysis of variance, the alpha-risk is the chance of obtaining an F-ratio of at least the magnitude indicated in the F-table when taking samples out of the same population (i.e. a group of items having the same mean and variance). Since the alpha-risk is chosen as a small value, 5 % for example, the experimenter would rather believe that two or more populations with different averages have been sampled when the F-ratio does exceed the tabulated critical F-ratio. Therefore, it is highly unlikely that samples from one population will exceed the tabulated critical F-ratio. It is much more likely that the samples were obtained from two or more populations. In other words, factor A has a *significant effect* on the results.

In the design of experiments, the alpha-error is the error of finding an insignificant factor as significant. Correspondingly, the beta-error is the error of finding a significant factor as insignificant. Of course both errors are inevitable due to experimental errors and it is important to minimize both errors. If an alpha-error is committed, an insignificant factor is regarded as significant. If a beta-error is committed, a significant factor is deemed insignificant and may be omitted in subsequent experimentation. In this case, committed the alpha error is less serious.

Experimental Testing			
		Experimental evidence	
		Insignificant factor	Significant factor
T r u e	Insignificant factor	No error	error (α)
	Significant factor	error (β)	No error

Figure 5:2.11   Alpha and beta errors in experimentation.

5.2.9 Self-assessment questions

1. The thickness of a moulded plastic disc is studied at two injection speeds: 50 and 75 g min^{-1}. The thicknesses of ten discs were measured and recorded to be:

50 g min^{-1}	5.0	5.1	5.2	5.0	5.3	4.9	5.1	5.2	5.0	5.1
75 g min^{-1}	5.3	5.1	5.3	5.3	5.2	5.0	5.2	5.0	5.1	5.5

Conduct a one-way analysis of variance and comment on the injection speed.

2. The thickness of a moulded plastic disc is studied at two injection pressures: the current pressure and a low pressure. The thicknesses of ten discs were measured and recorded to be:

Current	5.0	5.1	5.2	5.0	5.3	4.9	5.1	5.2	5.0	5.1
Low	5.0	5.8	5.5	6.1	6.0	5.9	5.8	6.0	5.8	5.9

Conduct a one-way analysis of variance and comment on the injection pressure.

5.2.10   Answers to self-assessment questions

1.   The thickness of a moulded plastic disc is studied at two injection speeds: 50 and 75 g min^{-1}. The thicknesses of ten discs were measured and recorded to be:

50 g min^{-1}	5.0	5.1	5.2	5.0	5.3	4.9	5.1	5.2	5.0	5.1
75 g min^{-1}	5.3	5.1	5.3	5.3	5.2	5.0	5.2	5.0	5.1	5.5

Conduct a one-way analysis of variance and comment on the injection speed.

Answer   The one-way analysis of variance is as follows.

Source	Sq	ν	Mq	F-ratio
A	0.06	1	0.06	3.12
error	0.35	18	0.02	1.00
St	0.41	19	0.02	-
Mean	529.42	1	-	-
ST	529.83	20	-	-

From the analysis of variance, the calculated F-ratio for factor A is 3.12. From F-tables, the tabulated F-ratio for a comparison of a factor (with 1 degree of freedom) against an experimental error (with 18 degrees of freedom), using 5 % confidence interval, is $F_{0.05,1,18} = 4.41$. That is $F_{Calculated} < F_{Tabulated}$. Hence, factor A is not a significant factor. In other words, allowing for an alpha-error (cf. a genuinely guilty person proven not guilty, or accepting a bad batch) 5 % of the time, we do not have sufficient experimental evidence that the injection speed affects the disc size.

2.   The thickness of a moulded plastic disc is studied at two injection pressures: the current pressure and a low pressure. The thicknesses of ten discs were measured and recorded to be:

Current	5.0	5.1	5.2	5.0	5.3	4.9	5.1	5.2	5.0	5.1
Low	5.0	5.8	5.5	6.1	6.0	5.9	5.8	6.0	5.8	5.9

Conduct a one-way analysis of variance and comment on the injection pressure.

Answer

The one-way analysis of variance is as follows.

Source	*Sq*	*v*	*Mq*	*F-ratio*
A	2.38	1	2.38	41.00
error	1.04	18	0.06	1.00
St	3.43	19	0.18	-
Mean	590.78	1	-	-
ST	594.21	20	-	-

Figure 5:2.17   One-way analysis of variance.

From the analysis of variance, the calculated F-ratio for factor A is 41.00. From F-tables, the tabulated F-ratio for a comparison of a factor (with 1 degree of freedom) against an experimental error (with 18 degrees of freedom), using 5 % confidence interval, is $F_{0.05,1,18} = 4.41$. That is $F_{Calculated} > F_{Tabulated}$. Hence, factor A is a significant factor. In other words, allowing for an alpha-error (cf. a genuinely guilty person proven not guilty, or accepting a bad batch) 5 % of the time, we do have sufficient experimental evidence that the injection pressure affects the disc size.

5.3 Two-way Analysis of Variance

5.3.1 Introduction
The no-way analysis of variance only established the variance due to the mean and error. When a specific factor was compared (at two levels at least), we used the one-way analysis of variance to establish the variance due to the mean, the factor and the error variance. In this section we introduce another factor (at two levels at least) to establish the variance due to the mean, the two factors and the error variance.

5.3.2 Two-way analysis of variance
Suppose that, unknown to us, the engineer had used two types of darts. The engineer believed that the Nomohit darts are not as good as Supahit darts and had actually conducted the experiment with five throws from each type of dart at the two temperatures. We can code this factor DART as B, where B1 refers to Nomohit and B2 refers to Supahit. The results have now been classified with the appropriate darts as shown in Figure 5:3.1. Of course, we could always assume that we started with the two factors at the beginning of the experiment.

	B1					B2				
A1	6	4	5	3	5	5	4	3	4	6
A2	4	2	4	3	4	3	2	1	2	2

Figure 5:3.1 Two-way classification of data.

Obviously, the data can be considered in two ways.
- Compare A1 (10 observations) against A2 (10 observations). In Figure 5:3.1, this comparison is a horizontal comparison. Notice also that this is exactly what was done in the one-way analysis of variance in the last section.
- Compare B1 (10 observations) against B2 (10 observations). In Figure 5:3.1, this comparison is a vertical comparison.

There is however, one other valid comparison.
- Compare the shaded (10 observations) against unshaded (10 observations) boxes as shown in Figure 5:3.2. From Figure 5:3.2, this comparison is diagonal. In fact, this is referred to as the A×B interaction.

For ease of reference, we can regard this experiment as consisting of the A1B1, A1B2, A2B1 and A2B2 combinations shown in Figure 5:3.3, each with five observations. The data can then be expressed using an $L_4(2^3)$ orthogonal array as shown in Figure 5:3.3.

	B1					B2				
A1										
A2										

Figure 5:3.2 Comparison of the A×B interaction effect.

Exp	A	B	A×B	Results				
1	1	1	1	6	4	5	3	5
2	1	2	2	5	4	3	4	6
3	2	1	2	4	2	4	3	4
4	2	2	1	3	2	1	2	2

Figure 5:3.3 Orthogonal array representation of experimental data.

The reader should compare Figures 5:3.1 and 5:3.3 and verify that the data classifications are identical. In the orthogonal array form, it is in fact much easier to visualize the factor effects for A, B and A×B. Indeed, as more factors are added, the two-way layout becomes very cumbersome and it becomes very difficult to visualize the data structure. The orthogonal array form however, would still allow us to visualize the data structure. For this reason, the orthogonal array form will be used henceforth. The appropriate decomposition diagram for factor A, factor B and interaction A×B is shown in Figure 5:3.4. Notice[4] that the sum of squares of A and B (SAB) is the sum of squares of A (SA), sum of squares of B (SB) and sum of squares of interaction A×B ($SA×B$). We shall now calculate the following quantities:

- ST – the total sum of squares,
- Sm – the sum of squares due to mean,
- SA – the factor A sum of squares,
- SB – the factor B sum of squares,
- $SA×B$ – the interaction A×B sum of squares,
- Se – the error sum of squares.

---

4   The reader's attention is brought to the syntax of writing the name of a factor or interaction and the associated sum of squares. For example, $SA×B$ could formally be written $S(A×B)$. However, since there can be no misunderstanding in reading $SA×B$ as SA times SB there is no advantage in writing $S(A×B)$.

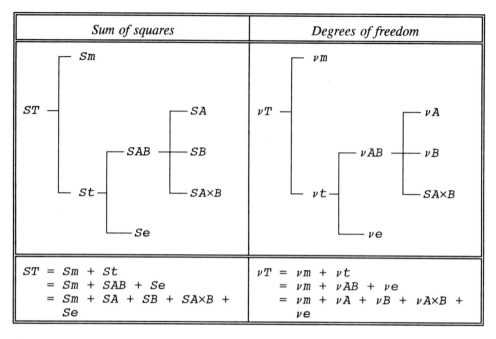

Sum of squares	*Degrees of freedom*

$$ST = Sm + St$$
$$= Sm + SAB + Se$$
$$= Sm + SA + SB + SA{\times}B + Se$$

$$\nu T = \nu m + \nu t$$
$$= \nu m + \nu AB + \nu e$$
$$= \nu m + \nu A + \nu B + \nu A{\times}B + \nu e$$

Figure 5:3.4 Conservation of sums of squares and degrees of freedom.

1.   *ST* – the total sum of squares
The total sum of squares is given by:

$$ST = \sum y^2$$
$$= 296.00$$

2.   *Sm* – the sum of squares due to mean
The sum of squares due to mean is given by:

$$Sm = n\,\bar{y}^2$$
$$= 259.20$$

3.   *SA* – the factor A sum of squares
The factor A sum of squares is given by:

$$SA = \frac{[Total\ of\ A1]^2}{n1} + \frac{[Total\ of\ A2]^2}{n2} - \frac{[Total\ of\ A]^2}{n1 + n2}$$

$$= \frac{45^2}{10} + \frac{27^2}{10} - \frac{(45 + 27)^2}{(10 + 10)}$$

$$= 16.20$$

The total for A1 is obtained by summing all observations for which factor A is in level 1, i.e. Exp rows 1 and 2 of Figure 5:3.3. Similarly, the total for A2 is obtained by summing all observations in Exp rows 3 and 4.

4.   *SB* – the factor B sum of squares
Similarly, the factor B sum of square is given by:

$$SB = \frac{[Total\ of\ B1]^2}{n1} + \frac{[Total\ of\ B2]^2}{n2} - \frac{[Total\ of\ B]^2}{n1 + n2}$$

$$= \frac{40^2}{10} + \frac{32^2}{10} - \frac{(40 + 32)^2}{(10 + 10)}$$

$$= 3.20$$

5.   *SA×B* – the interaction A×B sum of squares
Similarly, the A×B interaction sum of square is given by:

$$SA{\times}B = \frac{[Total\ of\ A{\times}B1]^2}{n1} + \frac{[Total\ of\ A{\times}B2]^2}{n2} - \frac{[Total\ of\ A{\times}B]^2}{n1 + n2}$$

$$= \frac{33^2}{10} + \frac{39^2}{10} - \frac{(33 + 39)^2}{(10 + 10)}$$

$$= 1.80$$

The reader should note the similarities in the formulas above arising directly from using the orthogonal array.

6.   *Se* – the error sum of squares
Henceforth, we denote e as error. The error sum of squares is given by:

$$ST = Sm + SA + SB + SA{\times}B + Se$$

$$Se = ST - Sm - SA - SB - SA{\times}B$$

$$= 296.00 - 259.20 - 16.20 - 3.20 - 1.80$$

$$= 15.60$$

The results of these calculations are tabulated in the analysis of variance shown in Figure 5:3.5. The mean sum of squares Mq, is calculated by dividing the sum of squares Sq, by the corresponding degrees of freedom ν. The F-ratio is calculated by dividing Mq by the mean sum of squares due to error (Mq of error, also called Ve).

Source	Sq	ν	Mq	F-ratio
A	16.20	1	16.20	16.62
B	3.20	1	3.20	3.28
A×B	1.80	1	1.80	1.85
e	15.60	16	0.98	1.00
St	36.80	19	1.94	-
Mean	259.20	1	-	-
ST	296.00	20	-	-

Figure 5:3.5 Two-way analysis of variance.

When considering the significance of a factor we need to refer to an F-table. For factor A, assuming an alpha-error of 5 %, $\alpha = 0.05$, $\nu1 = 1$ since the sum of squares of factor A was calculated with one degree of freedom, $\nu2 = 16$ since the sum of squares due to error was calculated with 16 degrees of freedom. Hence, we look-up for $F_{0.05,1,16}$ in an F-table and read 4.49. Since the calculated F-ratio of factor A is much larger than 4.49 we can infer that factor A is a significant factor. A factor that is significant at $\alpha = 0.05$ is usually marked with an asterisk (*). Additionally, we note that the F-values for factor B and interaction A×B must be the same if α remains the same (i.e. $\alpha = 0.05$) since $\nu1$ and $\nu2$ are the same as for factor A. However, the F-ratio of factor B is less than 4.49 and we do not have sufficient experimental evidence to show that factor B is significant. Similarly, we do not have sufficient experimental evidence to show that the interaction A×B is significant.

Assuming $\alpha = 0.05$ implies that we are willing to accept a 5 % chance of wrongly classifying an insignificant factor as significant. If want to reduce this type of error, we could make $\alpha = 0.01$. In that case, we are willing to accept a 1 % chance of wrongly classifying an insignificant factor as significant. Referring to an appropriate F-table, $F_{0.01,1,16}$ is 8.53. In our engineer's darts game, factor A is still significant. A factor that is significant at the $\alpha = 0.01$ is usually marked with a double asterisk (**).

5.3.3 Critique of the F-test

There are three limitations in the F-test of which we should be aware:
- assumption of equal error variance,
- alpha-risk,
- go or no-go dichotomy.

1.   Assumption of equal error variance

A basic assumption of analysis of variance is that the error variance is equal for all combinations of the various factor and levels. However, this may not be true. Because of the inherent averaging in analysis of variance, an opportunity to reduce variation by controlling levels of design parameters may go unrecognized. The quality loss function suggests that opportunities to reduce variation should be sought and utilized. Therefore, the F-test can be used only as a reference decision making tool to identify significant factors.

2.   Alpha-risk

Another limitation of the F-test (as discussed earlier) is that only the alpha-risk is addressed. If a truly significant factor is tested and found to be significant, there is no error. Similarly, when a truly insignificant factor is tested and found to be insignificant, there is no error. However, when a truly insignificant factor is tested and found to be significant a mistake occurs. This is the alpha-risk. Similarly, when a truly significant factor is tested and found to be insignificant a mistake occurs. This is the beta-risk. Since this risk is not assessed in an analysis of variance, the experimenter may not·know what chance there is of missing an important factor. The consequence of this risk is that a factor which is truly significant factor but was tested and found to be insignificant may not be included in subsequent experimentation. In society, when a guilty person is found guilty or an innocent person is found innocent, no mistake is made. Sometimes a guilty person may be found innocent and sometimes an innocent person may be found guilty. The chances of a guilty person being found innocent is the alpha-risk and the chances of an innocent person being found guilty is the beta-risk. Of course, we would rather free a guilty person (make an alpha-error) than to punish an innocent person (make a beta-error).

3.   Go or no-go dichotomy

According to the traditional F-test procedure, when an F-value calculated from experimental data exceeds the tabulated critical F-value, the effect is fully recognized as significant, otherwise the effect is deemed insignificant. Thus the F-test divides a continuum into two parts; go or no-go dichotomy.

Some of these limitations can be overcome by using the *percent contribution* suggested by Taguchi. In this method, rather than use the F-ratio to evaluate the significance of a factor, we calculate the factor's contribution to the total sum of squares. This method is discussed in the next section.

5.3.4 Self-assessment questions

1.   An experiment is conducted on two 2-level factors A, B and the associated interaction $A \times B$. The results of the experiment are given below.

Exp	A	B	A×B	Results			
1	1	1	1	1.1	2.3	2.5	1.2
2	1	2	2	5.1	6.9	6.5	4.3
3	2	1	2	0.8	2.2	1.0	1.4
4	2	2	1	4.3	5.6	3.3	4.7

Comment on the experimental findings based on an analysis of variance.

5.3.5 Answers to self-assessment questions

1. An experiment is conducted on two 2-level factors A, B and the associated interaction A×B. The results of the experiment are given below.

Exp	A	B	A×B	Results			
1	1	1	1	1.1	2.3	2.5	1.2
2	1	2	2	5.1	6.9	6.5	4.3
3	2	1	2	0.8	2.2	1.0	1.4
4	2	2	1	4.3	5.6	3.3	4.7

Comment on the experimental findings based on an analysis of variance.

Answer

Source	Sq	ν	Mq	F-ratio
A	2.72	1	2.72	3.31
B	49.70	1	49.70	60.46
A×B	0.64	1	0.64	0.78
e	9.86	12	0.82	1.00
St	62.93	15	4.20	-
Mean	176.89	1	-	-
ST	239.82	16	-	-

From F-tables, the tabulated F-ratio for a comparison of a factor (with 1 degree of freedom) against an experimental error (with 12 degrees of freedom), using 5 % confidence level is $F_{0.05,1,12} = 4.75$. Hence, B is a significant factor, while factor A and the interaction A×B are insignificant factors.

5.4 Percent Contribution

5.4.1 Introduction
When an analysis of variance has been performed on a set of data and the respective sums of squares have been calculated it is possible to use this information to apportion the corrected sums of squares to the appropriate factors. Comparing this value to the total sum of squares then gives us a percent contribution by each factor or source.

5.4.2 Percent contribution
In the last section, we discussed a method of conducting an analysis of variance. Using the F-ratio is a good method of identifying the significance of factors. However, the F-ratio only informs us of the significance of a factor. An extension to this method is to use the percent contribution (ρ) so that the contribution of a factor to the total sum of squares can be established. We develop this idea in the following.

So far, we have used the sum of squares Sq to mean the sum of squares of deviations from mean. We shall now introduce the sum of squares of deviations from target Sd. Suppose a set of observations: $\{y_1, y_2, y_3, \dots , y_n\}$. The mean of this set of data is $\bar{y}$ and the variance is σ^2. If the target value is m, then $(\bar{y} - m)$ is the offset or bias. The sum of squares of deviations from target for this set can be written as follows:

$$Sd = (y_1 - m)^2 + (y_2 - m)^2 + \dots + (y_n - m)^2$$

$$= \sum (y_i - m)^2 \qquad (5:4.1)$$

We can then introduce $\bar{y}$ into the equation and proceed as follows:

$$Sd = \sum (y_i - \bar{y} + \bar{y} - m)^2$$

$$= \sum ([y_i - \bar{y}] + [\bar{y} - m])^2$$

$$= \sum ([y_i - \bar{y}]^2 + 2 [y_i - \bar{y}][\bar{y} - m] + [\bar{y} - m]^2)$$

$$= \sum (y_i - \bar{y})^2 + 2 (\bar{y} - m) \sum (y_i - \bar{y}) + \sum (\bar{y} - m)^2 \qquad (5:4.2)$$

Notice that the middle term $2 (\bar{y} - m) \sum (y_i - \bar{y})$ is identically zero. This is because $\sum (y_i - \bar{y})$ evaluates to zero following from a basic definition of the mean; that the sum of deviations from the mean must be equal to zero. Therefore,

$$Sd = \sum (y_i - \bar{y})^2 + \sum (\bar{y} - m)^2$$
$$= n\, \sigma^2 + n\, (\bar{y} - m)^2 \qquad (5{:}4.3)$$

Thus, it can be seen that:

$$Sd = n\, \sigma^2 + n\, (\bar{y} - m)^2$$
$$= n\, (\bar{y} - m)^2 + n\, \sigma^2$$
$$= Sd' + n\, \sigma^2 \qquad (5{:}4.4)$$

where Sd' is the *pure sum of squares of deviation from mean*. This quantity is also variously referred to as *true effect*, *net variation* or *pure variation* of a source. In any case,

$$Sd = Sd' + n\, \sigma^2$$
$$\therefore\ Sd' = Sd - n\, \sigma^2 \qquad (5{:}4.5)$$

Hence, the *pure sum of squares of deviations from the mean* is the *sum of squares of deviations from target minus n times the error variance, where n is the degrees of freedom associated with the sum of squares*. Indeed, this is true for any sum of squares of deviation and we generalise:

$$SA' = SA - vA\ Ve \qquad (5{:}4.6)$$

where SA is the sum of squares of deviations from the target, SA' is the pure sum of squares of factor A, vA is the degrees of freedom of factor A and Ve is the variance (σ^2). The portion of the sum of squares $vA\ Ve$ must be added to the sum of squares due to error in order to ensure that the total sum of squares is accounted for.

Additionally, we define ρ (rho) as the percentage of the pure sum of squares of a source to the total sum of squares, St:

$$\rho A = \frac{SA'}{St} \times 100 \% \qquad (5:4.7)$$

Any portion of error subtracted from a sum of squares of deviations for a source must be added to the sum of squares to conserve the total sum of squares St. Hence, the percent contributions of all the sources (including error) must be 100 %.

Continuing from Figure 5:3.5, we now add two more columns to the analysis of variance for the engineer's game of darts as shown in Figure 5:4.1.

Source	Sq	v	Mq	F-ratio	Sq'	rho %
A	16.20	1	16.20	16.62	15.22	41.37
B	3.20	1	3.20	3.28	2.22	6.05
A×B	1.80	1	1.80	1.85	0.82	2.24
e	15.60	16	0.98	1.00	18.54	50.34
St	36.80	19	1.94	-	36.80	100.00
Mean	259.20	1	-	-	-	-
ST	296.00	20	-	-	-	-

Figure 5:4.1 Two-way analysis of variance.

We now follow with calculation of the pure sums of squares for:
- factor A,
- factor B,
- interaction A×B,
- error.

1.   Factor A

$$SA' = SA - vA\ Ve$$

$$= 16.20 - 1 \times 0.98$$

$$= 15.22$$

2.   **Factor B**

$$SB' = SB - \nu B\ Ve$$

$$= 3.20 - 1 \times 0.98$$

$$= 2.22$$

3.   **Interaction A×B**

$$SA{\times}B' = SA{\times}B - \nu A{\times}B\ Ve$$

$$= 1.80 - 1 \times 0.98$$

$$= 0.82$$

4.   **Error**

$$Se = St - SA' - SB' - SA{\times}B'$$

$$= 36.80 - 15.22 - 2.22 - 0.82$$

$$= 18.54$$

Next, we calculate the percent contributions for:
- factor A,
- factor B,
- interaction A×B,
- error.

1.   **Factor A**

$$\rho A = \frac{SA'}{St} \times 100\ \%$$

$$= \frac{15.22}{36.80} \times 100\ \%$$

$$= 41.37\ \%$$

2.   **Factor B**

$$\rho B = \frac{SB'}{St} \times 100 \ \%$$

$$= \frac{2.22}{36.80} \times 100 \ \%$$

$$= 6.05 \ \%$$

3.   **Interaction A×B**

$$\rho A \times B = \frac{SA \times B'}{St} \times 100 \ \%$$

$$= \frac{0.82}{36.80} \times 100 \ \%$$

$$= 2.24 \ \%$$

4.   **Error**

$$\rho e = \frac{Se}{St} \times 100 \ \%$$

$$= \frac{18.54}{36.80} \times 100 \ \%$$

$$= 50.34 \ \%$$

The percent contribution due to error provides an estimate of the adequacy of the experiment. Since *error* refers to *unknown and uncontrolled* factors, the percent contribution due to error suggests the sufficiency (or insufficiency) of the experiment. As a rule of thumb, if the percent contribution due to error is low (15 % or less), then it can be assumed that no important factors have been omitted from the experiment. If the percent contribution due to error is high (50 % or more), then it can be assumed that some important factors have been omitted, conditions were not well controlled or there was a large measurement error.

We shall now consider the analysis of variance for data from an $L_8(2^7)$ orthogonal array experiment. We do this in three stages:

1.   Construct an analysis of variance,
2.   Calculate the percent contribution,
3.   Pooling of insignificant factors.

5.4.3 Construct an analysis of variance

Consider an $L_8(2^7)$ experiment on the concentricity[5] of lenses given below in Figure 5:4.2, together with the experimental results. The response table is given in Figure 5:4.3.

	A	B	C	D	E	F	G	Results			
1	1	1	1	1	1	1	1	5.00	5.10	5.50	4.70
2	1	1	1	2	2	2	2	6.60	6.80	6.70	6.50
3	1	2	2	1	1	2	2	5.50	5.30	5.10	5.00
4	1	2	2	2	2	1	1	7.00	7.30	7.50	7.40
5	2	1	2	1	2	1	2	6.10	6.30	6.10	6.00
6	2	1	2	2	1	2	1	2.80	2.50	3.00	3.50
7	2	2	1	1	2	2	1	5.70	5.40	5.20	6.00
8	2	2	1	2	1	1	2	6.10	6.30	6.10	6.00

Figure 5:4.2 $L_8(2^7)$ Orthogonal Array.

Our intent now is to construct an analysis of variance. We will provide a step-by-step method to calculate:

- the average response for each experiment,
- the overall experimental average,
- the response table,
- the total sum of squares,
- the sum of squares due to the mean,
- the sum of squares due to factors,
- the sum of squares due to error,
- the mean sum of squares,
- the F-ratio.

1.   The average response for each experiment
The average of experiment 1 is:

$$\bar{y}_1 = \frac{5.00 + 5.10 + 5.50 + 4.70}{4}$$

$$= 5.08$$

2.   The overall experimental average
The overall experimental average is the average of all the experimental data:

$$\bar{y} = \frac{\sum y}{n}$$

$$= \frac{5.0 + 5.1 + 5.5 + 4.7 + ... + 6.1 + 6.0}{32}$$

$$= 5.63$$

3.   The response table

	A	B	C	D	E	F	G
Level 1	6.06	5.20	5.86	5.50	4.84	6.15	5.23
Level 2	5.19	6.05	5.40	5.76	6.41	5.10	6.03
Difference	0.87	0.86	0.46	0.26	1.57	1.06	0.81
Rank	3	4	6	7	1	2	5

Figure 5:4.3 Response table of factor effects.

4.   The total sum of squares
The total sum of squares is:

$$ST = \sum y^2$$

$$= 5.0^2 + 5.1^2 + 5.5^2 + 4.7^2 + ... + 6.1^2 + 6.0^2$$

$$= 1063.19$$

5.   The sum of squares due to the mean
The sum of squares due to the mean is:

$$Sm = n \, \bar{y}^2$$

$$= 32 \times 5.63^2$$

$$= 1013.63$$

6. The sum of squares due to factors

Sum of squares of deviation from target for factor A:

$$SA = \frac{[Total \ of \ A1]^2}{n_{A1}} + \frac{[Total \ of \ A2]^2}{n_{A2}} - \frac{[Total \ of \ A]^2}{n_{A}}$$

Since, the response table gives us the mean values of each factor in level 1 and level 2, we reformulate SA as follows:

$$SA = \frac{[Total \ of \ A1]^2}{n_{A1}} + \frac{[Total \ of \ A2]^2}{n_{A2}} - \frac{[Total \ of \ A]^2}{n_{A}}$$

$$= n_{A1} \times \frac{[Total \ of \ A1]^2}{n_{A1} \, n_{A1}} + n_{A2} \times \frac{[Total \ of \ A2]^2}{n_{A2} \, n_{A2}} - n_{A} \times \frac{[Total \ of \ A]^2}{n_{A} \, n_{A}}$$

$$= n_{A1} \times \overline{A1}^2 + n_{A2} \times \overline{A2}^2 - n_{A} \times \overline{A}^2$$

The last term in the equation above, $n_{A} \times \overline{A}^2$, is really the sum of squares due to the mean (Sm) and is usually the same for all factors and interactions. Generalizing for a factor A, we may rewrite:

$$SA = n_{A1} \times \overline{A1}^2 + n_{A2} \times \overline{A2}^2 - Sm$$

Specifically, for factor A, and recalling that the average of A1 and A2 are made over 16 observations each,

$$SA = n_{A1} \times \overline{A1}^2 + n_{A2} \times \overline{A2}^2 - Sm$$

$$= 16 \times 6.06^2 + 16 \times 5.19^2 - 1013.63$$

$$= 6.04$$

The sums of squares due to factors B, C, D, E, F and G are calculated similarly.

7.   The sum of squares due to error
Se is then calculated as follows:

$$Se = ST - Sm - SA - SB - SC - SD - SE - SF - SG$$

$$= 1063.19 - 1013.63 - 6.04 - 5.87 - 1.67 - 0.53 - 19.69 - 8.93 - 5.20$$

$$= 1.66$$

8.   The mean sum of squares
The mean sum of squares is calculated by dividing the sum of squares by the degrees of freedom. For factor A,

$$Mq_A = \frac{Sq_A}{vA}$$

$$= \frac{6.04}{1}$$

$$= 6.04$$

The mean sums of squares of the remaining factors are calculated similarly.

9.   The F-ratio
The F-ratio is calculated by dividing the mean sum of squares by the error sum of squares. For factor A,

$$F_A = \frac{Mq_A}{Se}$$

$$= \frac{6.04}{0.07}$$

$$= 87.43$$

The results of the calculations above are used to draw the analysis of variance as shown in Figure 5:4.4. The calculation of Sq' and percent contribution is shown in the next section.

Source	Sq	ν	Mq	F-ratio	Sq'	rho %
A	6.04	1	6.04	87.43	5.97	12.04
B	5.87	1	5.87	84.93	5.80	11.69
C	1.67	1	1.67	24.11	1.60	3.22
D	0.53	1	0.53	7.61	0.46	0.92
E	19.69	1	19.69	285.07	19.62	39.58
F	8.93	1	8.93	129.24	8.86	17.87
G	5.20	1	5.20	75.30	5.13	10.35
e	1.66	24	0.07	1.00	2.14	4.32
St	49.56	31	1.60	-	49.56	100.00
Mean	1013.63	1	-	-	-	-
ST	1063.19	32	-	-	-	-

Figure 5:4.4 Analysis of variance for the concentricity data.

5.4.4 Calculate the percent contribution
In order to calculate the percent contribution of the various sources in an analysis of variance we need to calculate the pure sum of squares and divide by the total sum of squares. We show some examples:
- factor A,
- factor B,
- error.

   1.   **Factor A**

$$SA' = SA - \nu A \; Ve$$

$$= 6.04 - 1 \times 0.07$$

$$= 5.97$$

$$\rho A = \frac{SA'}{St} \times 100 \%$$

$$= \frac{5.97}{49.56} \times 100 \%$$

$$= 12.04 \%$$

2.   Factor B

$$SB' = SB - vB\ Ve$$

$$= 5.87 - 1 \times 0.07$$

$$= 5.80$$

$$\rho B = \frac{SB'}{St} \times 100 \%$$

$$= \frac{5.80}{49.56} \times 100 \%$$

$$= 11.69 \%$$

A similar method is used to calculate the pure sums of squares and the percent contributions for factors C to G.

3.   For error

$$Se = St - SA' - SB' - SC' - SD' - SE' - SF' - SG'$$

$$= 49.56 - 5.97 - 5.80 - 1.60 - 0.46 - 19.62 - 8.86 - 5.13$$

$$= 2.14$$

$$\rho e = \frac{Se}{St} \times 100 \%$$

$$= \frac{2.14}{49.56} \times 100 \%$$

$$= 4.32 \%$$

5.4.5 Pooling of insignificant factors

From an F-table, $F_{0.05,1,24} = 4.26$. This suggests that all factors A to G are significant. Thus, the analysis of variance shown in Figure 5:4.4 only shows us the significance of factors and is not very useful in an engineering sense. In order to avoid over-estimation, it is recommended that we use only about half the degrees of freedom of the orthogonal array used in the experiment. Since this experiment is an $L_8(2^7)$ orthogonal array experiment we may only take about 3 (or 4) main effects for estimations. To do this, we *pool* the 4 (or 3) factors with the smallest F-ratios into the error. To show this in the analysis of variance, we introduce a column called *Pool* and a row called *Pooled e*, as shown in Figure 5:4.5. Initially we pool the smallest variance (error variance in this case) and mark a Y (for yes) in column Pool to indicate that that source has been pooled into Pooled e (pooled error) as shown in Figure 5:4.5. Next, we calculate the F-ratio based on the Pooled e. In this case, there is no difference to the analysis of variance since the pooled error is the same as the error variance.

Next, we pool a factor with the next largest variance (column *Mq*) into Pooled e. By pooling an insignificant factor we are really treating that factor as if it was not included in the experiment and that its sum of squares was part of the sum of squares due to error *(Se)*. We can now pool as many insignificant factors and calculate the pure sum of squares for the remaining factors and complete the percent contribution. We start by pooling the factor with the smallest sum of squares, i.e. factor D. Thus, we pool factor D by marking a Y against it and adding its sum of squares to Pooled e as shown in Figure 5:4.6. Doing so changes the pooled error and hence the pooled error variance as follows.

$$S(Pooled\ e) = Se + SD$$

$$= 1.66 + 0.53$$

$$= 2.18$$

$$v(Pooled\ e) = ve + vD$$

$$= 24 + 1$$

$$= 25$$

$$M(Pooled\ e) = \frac{S(Pooled\ e)}{v(Pooled\ e)}$$

$$= \frac{2.18}{25}$$

$$= 0.09$$

Source	Pool	Sq	ν	Mq	F-ratio	Sq'	rho %
A		6.04	1	6.04	87.43	5.97	12.04
B		5.87	1	5.87	84.93	5.80	11.69
C		1.67	1	1.67	24.11	1.60	3.22
D		0.53	1	0.53	7.61	0.46	0.92
E		19.69	1	19.69	285.07	19.62	39.58
F		8.93	1	8.93	129.24	8.86	17.87
G		5.20	1	5.20	75.30	5.13	10.35
e	Y	1.66	24	0.07	-	-	-
Pooled e		1.66	24	0.07	1.00	2.14	4.32
St		49.56	31	1.60	-	49.56	100.00
Mean		1013.63	1	-	-	-	-
ST		1063.19	32	-	-	-	-

Figure 5:4.5 Analysis of variance.

Consequently, *Sq'* and *rho* % will also change for all the factors. However, the method of recalculation is similar to that shown in Section 5.4.4 and is repeated here only for factor A.

$$SA' = SA - \nu A \ Ve$$

$$= 6.04 - 1 \times 0.09$$

$$= 5.95$$

$$\rho A = \frac{SA'}{St} \times 100 \ \%$$

$$= \frac{5.95}{49.56} \times 100 \ \%$$

$$= 12.01 \ \%$$

Source	Pool	Sq	ν	Mq	F-ratio	Sq'	rho %
A		6.04	1	6.04	69.15	5.95	12.01
B		5.87	1	5.87	67.18	5.78	11.66
C		1.67	1	1.67	19.07	1.58	3.18
D	Y	0.53	1	0.53	-	-	-
E		19.69	1	19.69	225.49	19.60	39.55
F		8.93	1	8.93	102.22	8.84	17.83
G		5.20	1	5.20	59.56	5.11	10.32
e	Y	1.66	24	0.07	-	-	-
Pooled e		2.18	25	0.09	1.00	2.71	5.46
St		49.56	31	1.60	-	49.56	100.00
Mean		1013.63	1	-	-	-	-
ST		1063.19	32	-	-	-	-

Figure 5:4.6 Analysis of variance.

For the pooled error,

$$Se' = St - SA' - SB' - SC' - SE' - SF' - SG'$$

$$= 49.56 - 5.95 - 5.78 - 1.58 - 19.60 - 8.84 - 5.11$$

$$= 2.71$$

$$\rho e = \frac{Se'}{St} \times 100 \ \%$$

$$= \frac{2.71}{49.56} \times 100 \ \%$$

$$= 5.46 \ \%$$

The reader should be able to pool factors C, G and B and complete the remaining calculations to obtain the analysis of variance, as shown in Figure 5:4.7.

Source	Pool	Sq	ν	Mq	F-ratio	Sq'	rho %
A		6.04	1	6.04	11.34	5.51	11.11
B	Y	5.87	1	5.87	-	-	-
C	Y	1.67	1	1.67	-	-	-
D	Y	0.53	1	0.53	-	-	-
E		19.69	1	19.69	36.96	19.16	38.65
F		8.93	1	8.93	16.76	8.39	16.93
G	Y	5.20	1	5.20	-	-	-
e	Y	1.66	24	0.07	-	-	-
Pooled e		14.91	28	0.53	1.00	16.51	33.31
St		49.56	31	1.60	-	49.56	100.00
Mean		1013.63	1	-	-	-	-
ST		1063.19	32	-	-	-	-

Figure 5:4.7 Analysis of variance.

Of course, the more degrees of freedom for error sum of squares, the better the estimate of error sum of squares. When pooling factors, we use the *pooling-up technique*, which is simply pooling by starting with the factor with the lowest sum of squares and pooling the next lowest factor, etc. Hence, in this case, we would pool factor C next. Generally, we would pool about half the degrees of freedom in the $L_8(2^7)$ orthogonal array experiment. Therefore, we may pool other factors such as C, G and B as well. The recalculated analysis of variance is now shown in the Figure 5:4.7. From the percent contribution from this figure, we note that the significant factors are, in descending order, E, F and A. Between them, they account for 66.69 % of the total sum of squares. The pooled error sum of squares contributes to 33.31 % of the total sum of squares. As a general rule, the pooled error sum of squares may be up to 50 % of the total sum of squares for half the degrees of freedom in the orthogonal array. The final analysis of variance could in fact be shown as Figure 5:4.8.

5.4.6 Advantage of using analysis of variance
In the previous chapter, factor selection was based on the response table. The response table for the data in the present example is shown in Figure 5:4.9.

Source	Pool	Sq	ν	Mq	F-ratio	Sq'	rho %
A		6.04	1	6.04	11.34	5.51	11.11
E		19.69	1	19.69	36.96	19.16	38.65
F		8.93	1	8.93	16.76	8.39	16.93
Pooled e		1.66	28	0.53	1.00	16.51	33.31
St		49.56	31	1.60	-	49.56	100.00
Mean		1013.63	1	-	-	-	-
ST		1063.19	32	-	-	-	-

Figure 5:4.8 Analysis of variance.

	A	B	C	D	E	F	G
Level 1	6.06	5.20	5.86	5.50	4.84	6.16	5.23
Level 2	5.19	6.06	5.40	5.76	6.41	5.10	6.03
Difference	0.87	0.86	0.46	0.26	1.57	1.06	0.81
Rank	3	4	6	7	1	2	5

Figure 5:4.9 Response graph of factor effects.

If we compare the rank of the factors by the response table against rank of factors by the sum of squares (or percent contribution) in the analysis of variance (Figure 5:4.6), we get the results shown in Figure 5:4.10. Hence, we might ask *What then is the advantage of analysis of variance?*

	A	B	C	D	E	F	G
Response table	3	4	6	7	1	2	5
Analysis of variance	3	4	6	7	1	2	5

Figure 5:4.10   Comparison of factor selection.

In performing an analysis of variance, we account for the sum of squares of each factor. If the error sum of squares is large compared to the control factors in the experiment, then the analysis of variance together with the percent contributions will suggest that there is little to be gained by selecting optimum conditions. In fact, there may be nothing to be gained at all. This information is not available from the response table. Although it is understandable that analysis of variance calculations are tedious, they are nevertheless well defined and easily programmed onto a spreadsheet.

5.4.7 Discussion on the pooling-up technique

When using saturated orthogonal arrays with only one set of results, there is no degree of freedom for error. Hence, it will be necessary to use the pooling technique to estimate error. The pooling-up technique starts by regarding the factor with the smallest variance for the error variance. The remaining factors are now F-tested against this error variance. If no significant factor exists, the factor with the smallest F-ratio is pooled into the error. This will tend to increase the sum of squares for error and the degrees of freedom for error and will therefore improve the estimate of error variance. The remaining factors are again F-tested until we have pooled about half the degrees of freedom for the orthogonal array.

Another method of pooling is the *pooling-down* technique. This technique starts by testing the largest factor variance against the pooled variance of all the remaining factors. If that factor is significant, the next largest factor is removed from the pool and the F-test is repeated until some insignificant F-ratio is obtained.

Since there are two techniques of pooling available, which one should an engineer use? In the pooling-up technique we tend to maximize the alpha-error, whereas the pooling-down technique tends to maximize the beta-error. The tendency to maximize an alpha-error may cause us to regard insignificant factors as significant. Doing so will result in thinking that some factor may improve our response when in fact it may not. But there is no harm in doing so. However, a beta-error would cause us to regard a significant factor as insignificant – in which case we may disregard the factor in subsequent experimentation. Doing so will result in thinking that some factor will not improve our response when in fact it might. Therefore, it is better to use the pooling-up technique for F-testing.

5.4.8 Self-assessment questions

1.   Perform a suitable analysis of variance for the following experimental data.

Exp	Control factors							Results	
	A	B	C	D	E	F	G	P	Q
1	1	1	1	1	1	1	1	5.0	5.9
2	1	1	1	2	2	2	2	6.6	6.8
3	1	2	2	1	1	2	2	5.5	5.3
4	1	2	2	2	2	1	1	10.5	10.3
5	2	1	2	1	2	1	2	6.1	6.5
6	2	1	2	2	1	2	1	1.8	1.9
7	2	2	1	1	2	2	1	5.7	5.4
8	2	2	1	2	1	1	2	6.1	6.5

5.4.9 Answers to self-assessment questions

1.   Perform a suitable analysis of variance for the following experimental data.

Exp	Control factors							Results	
	A	B	C	D	E	F	G	P	Q
1	1	1	1	1	1	1	1	5.0	5.9
2	1	1	1	2	2	2	2	6.6	6.8
3	1	2	2	1	1	2	2	5.5	5.3
4	1	2	2	2	2	1	1	10.5	10.3
5	2	1	2	1	2	1	2	6.1	6.5
6	2	1	2	2	1	2	1	1.8	1.9
7	2	2	1	1	2	2	1	5.7	5.4
8	2	2	1	2	1	1	2	6.1	6.5

Answer

1.   **The average response for each experiment**

The average of experiment 1 is:

$$\bar{y}_1 = \frac{5.00 + 5.90}{2}$$

$$= 5.45$$

2.   **The overall experimental average is**

$$\bar{y} = \frac{\sum y}{n}$$

$$= \frac{5.0 + 5.9 + 6.6 + 6.8 + ... + 6.1 + 6.5}{16}$$

$$= 5.99$$

3.   **The response table**

	A	B	C	D	E	F	G
Level 1	6.99	5.08	6.00	5.68	4.75	7.11	5.81
Level 2	5.00	6.91	5.99	6.31	7.24	4.88	6.18
Difference	1.99	1.84	0.01	0.64	2.49	2.24	0.36
Rank	3	4	7	5	1	2	6

4.   **The total sum of squares**

$$ST = \sum y^2$$

$$= 5.0^2 + 5.9^2 + 6.6^2 + 6.8^2 + ... + 6.1^2 + 6.5^2$$

$$= 651.71$$

5.   **The sum of squares due to the mean**

$$Sm = n \, \bar{y}^2$$

$$= 16 \times 5.99^2$$

$$= 574.80$$

6.   **Sum of squares of due to factors**

$$SA = n_{A1} \times \overline{A1}^2 + n_{A2} \times \overline{A2}^2 - Sm$$

$$= 8 \times 6.99^2 + 8 \times 5.00^2 - 574.80$$

$$= 15.80$$

The sums of squares for factors B, C, D, E, F and G are similarly calculated and the results are then tabulated into an analysis of variance as shown in Figure 5:4.11.

Source	Pool	Sq	ν	Mq	F-ratio	Sq'	rho %
A		15.80	1	15.80			
B		13.51	1	13.51			
C		0.00	1	0.00			
D		1.63	1	1.63			
E		24.75	1	24.75			
F		20.03	1	20.03			
G		0.53	1	0.53			
e		0.68	8	0.08			
Pooled e							
St		76.91	15	5.13			
Mean		574.80	1				
ST		651.71	16				

Figure 5:4.11   Analysis of variance.

7. **Sum of squares due to error**

The next step is to pool factors with small sums of squares. Of course, the more degrees of freedom for the error sum of squares, the better will be the estimate of the error sum of squares. Therefore, we may pool about half the number of factors. When pooling factors, we use the pooling-up technique, which is simply pooling by starting with the factor with the lowest sum of squares and pooling the next lowest factor, etc. Hence, in this case, we would pool the error e and the factors C, G and D.

$$S(Pooled\ e) = Se + SC + SD + SG$$

$$= 0.68 + 0.00 + 1.63 + 0.53$$

$$= 2.83$$

8.   **Calculation of F-ratio**
Having pooled the insignificant factors we need to calculate the
F-ratios. We calculate the F-ratios for the factors by taking the
mean sum of squares Mq and dividing by the pooled error sum
of squares Se.

$$F_A = \frac{Mq_A}{Se}$$

$$= \frac{15.80}{0.26}$$

$$= 61.48$$

9.   **Calculation of pure sum of squares**
We then calculate the pure sums of squares for the factors. The
pure sum of squares for a factor, say A, is given by the formula:

$$SA' = SA - vA\ Ve$$

Since SA is known, degree of freedom for factor A, $vA = 1$ and
Ve is 0.26 and:

$$SA' = SA - vA\ Ve$$

$$= 15.80 - 1 \times 0.26$$

$$= 15.54$$

The portion of the sum of squares, $vA\ Ve = 0.26$ must be added
to the pooled error sum of squares in order to conserve the total
sum of squares.

10.   **Calculation of percent contribution**
We then calculate the percent contribution for the factors. The
percent contribution for a factor, say A, is given by the formula:

$$pA = \frac{SA'}{St} \times 100\ \%$$

$$= \frac{15.54}{76.91} \times 100\ \%$$

$$= 20.21\ \%$$

The remaining calculations are then completed and the results are as shown in Figure 5:4.12.

Source	Pool	Sq	ν	Mq	F-ratio	Sq'	rho %
A		15.80	1	15.80	61.48	15.54	20.21
B		13.51	1	13.51	52.55	13.25	17.23
C	Y	0.00	1	0.00	-	-	-
D	Y	1.63	1	1.63	-	-	-
E		24.75	1	24.75	96.31	24.49	31.85
F		20.03	1	20.03	77.92	19.77	25.70
G	Y	0.53	1	0.53	-	-	-
e	Y	0.68	8	0.08	-	-	-
Pooled e		2.83	11	0.26	1.00	3.85	5.01
St		76.91	15	5.13	-	76.91	100.00
Mean		574.80	1	-	-	-	-
ST		651.71	16	-	-	-	-

Figure 5:4.12   Analysis of variance.

From the analysis of variance, factors E, F, A and B are significant and account for 94.99 % of the variation (Figure 5:4.12).

5.5 Confidence Intervals

5.5.1 Introduction

After an experiment has been conducted, it is important to establish the process average at the predicted condition. This prediction is usually a point estimate. To improve the situation we need to know for instance that 95 % (confidence level) of the confirmation test results must be within $\pm x$ units (confidence interval) of the predicted mean. To most engineers the calculation of confidence intervals is very bothersome. However, whenever a prediction is made, unless we establish the confidence interval within which the observed data must lie we cannot infer that our experimental result would be reproducible.

5.5.2 Confidence intervals

Experimental data are used to make a number of estimates. We generally make estimates of the factor levels and the predicted optimum process average. However, these are point estimates based on the averages of results obtained from the experiment. For an estimate of a factor level such as A1, statistically, there is a 50 % chance of the true mean (μ_{A1}) being greater than this estimate and a 50 % chance of the true average being less than this estimate. To improve the situation, we calculate a confidence interval between which the estimate must lie at some stated confidence level.

A high confidence level may be chosen to reduce risk, but a high confidence level results in a wider confidence interval, as illustrated in Figure 5:5.1. Consider for example, the confidence level that you can finish reading this book by a certain time:

Can I finish reading this book within:	Response	Implied confidence level	Implied confidence interval
5 years	Certainly	Better than 99 %	0 to 1750 days
5 months	Positive	At least 95 %	0 to 150 days
5 weeks	Possible	About 50 %	0 to 35 days
5 days	Unlikely	About 1 %	0 to 5 days

Figure 5:5.1 Confidence levels and confidence intervals.

There are three cases where we may need to calculate confidence intervals:

1.   for a factor level,
2.   for a predicted mean,
3.   for a confirmation experiment.

5.5.3 Confidence interval – for a factor level

We show the calculation of confidence intervals for a factor level using the concentricity data (Figure 5:4.2) once again. The response table is given here for convenience.

	A	B	C	D	E	F	G
Level 1	6.06	5.20	5.86	5.50	4.84	6.16	5.23
Level 2	5.19	6.06	5.40	5.76	6.41	5.10	6.03
Difference	0.87	0.86	0.46	0.26	1.57	1.06	0.81
Rank	3	4	6	7	1	2	5

Figure 5:5.2 Response table of factor effects.

The method of calculating the confidence interval for a factor level is to use the formula:

$$CI = \sqrt{F_{\alpha,v1,v2} \times V_e \times \left[\frac{1}{n}\right]} \qquad (5:5.1)$$

where

$F_{\alpha,v1,v2}$	=	the tabulated F-ratio[6],
α	=	risk. The confidence level = 1 − risk,
$v1$	=	the degrees of freedom for the numerator associated with a mean and is always 1 for a confidence interval,
$v2$	=	the degrees of freedom for the denominator associated with the degrees of freedom for the pooled error variance,
Ve	=	is the pooled error variance,
n	=	number of observations used to calculate the mean.

---

6   The tabulated F-ratio may be obtained from Murdoch and Barnes, *Statistical Tables for Science, Engineering, Management and Business Studies*, MacMillan Education Ltd.

Hence, if the true mean is $\mu_{\overline{A1}}$,

$$\mu_{\overline{A1}} = \overline{A1} \pm CI$$

$$\overline{A1} - CI \le \mu_{\overline{A1}} \le \overline{A1} + CI$$

The mean value of A1 is 6.06. The associated confidence interval is calculated as follows:

$$CI = \sqrt{F_{\alpha,v1,v2} \times V_e \times \left[\frac{1}{n}\right]}$$

where: $\alpha = 0.05$, $v1 = 1$, $v2 = 28$, $Ve = 0.53$ (from the analysis of variance shown in Figure 5:4.7), $n = 16$, $F_{\alpha,v1,v2} = F_{0.05,1,28} = 4.20$ and,

$$CI = \sqrt{F_{0.05,1,28} \times V_e \times \left[\frac{1}{n}\right]}$$

$$= \sqrt{4.20 \times 0.53 \times \left[\frac{1}{16}\right]}$$

$$= \pm 0.37$$

Thus the confidence interval for the average of A1 is:

$$\mu_{\overline{A1}} = \overline{A1} \pm CI$$

$$\overline{A1} - CI \le \mu_{\overline{A1}} \le \overline{A1} + CI$$

$$6.06 - 0.37 \le \mu_{\overline{A1}} \le 6.06 + 0.37$$

$$5.69 \le \mu_{\overline{A1}} \le 6.44$$

and the confidence interval for the average of A2 is:

$$\overline{A2} - CI \le \mu_{\overline{A2}} \le \overline{A2} + CI$$

$$5.19 - 0.37 \le \mu_{\overline{A2}} \le 5.19 + 0.37$$

$$4.82 \le \mu_{\overline{A2}} \le 5.57$$

The confidence interval for the remaining factors are similarly calculated and shown as in Figure 5:5.3.

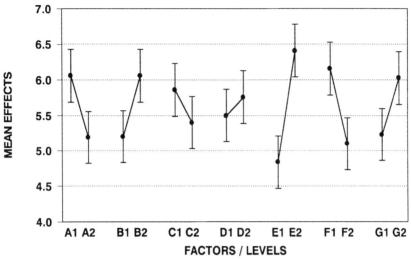

Figure 5:5.3 Confidence intervals for factor levels.

5.5.4 Confidence interval – for a predicted mean
To calculate the confidence interval for a predicted optimum process mean, we use the following formula:

$$CI = \sqrt{F_{\alpha,v1,v2} \times V_e \times \left[\frac{1}{n_{eff}}\right]}$$
(5:5.2)

where n_{eff} is the effective number of observations,

$$n_{eff} = \frac{total\ number\ of\ experiments}{sum\ of\ degrees\ of\ freedom\ used\ in\ estimate\ of\ mean}$$
(5:5.3)

The reader should note that the denominator in Equation 5:5.3 must *include the degrees of freedom for the overall mean.*

For the example of the concentricity data, the predicted process mean ($\mu_{Predicted}$) can be calculated as:

$$\mu_{Predicted} = \bar{y} + (\overline{E1} - \bar{y}) + (\overline{F2} - \bar{y}) + (\overline{A2} - \bar{y})$$

$$= \overline{E1} + \overline{F2} + \overline{A2} - 2 \times \bar{y}$$

$$= 4.84 + 5.10 + 5.19 - 2 \times 5.63$$

$$= 3.88$$

Since three factors with 1 degree of freedom each were used to establish the predicted process mean,

$$n_{eff} = \frac{total\ number\ of\ experiments}{sum\ of\ degrees\ of\ freedom\ used\ in\ estimate\ of\ mean}$$

$$= \frac{8 \times 4}{v_\mu + v_E + v_F + v_A}$$

$$= \frac{32}{1 + 1 + 1 + 1} = 8$$

and the confidence interval is therefore:

$$CI = \sqrt{F_{0.05,1,28} \times V_e \times \left[\frac{1}{n_{eff}}\right]}$$

$$= \sqrt{4.20 \times 0.53 \times \left[\frac{1}{8}\right]}$$

$$= \pm\ 0.53$$

Note that n_{eff} depends on the number of degrees of freedom used to calculate the predicted optimum process mean and does not depend on which factor level is used. Thus, all factors and interaction terms used in calculating the predicted mean must be included in the degrees of freedom for calculating n_{eff}. The confidence interval for this optimum process mean is therefore:

$$\mu_{Predicted} - CI \le \mu_{Predicted} \le \mu_{Predicted} + CI$$

$$3.88 - 0.53 \le \mu_{Predicted} \le 3.88 + 0.53$$

$$3.35 \le \mu_{Predicted} \le 4.41$$

5.5.5 Confidence interval – for a confirmation experiment

The confirmation experiment is used to verify that the predicted mean for the factors and levels chosen from an orthogonal array experiment is valid. If too few samples are taken, then it would be difficult to establish the validity of the predicted mean. Hence, we shall provide a formula to calculate the confidence intervals for a confirmation experiment as follows:

$$CI = \sqrt{F_{\alpha,v1,v2} \times V_e \times \left[\frac{1}{n_{\textit{eff}}} + \frac{1}{r}\right]} \qquad (5:5.4)$$

where r is the sample size (number of replicates) for the confirmation experiment ($r \neq 0$). If r approached a very large number (infinity), then $1/r$ approaches zero and the formula is reduced to that of the confidence interval around a predicted mean. As r becomes smaller, $1/r$ becomes larger and the confidence interval increases until it is a maximum at 1. Of course, r cannot be less than one.

For the concentricity data, suppose a confirmation experiment is run 10 times with a mean result of 3.52. With $r = 10$, the confidence interval is:

$$CI = \sqrt{F_{0.05,1,28} \times V_e \times \left[\frac{1}{n_{\textit{eff}}} + \frac{1}{r}\right]}$$

$$= \sqrt{4.20 \times 0.53 \times \left[\frac{1}{8} + \frac{1}{10}\right]}$$

$$= \pm\ 0.71$$

The confidence interval is therefore

$$\mu_{Confirmation} - CI \leq \mu_{Confirmation} \leq \mu_{Confirmation} + CI$$

$$3.52 - 0.71 \leq \mu_{Confirmation} \leq 3.52 + 0.71$$

$$2.81 \leq \mu_{Confirmation} \leq 4.23$$

To interpret the confidence interval for the predicted optimum process mean and the confirmation experiment, we shall refer to Figure 5:5.4, showing both means and confidence intervals. In A and C there is a good overlap between the predicted and confirmation confidence intervals. In C, there is no overlap between the predicted and confirmation confidence intervals. The confirmation experiment would therefore show that the predicted value is not acceptable and the experiment is unlikely to be reproducible. The importance of the confirmation experiment cannot be overemphasized. The fundamental concept of using orthogonal arrays is based on the assumption that the predicted optimum condition *will be tested* to ensure the additivity (correctness of model) of the quality characteristic.

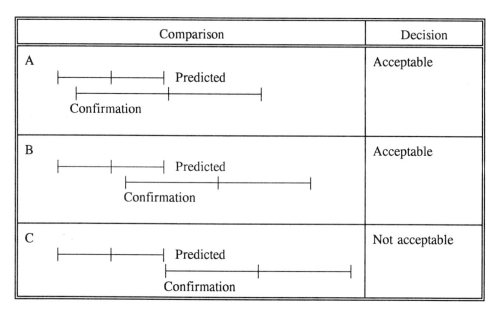

Comparison	Decision

Figure 5:5.4 Comparison of confidence intervals for predicted mean and confirmation experiment.

A priori, we have no way of establishing whether additivity is present in an experiment. However, if the confidence interval of a confirmation experiment overlaps with the confidence interval of the predicted mean, then we may accept that the results are additive. Additivity implies that the effect of a factor and the effect of another factor can be numerically added (or subtracted) and the effect of the better levels of the factors will be the best. If the confidence interval of a confirmation experiment does not overlap with the confidence interval of the predicted mean, then the result is not additive or is full of interactions. Interactions imply that the effect of a factor and the effect of another factor cannot be added numerically. Therefore, the effect of the better levels of the factors may be poorer than either.

For the concentricity data above, the confidence interval of the predicted mean [3.35, 4.41] overlaps fairly well with the confidence interval of the confirmation experiment [2.81, 4.23]. Thus we may infer that the experimental results will be reproducible.

5.5.6 Industrial experiment datasheet
From this and the previous chapters, it is clear that there is a methodical approach to the orthogonal array experiment analysis. For an $L_8(2^7)$ orthogonal array in a direct product design with an $L_4(2^3)$ noise array, we may suggest a datasheet (Figure 5:5.5) on which to fill in the results and the various analyses. The following example shows a direct-product design for an experiment in plastic moulding. Notice that in this example, we have 32 experiments since there are four measurements in the noise factor array for each of the eight control factor array experiment. For convenience, control factors have been labelled A, B, ..., F and G, while the noise factors have been labelled H, I and J. The results of the 32 experiments are given, together with a complete analysis of the data.

For the benefit of the interested reader, a number of calculations are repeated in order to help better understanding.

1.   **Average response of each experiment**
      The average of experiment 1is:

$$\bar{y}_1 = \frac{5.50 + 5.1 + 5.5 + 4.7}{4} = \frac{20.3}{4} = 5.08$$

2.   **The overall experimental average**

$$\bar{y} = \frac{\sum y}{n}$$

$$= \frac{5.0 + 5.1 + 5.5 + 4.7 + ... + 6.1 + 6.0}{32}$$

$$= 5.63$$

3.   **Average effect of factors**
      The mean effect of a factor is calculated by averaging all experiments in the same level, e.g. factor A1:

$$\overline{A1} = \frac{5.08 + 6.65 + 5.23 + 7.30}{4} = \frac{24.26}{4} = 6.06$$

4.   **Complete the response table**
      Difference is calculated by taking the difference of the factor levels for each factor. The factors are then ranked in descending order of difference.

									N	1	2	3	4			
									A	1	1	2	2			
									B	1	2	1	2			
									C	1	2	2	1	$\bar{y}$	σ	η

C	A	B	C	D	E	F	G		1	2	3	4	$\bar{y}$		
1	1	1	1	1	1	1	1		5.0	5.1	5.5	4.7	5.08		
2	1	1	1	2	2	2	2		6.6	6.8	6.7	6.5	6.65		
3	1	2	2	1	1	2	2		5.5	5.3	5.1	5.0	5.23		
4	1	2	2	2	2	1	1		7.0	7.3	7.5	7.4	7.30		
5	2	1	2	1	2	1	2		6.1	6.3	6.1	6.0	6.13		
6	2	1	2	2	1	2	1		2.8	2.5	3.0	3.5	2.95		
7	2	2	1	1	2	2	1		5.7	5.4	5.2	6.0	5.58		
8	2	2	1	2	1	1	2		6.1	6.3	6.1	6.0	6.13		

	A	B	C	D	E	F	G
Level 1	6.06	5.20	5.86	5.50	4.84	6.16	5.23
Level 2	5.19	6.06	5.40	5.76	6.41	5.10	6.03
Difference	0.87	0.86	0.46	0.26	1.57	1.96	0.81
Rank	3	4	6	7	1	2	5

Characteristic type: Smaller-the-better	Experimental average: 5.63	Predicted mean: 3.88	Confirmation: 3.52

Source	Pool	Sq	ν	Mq	F-ratio	Sq'	rho %
A		6.04	1	6.04	11.34	5.51	11.11
B	Y	5.87	1	5.87	-	-	-
C	Y	1.67	1	1.67	-	-	-
D	Y	0.53	1	0.53	-	-	-
E		19.69	1	19.69	36.96	19.16	38.65
F		8.93	1	8.93	16.76	8.39	16.93
G	Y	5.20	1	5.20	-	-	-
e	Y	1.66	24	0.07	-	-	-
Pooled e		14.91	28	0.53	1.00	16.51	33.31
St		49.56	31	1.60	-	49.56	100.00
Mean		1013.60	1	-	-	-	-
ST		1063.16	32	-	-	-	-

Confidence intervals	Factor	Predicted mean	Confirmation experiment
	± 0.37	± 0.53	± 0.71

Figure 5:5.5 Industrial experiment datasheet.

5. The total sum of squares is calculated as follows

$$ST = \sum y^2$$

$$= 5.0^2 + 5.1^2 + 5.5^2 + 4.7^2 + ... + 6.1^2 + 6.0^2$$

$$= 1063.19$$

6. The sum of squares due to the mean is calculated as follows

$$Sm = n\,\bar{y}^2$$

$$= 8 \times 5.63^2$$

$$= 1013.63$$

7. The sum of squares for factor A is calculated as follows

$$SA = n_{A1} \times \overline{A1}^2 + n_{A2} \times \overline{A2}^2 - Sm$$

$$= 16 \times 6.06^2 + 16 \times 5.19^2 - 1013.63$$

$$= 6.04$$

8. **Calculate the mean sum of squares**
Mq is calculated by dividing *Sq* by *v*. The degrees of freedom *v* for a 2-level factor is 1.

9. **Pool insignificant errors**
The error e and factors D, C, G and B are pooled.

10. **Calculate the pure sum of squares**
The pure sum of squares for factor A is calculated as follows:

$$SA' = SA - vA\ Ve$$

$$= 6.04 - 1 \times 0.53$$

$$= 5.51$$

11. **Calculate the percent contribution**
The percent contribution for factor A is calculated as follows:

$$pA = \frac{SA'}{St} \times 100 \%$$

$$= \frac{5.51}{49.56} \times 100 \%$$

$$= 11.11 \%$$

12. **Predict the process average**
    The predicted process average is calculated as follows:

$$\mu_{Predicted} = \bar{y} + (\overline{E1} - \bar{y}) + (\overline{F2} - \bar{y}) + (\overline{A2} - \bar{y})$$

$$= \overline{E1} + \overline{F2} + \overline{A2} - 2 \times \bar{y}$$

$$= 4.84 + 5.10 + 5.19 - 2 \times 5.63$$

$$= 3.88$$

Note that we have used 3 degrees of freedom for estimating the process average.

13. **Calculate the confidence interval for factors**
    The confidence interval of a factor level is calculated as follows:

$$CI = \sqrt{F_{0.05,1,28} \times V_e \times \left[\frac{1}{n}\right]}$$

$$= \sqrt{4.20 \times 0.53 \times \left[\frac{1}{16}\right]}$$

$$= \pm 0.37$$

14. **Calculate confidence interval for the predicted mean**
    The confidence interval of the predicted mean is calculated as follows:

$$CI = \sqrt{F_{0.05,1,28} \times V_e \times \left[\frac{1}{n_{eff}}\right]}$$

$$= \sqrt{4.20 \times 0.53 \times \left[\frac{1}{8}\right]}$$

$$= \pm \, 0.53$$

where $n_{eff} = 32/4$.

15.   **Calculate the confidence interval for the confirmation experiment**
If ten observations are made for the confirmation experiment, the confidence interval is:

$$CI = \sqrt{F_{0.05,1,28} \times V_e \times \left[\frac{1}{n_{eff}} + \frac{1}{r}\right]}$$

$$= \sqrt{4.20 \times 0.53 \left[\frac{1}{8} + \frac{1}{10}\right]}$$

$$= \pm \, 0.71$$

The confidence intervals for factor levels can be used to enhance the response graph of factor effects by indicating the interval of the factor levels as shown in Figure 5:5.3. The industrial experiment datasheet can thus be used as a standard summary sheet for an $L_8(2^7)$ experiment.

5.5.7 Self-assessment questions

1. Using the datasheet for the concentricity data below, complete the calculations using SN ratio and make your inference.

N	1	2	3	4
A	1	1	2	2
B	1	2	1	2
C	1	2	2	1

C	A	B	C	D	E	F	G					$\bar{y}$	η
1	1	1	1	1	1	1	1	5.0	5.1	5.5	4.7		
2	1	1	1	2	2	2	2	6.6	6.8	6.7	6.5		
3	1	2	2	1	1	2	2	5.5	5.3	5.1	5.0		
4	1	2	2	2	2	1	1	7.0	7.3	7.5	7.4		
5	2	1	2	1	2	1	2	6.1	6.3	6.1	6.0		
6	2	1	2	2	1	2	1	2.8	2.5	3.0	3.5		
7	2	2	1	1	2	2	1	5.7	5.4	5.2	6.0		
8	2	2	1	2	1	1	2	6.1	6.3	6.1	6.0		

Level 1													
Level 2													
Difference													
Rank													
Optimum													

Characteristic type:	Experimental average:	Predicted mean:	Confirmation:

Source	Pool	Sq	ν	Mq	F-ratio	Sq'	rho %
A							
B							
C							
D							
E							
F							
G							
e							
Pooled e							
St							
Mean							
ST							

Confidence intervals	Factor	Predicted mean	Confirmation experiment

5.5.8 Answers to self-assessment questions

1.   Using the datasheet for the concentricity data below, complete the calculations using SN ratio and make your inference.

N	1	2	3	4
A	1	1	2	2
B	1	2	1	2
C	1	2	2	1

C	A	B	C	D	E	F	G	1	2	3	4	$\bar{y}$	η
1	1	1	1	1	1	1	1	5.0	5.1	5.5	4.7		
2	1	1	1	2	2	2	2	6.6	6.8	6.7	6.5		
3	1	2	2	1	1	2	2	5.5	5.3	5.1	5.0		
4	1	2	2	2	2	1	1	7.0	7.3	7.5	7.4		
5	2	1	2	1	2	1	2	6.1	6.3	6.1	6.0		
6	2	1	2	2	1	2	1	2.8	2.5	3.0	3.5		
7	2	2	1	1	2	2	1	5.7	5.4	5.2	6.0		
8	2	2	1	2	1	1	2	6.1	6.3	6.1	6.0		

Level 1													
Level 2													
Difference													
Rank													
Optimum													

Characteristic type:	Experimental average:	Predicted mean:	Confirmation:

Source	Pool	Sq	v	Mq	F-ratio	Sq'	rho %
A							
B							
C							
D							
E							
F							
G							
e							
Pooled e							
St							
Mean							
ST							

Confidence intervals	Factor	Predicted mean	Confirmation experiment

Answer

For the benefit of the reader, the following calculations are shown in order to help better understanding.

1. **Average response of each experiment**
   To calculate the SN ratio (for smaller-the-better) for experiment 1 we use the formula:

$$\eta_1 = -10 \log_{10} \left[\frac{y_1^2 + y_2^2 + ... + y_n^2}{n} \right]$$

$$= -10 \log_{10} \left[\frac{5.0^2 + 5.1^2 + 5.5^2 + 4.7^2}{4} \right]$$

$$= -10 \log_{10} (25.84)$$

$$= -14.12 \text{ dB}$$

The remaining SN ratio values (η_2 to η_8) are similarly calculated.

2. **The overall experimental average**

$$\bar{y} = \frac{\Sigma y}{n}$$

$$= \frac{-14.12 - 16.46 - 14.37 - 17.27 - 15.74 - 9.46 - 14.94 - 15.74}{8}$$

$$= -14.76 \text{ dB}$$

3. **Average effect of factors**
   The mean effect of a factor is calculated by averaging all experiments in the same level, e.g. factor A:

$$\overline{A1} = \frac{-14.12 - 16.46 - 14.37 - 17.27}{4} = \frac{-62.22}{4} = -15.55 \text{ dB}$$

$$\overline{A2} = \frac{-15.74 - 9.46 - 14.94 - 15.74}{4} = \frac{-55.88}{4} = -13.97 \text{ dB}$$

								N	1	2	3	4			
								A	1	1	2	2			
								B	1	2	1	2			
								C	1	2	2	1		$\bar{y}$	n
C	A	B	C	D	E	F	G								
1	1	1	1	1	1	1	1		5.0	5.1	5.5	4.7			−14.12
2	1	1	1	2	2	2	2		6.6	6.8	6.7	6.5			−16.46
3	1	2	2	1	1	2	2		5.5	5.3	5.1	5.0			−14.37
4	1	2	2	2	2	1	1		7.0	7.3	7.5	7.4			−17.27
5	2	1	2	1	2	1	2		6.1	6.3	6.1	6.0			−15.74
6	2	1	2	2	1	2	1		2.8	2.5	3.0	3.5			−9.46
7	2	2	1	1	2	2	1		5.7	5.4	5.2	6.0			−14.94
8	2	2	1	2	1	1	2		6.1	6.3	6.1	6.0			−15.74

	A	B	C	D	E	F	G
Level 1	−15.55	−13.95	−15.32	−14.79	−13.42	−15.72	−13.95
Level 2	−13.97	−15.58	−14.21	−14.73	−16.10	−13.81	−15.58
Difference	1.58	1.63	1.11	0.06	2.68	1.91	1.63
Rank	5	3	6	7	1	2	4
Optimum	A2	B1	C2	D2	E1	F2	G1

Characteristic type: Larger-the-better	Experimental average: −14.76 dB	Predicted mean: −10.84 dB	Confirmation: -9.20 dB

Source	Pool	Sq	ν	Mq	F-ratio	Sq'	rho %
A	Y	5.01	1	5.01	-	-	-
B		5.33	1	5.33	2.15	2.85	7.16
C	Y	2.44	1	2.44	-	-	-
D	Y	0.01	1	0.01	-	-	-
E		14.34	1	14.34	5.77	11.86	29.82
F		7.32	1	7.32	2.95	4.84	12.16
G		5.31	1	5.31	2.14	2.83	7.11
e		0.00	0	-	-	-	-
Pooled e		7.46	3	2.49	1.00	17.40	43.75
St		39.77	7	5.68	-	39.77	100.00
Mean		1743.56	1	-	-	-	-
ST		1783.34	8	-	-	-	-

Confidence intervals	Factor	Predicted mean	Confirmation experiment
	±2.51 dB	± 3.97 dB	± 4.27 dB

4.   **Complete the response table**
      Difference is calculated by taking the difference of the factor levels for each
      factor. The factors are then ranked in descending order of difference.

5.   **The total sum of squares is calculated as follows**

$$ST = \sum y^2$$

$$= -14.12^2 - 16.46^2 - 14.37^2 - 17.27^2 - 15.74^2 - 9.46^2 - 14.94^2 - 15.74^2$$

$$= 1783.34$$

6.   **The sum of squares due to the mean is calculated as follows**

$$Sm = n\,\bar{y}^2$$

$$= 8 \times (-14.76)^2$$

$$= 1743.56$$

7.   **The sum of squares for factor A is calculated as follows**

$$SA = n_{A1} \times \overline{A1}^2 + n_{A2} \times \overline{A2}^2 - Sm$$

$$= 4 \times (-15.55)^2 + 4 \times (-13.97)^2 - 1743.56$$

$$= 5.01$$

8.   **Calculate the mean sum of squares**
      Mq is calculated by dividing Sq by v. The degrees of freedom v for a 2-level
      factor is 1.

9.   **Pool insignificant factors**
      Factors D, C and A are pooled.

10.   **Calculate the pure sum of squares**
      The pure sum of squares for factor B is calculated as follows:

$$SB' = SB - vB\ Ve$$

$$= 5.33 - 1 \times 2.49$$

$$= 2.85$$

11.   Calculate the percent contribution

The percent contribution for factor B is calculated as follows:

$$\rho B = \frac{SB'}{St} \times 100\ \%$$

$$= \frac{2.85}{39.77} \times 100\ \%$$

$$= 7.16\ \%$$

12.   Predict the process average

The predicted process average is calculated as follows:

$$\mu_{Predicted} = \bar{y} + (\overline{E1} - \bar{y}) + (\overline{F2} - \bar{y}) + (\overline{B1} - \bar{y}) + (\overline{G1} - \bar{y})$$

$$= \overline{E1} + \overline{F2} + \overline{B1} + \overline{G1} - 3 \times \bar{y}$$

$$= -13.42 - 13.81 - 13.95 - 13.95 - 3 \times (-14.76)$$

$$= -10.84\ dB$$

Note that we have used 4 degrees of freedom for estimating the process average.

13.   Calculate the confidence interval for factors

The confidence interval of a factor level is calculated as follows:

$$CI = \sqrt{F_{0.05,1,3} \times V_e \times \left[\frac{1}{n}\right]}$$

$$= \sqrt{10.13 \times 2.49 \times \left[\frac{1}{4}\right]}$$

$$= \pm\ 2.51\ dB$$

14. **Calculate the confidence interval for the predicted mean**
     The confidence interval of the predicted mean is calculated as follows:

$$CI = \sqrt{F_{0.05,1,3} \times V_e \times \left[\frac{1}{n_{eff}}\right]}$$

$$= \sqrt{10.13 \times 2.49 \times \left[\frac{5}{8}\right]}$$

$$= \pm\ 3.97 \text{ dB}$$

where n_{eff} is the effective number of degrees of freedom and

$$n_{eff} = \frac{total\ number\ of\ experiments}{sum\ of\ degrees\ of\ freedom\ used\ in\ estimate\ of\ mean}$$

$$= \frac{8}{v_\mu + v_E + v_F + v_B + v_G}$$

$$= \frac{8}{1 + 1 + 1 + 1 + 1}$$

$$= \frac{8}{5}$$

15. **Calculate the confidence interval for the confimation experiment**
     If ten observations are made for the confirmation experiment, the confidence
     interval is:

$$CI = \sqrt{F_{\alpha,1,v2} \times V_e \times \left[\frac{1}{n_{eff}} + \frac{1}{r}\right]}$$

$$= \sqrt{10.13 \times 2.49 \times \left[\frac{5}{8} + \frac{1}{10}\right]}$$

$$= \pm\ 4.27 \text{ dB}$$

The confidence intervals for factor levels can be used to enhance the response
graph of factor effects by indicating the interval of the factor levels as shown
in the following figure, from which it is clear that there is a large noise effect

in the data. The industrial experiment datasheet can thus be used as a standard summary sheet for an $L_8(2^7)$ experiment.

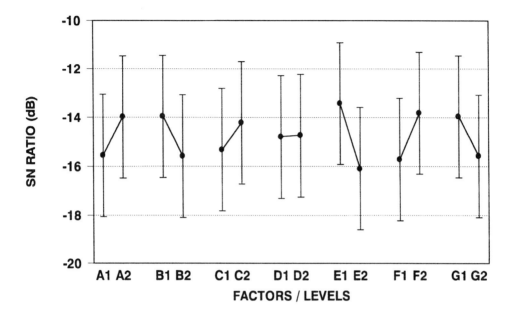

5.6   Analysis of Variance for Attribute Data

5.6.1 Introduction
Attribute data analysis was introduced in Section 4.1.8 where the Omega (Ω) transformation was introduced as an objective characteristic. However, the selection of important factors was based on a simple numerical analysis based on the response table. While such a numeric analysis is relatively simple to perform, a better method, the attribute accumulation analysis, is introduced in this chapter. Attribute accumulation analysis uses analysis of variance and the contribution ratio to establish significant factors. In fact, only the analysis of variance is given here, since the response analysis has been covered in Chapter 4.3.

5.6.2 Attribute accumulation
Attribute accumulation is used when the experimental data can be ranked or categorized. Examples of such data are those that can be categorized into good, fair or bad. Continuous data that can be classified can also be analyzed using attribute accumulation. An example of such data are 0 % – 39 % : E, 40 % – 49 % : D, 50 % – 59 % : C, 60 % – 69 % : B, 71 % – 100 % : A.

5.6.3 Binomial distribution
Accumulation analysis needs some understanding of the binomial distribution. In the binomial distribution, if the fraction defective is p, then the corresponding variance is:

$$\sigma^2 = p \times (1 - p) \qquad\qquad (5{:}6.1)$$

This implies that the variance (σ^2) depends on p. However, if we are to compare two distributions (corresponding to classes or categories in an experiment), we can only make a fair comparison if the variances are (at least approximately) the same. Since, the sums of squares of different classes in accumulation analysis would have different bases, it is important to normalize these bases. This is done by *dividing the sum of squares of each class by its variance*. This procedure is frequently called weighting. Only when the classes have been weighted appropriately, can the sums of squares of those classes be added. For convenience we define the weight (ω) as follows:

$$\omega = \frac{1}{\sigma^2} \qquad\qquad (5{:}6.2)$$

5.6.4 Example of attribute accumulation analysis
Recall the example of fraction defective analysis from Chapter 4.3, the results of which
are given in Figure 5:6.1 for ease of reference.

Exp	A	B	C	D	E	F	G	None	Some	Severe	Total
1	1	1	1	1	1	1	1	15	4	1	20
2	1	1	1	2	2	2	2	5	13	2	20
3	1	2	2	1	1	2	2	8	12	0	20
4	1	2	2	2	2	1	1	2	11	7	20
5	2	1	2	1	2	1	2	17	2	1	20
6	2	1	2	2	1	2	1	4	15	1	20
7	2	2	1	1	2	2	1	7	13	0	20
8	2	2	1	2	1	1	2	3	11	6	20

Figure 5:6.1 Orthogonal array and results for accumulation analysis.

The method of conducting the analysis of variance is outlined as follows:

1.   **Create a table of data with the results**
       See Figure 5:6.1.

2.   **Establish the cumulative frequencies**
       For reasons of space, let None = I, Some = II and Severe = III. The
       cumulative frequencies (I) = I, (II) = I + II and (III) = I + II + III. This is
       shown in Figure 5:6.2.

3.   **Calculate the total cumulative frequency in each class**

$$f_{(I)} = 61$$

$$f_{(II)} = 142$$

$$f_{(III)} = 160$$

Exp	Control Factors							Frequencies			Cumulative Frequencies		
	A	B	C	D	E	F	G	I	II	III	(I)	(II)	(III)
1	1	1	1	1	1	1	1	15	4	1	15	19	20
2	1	1	1	2	2	2	2	5	13	2	5	18	20
3	1	2	2	1	1	2	2	8	12	0	8	20	20
4	1	2	2	2	2	1	1	2	11	7	2	13	20
5	2	1	2	1	2	1	2	17	2	1	17	19	20
6	2	1	2	2	1	2	1	4	15	1	4	19	20
7	2	2	1	1	2	2	1	7	13	0	7	20	20
8	2	2	1	2	1	1	2	3	11	6	3	14	20
Column totals								61	81	18	61	142	160

Figure 5:6.2 Orthogonal array and results.

4.   Calculate the fraction defective of each class

$$p_I = \frac{f_I}{f_I + f_{II} + f_{III}} = \frac{f_I}{f_{(III)}}$$

$$p_{II} = \frac{f_{II}}{f_I + f_{II} + f_{III}} = \frac{f_{II}}{f_{(III)}}$$

$$p_{III} = \frac{f_{III}}{f_I + f_{II} + f_{III}} = \frac{f_{III}}{f_{(III)}} \qquad (5:6.3)$$

from which,

$$p_I = 0.38125$$

$$p_{II} = 0.50625$$

$$p_{III} = 0.11250$$

Note also that, $p_{(I)} = 0.38125$, $p_{(II)} = 0.88750$, $p_{(III)} = 1.0000$.

5.   Calculate the weight of each class

$$\omega_I = \frac{1}{\sigma_I^2} = \frac{1}{p_I \times (1 - p_I)} \tag{5:6.4}$$

For ease of calculation, it may be better to use:

$$\omega_I = \frac{1}{p_I \times (1 - p_I)}$$

$$= \frac{1}{\dfrac{f_I}{f_{(III)}} \times \left(1 - \dfrac{f_I}{f_{(III)}}\right)}$$

$$= \frac{1}{\dfrac{f_I}{f_{(III)}} \times \left(\dfrac{f_{(III)} - f_I}{f_{(III)}}\right)}$$

$$= \frac{f_{(III)}^2}{f_I \times (f_{(III)} - f_I)} \tag{5:6.5}$$

Hence,

$$\omega_I = \frac{f_{(III)}^2}{f_I \times (f_{(III)} - f_I)}$$

$$= \frac{160^2}{61 \times (160 - 61)}$$

$$= 4.24$$

$$\omega_{II} = \frac{f_{(III)}^2}{f_{II} \times (f_{(III)} - f_{II})} \tag{5:6.6}$$

$$= \frac{160^2}{142 \times (160 - 142)}$$

$$= 10.02$$

6.   **Calculate the total sum of squares due to class I**

$$S_I = total\ sum\ of\ squares\ of\ class\ I$$

$$= \left(f_I - \frac{f_I^2}{f_{(III)}} \right) \omega_I$$

$$= \frac{f_I f_{(III)} - f_I^2}{f_{(III)}} \times \frac{f_{(III)}^2}{f_I \times (f_{(III)} - f_I)}$$

$$= \frac{f_I \times (f_{(III)} - f_I)}{f_{(III)}} \times \frac{f_{(III)}^2}{f_I \times (f_{(III)} - f_I)}$$

$$= f_{(III)} \tag{5:6.7}$$

Similarly, the total sum of squares of class II is also $f_{(III)}$.

7.   **Calculate the overall total sum of squares due to class I and class II**

$$ST = total\ sum\ of\ squares\ of\ class\ I\ and\ class\ II$$

$$= f_{(III)} + f_{(III)} \tag{5:6.8}$$

In general:

$$ST = total\ sum\ of\ squares\ of\ class\ I\ and\ class\ II$$

$$= (total\ number\ of\ measurements) \times (number\ of\ classes - 1)$$

In this case,

$$ST = (total\ number\ of\ measurements) \times (number\ of\ classes - 1)$$

$$= 160 \times (3 - 1)$$

$$= 320$$

8.   **Calculate the degrees of freedom due to class I**
The degrees of freedom of class I is νI.

$$v_I = f_{(III)} - 1 \qquad (5:6.9)$$

Similarly, the degrees of freedom of class II is also $f_{(III)} - 1$.

9. **Calculate the total degrees of freedom**
    The total degrees of freedom due to class I and class II.

$$vT = \text{total degrees of freedom of class I and class II}$$
$$= (f_{(III)} - 1) + (f_{(III)} - 1) \qquad (5:6.10)$$

In general:

vT = *total degrees of freedom of class I and class II*

  = *(total number of measurements - 1) × (number of classes - 1)*

In this case,

vT= *(total number of measurements - 1) × (number of classes - 1)*

  = (160 - 1) × (3 - 1)

  = 318

10. **Calculate the sum of squares due to the mean of each class**

$$Sm_1 = \text{sum of squares due to mean of class I}$$
$$= \frac{f_I^2}{f_{(III)}} \omega_I \qquad (5:6.11)$$

In this case,

$$Sm_I = \frac{f_I^2}{f_{(III)}}\, \omega_I$$

$$= \frac{61^2}{160} \times 4.24$$

$$= 98.59$$

$$Sm_{II} = \frac{142^2}{160} \times 10.02$$

$$= 1262.22$$

11.   **Calculate the total sum of squares due to the mean**

Sm = *sum of squares due to mean of class I and class II*

$$= \frac{f_I^2}{f_{(III)}}\, \omega_I + \frac{f_{II}^2}{f_{(III)}}\, \omega_{II} \qquad\qquad (5{:}6.12)$$

$$= Sm_I + Sm_{II}$$

In this case,

$$Sm = Sm_I + Sm_{II}$$

$$= 1360.81$$

12.   **Calculate the sum of squares due to a factor**
This is best done by first calculating the response table for the factors (or interactions) using the frequencies in Figure 5:6.3. Note that *totals* are used here.

		A	B	C	D	E	F	G
(I)	Level 1	30	41	30	47	30	37	28
	Level 2	31	20	31	14	31	24	33
(II)	Level 1	40	34	41	31	42	28	43
	Level 2	41	47	40	50	39	53	38

Figure 5:6.3 Response table of factor effects.

Note that if there are n observations in each of the $L_8(2^7)$ experiments, then there would be $4n$ observations at each factor level and $8n$ observations in total.

SA = *sum of squares due to factor A*

$$= \left(\frac{f_{I:A1}^2}{n_{I:A1}} + \frac{f_{I:A2}^2}{n_{I:A2}} - \frac{f_I^2}{n_{I:A}} \right) \omega_I + \left(\frac{f_{II:A1}^2}{n_{II:A1}} + \frac{f_{II:A2}^2}{n_{II:A2}} - \frac{f_{II}^2}{n_{II:A}} \right) \omega_{II}$$

$$= \frac{(f_{I:A1}^2 + f_{I:A2}^2) \times \omega_I + (f_{II:A1}^2 + f_{II:A2}^2) \times \omega_{II}}{4n} - \left(\frac{f_I^2}{8n} \omega_I + \frac{f_{II}^2}{8n} \omega_{II} \right)$$

$$= \frac{(f_{I:A1}^2 + f_{I:A2}^2) \times \omega_I + (f_{II:A1}^2 + f_{II:A2}^2) \times \omega_{II}}{4n} - Sm \qquad (5:6.13)$$

In this case,

$$SA = \frac{(f_{I:A1}^2 + f_{I:A2}^2) \times \omega_I + (f_{II:A1}^2 + f_{II:A2}^2) \times \omega_{II}}{4n} - Sm$$

$$= \frac{(30^2 + 31^2) \times 4.24 + (40^2 + 41^2) \times 10.02}{80} - 1360.81$$

$$= 0.28$$

$$SB = \frac{(f_{I:B1}^2 + f_{I:B2}^2) \times \omega_I + (f_{II:B1}^2 + f_{II:B2}^2) \times \omega_{II}}{4n} - Sm$$

$$= \frac{(41^2 + 20^2) \times 4.24 + (34^2 + 47^2) \times 10.02}{80} - 1360.81$$

$$= 15.69$$

Similarly, the sums of squares due to the remaining factors (or interactions) are calculated.

13.   **Calculate the degrees of freedom for a factor**
The degrees of freedom for a factor, say factor A, is:

$$vA = (number\ of\ classes - 1) \times (number\ of\ levels - 1)$$

$$= (3 - 1) \times (2 - 1) \qquad (5:6.14)$$

$$= 2$$

The degrees of freedom of the remaining factors (or interactions) are similarly calculated.

14.   **Calculate the error sum of squares**
Since more than one measurement is made in each experiment, it is necessary to calculate the error sum of squares, *Se*.

$$Se = ST - (SA + SB + SC + SD + SE + SF + SG) \qquad (5:6.15)$$

The degrees of freedom for *ve* is:

$$ve = vT - (vA + vB + vC + vD + vE + vF + vG) \qquad (5:6.16)$$

15.   **Draw the analysis of variance**

Source	Pool	Sq	ν	Mq	F-ratio	Sq'	rho %
A	Y	0.28	2	0.14	-	-	-
B		15.69	2	7.85	-	14.09	4.40
C	Y	0.03	2	0.01	-	-	-
D		41.02	2	20.56	-	39.52	12.35
E	Y	0.28	2	0.14	-	-	-
F		13.49	2	6.75	-	11.89	3.72
G	Y	0.66	2	0.33	-	-	-
e	Y	248.45	304	0.82	-	-	-
Pooled e		249.70	312	0.80	-	254.50	79.53
ST		320.00	318	-	-	320.00	100.00

Figure 5:6.4 Analysis of variance.

Notice that we have not used the F-ratio in these calculations. Pooling has been done based on the smaller variances (*Mq*).

16.   Calculate the predicted optimum process
From the analysis of variance, factors D, B and F are important factors. Since we want to minimize the Severe category we have to maximize cumulative category II. We therefore select factor levels D1, B1 and F2 and characterize the predicted process means for categories I and II using the method of Omega transformation as shown in Section 4.3.5.

For μ_I,

$$\Omega_I = \bar{\Omega}_I + (\Omega_{D1} - \bar{\Omega}_I) + (\Omega_{B1} - \bar{\Omega}_I) + (\Omega_{F2} - \bar{\Omega}_I)$$

$$= \Omega_{D1} + \Omega_{B1} + \Omega_{F2} - 2\,\bar{\Omega}_I$$

$$= 1.54 + 0.22 - 3.68 - 2 \times (-2.10)$$

$$= 2.28 \text{ dB}$$

$$\mu_I = \cfrac{1}{1 + 10^{\frac{\Omega}{-10}}}$$

$$= \cfrac{1}{1 + 10^{\frac{2.28}{-10}}}$$

$$= 0.6283$$

For μ_{II},:

$$\Omega_{II} = \bar{\Omega}_{II} + (\Omega_{D1} - \bar{\Omega}_{II}) + (\Omega_{B1} - \bar{\Omega}_{II}) + (\Omega_{F2} - \bar{\Omega}_{II})$$

$$= \Omega_{D1} + \Omega_{B1} + \Omega_{F2} - 2\,\bar{\Omega}_{II}$$

$$= 15.91 + 11.76 + 14.09 - 2 \times 8.97$$

$$= 23.82 \text{ dB}$$

$$\mu_{II} = \cfrac{1}{1 + 10^{\frac{23.82}{-10}}}$$

$$= \cfrac{1}{1 + 10^{\frac{\Omega}{-10}}}$$

$$= 0.9959$$

17.   Calculate the confidence interval of a predicted mean

For attribute accummulation analysis we use the following approximate formula[7]:

$$CI = \sqrt{\left(F_{\alpha,v1,v2} \times V_e \times \mu \times (1 - \mu) \times \frac{1}{n_{eff}} \right)}$$ (5:6.17)

where:

$$n_{eff} = \frac{total\ number\ of\ degrees\ of\ freedom}{sum\ of\ degrees\ of\ freedom\ used\ in\ estimate\ of\ mean}$$

$$= \frac{v_T}{v_\mu + v_D + v_B + v_F}$$ (5:6.18)

$$= \frac{318}{(2 + 2 + 2 + 2)}$$

$$= 39.75$$

The reader should note the difference in the **numerators** of Equations 5:5.3 and 5:6.18, i.e. *total number of experiments* in the former and *total number of degrees of freedom* in the latter.

Therefore,

$$CI_I = \sqrt{\left(F_{0.05,1,312} \times V_e \times \mu_I \times (1 - \mu_I) \times \frac{1}{n_{eff}} \right)}$$

$$= \sqrt{\left(3.84 \times 0.80 \times 0.6283 \times (1 - 0.6283) \times \frac{1}{39.75} \right)}$$

$$= \pm\ 0.1343$$

$$CI_{(II)} = \sqrt{\left(3.84 \times 0.80 \times 0.9959 \times (1 - 0.9959) \times \frac{1}{39.75} \right)}$$

$$= \pm\ 0.0178$$

---

7   Genichi Taguchi, *System of Experimental Design: Engineering Methods to Optimize Quality and Minimize Costs*, UNIPUB/Kraus, 1924.

The lower and upper confidence limits for class II are [0.9781, 1.0137]. Clearly a confidence limit of > 1.0000 is not sensible and it is therefore recommended to use the Omega transformation for the confidence interval. We will show this for class II as follows:

●   convert the optimum process average from fraction to dB scale,
●   convert the confidence interval from fraction to dB scale,
●   calculate the confidence limits in the dB scale,
●   transform the confidence limits back to the fraction scale.

1.   **Convert the optimum process average from fraction to dB scale**

$$dB \text{ value of mean, } \mu_{II} = -10 \log_{10}\left(\frac{1}{p} - 1\right)$$

$$= -10 \log_{10}\left(\frac{1}{0.9959} - 1\right) dB \qquad (5{:}6.19)$$

$$= 23.8249 \text{ dB}$$

2.   **Convert the confidence interval from fraction to dB scale**
Taguchi[8] has shown that:

$$dB \text{ value of } CI = \pm \frac{4.343 \times \sqrt{F_{\alpha,v1,v2} \times V_e \times \dfrac{1}{n_{eff}}}}{\overline{T}\,(1 - \overline{T})} \qquad (5{:}6.20)$$

where $\overline{T}$ is the fraction (corresponding to cumulative p) of the class. Hence, for class II,

$$dB \text{ value of } CI_{II} = \pm \frac{4.343 \times \sqrt{F_{\alpha,v1,v2} \times V_e \times \dfrac{1}{n_{eff}}}}{P_{(II)}\,(1 - P_{(II)})}$$

$$= \pm \frac{4.343 \times \sqrt{3.84 \times 0.8000 \times \dfrac{1}{39.75}}}{0.8875\,(1 - 0.8875)}$$

$$= \pm 12.0947 \text{ dB}$$

---

8   Genichi Taguchi, *System of Experimental Design: Engineering Methods to Optimize Quality and Minimize Costs*, UNIPUB/Kraus, 1924.

3.   **Calculate the confidence limits in the dB scale**

$$Lower\ limit = (\mu_{II} - CI_{II})\ dB$$

$$= (23.8249 - 12.0947)\ dB$$

$$= 11.7302\ dB$$

$$Upper\ limit = (\mu_{II} + CI_{II})\ dB$$

$$= 35.9196\ dB$$

4.   **Transform the confidence limits back to the fraction scale**

$$Lower\ limit = \frac{1}{1 + 10^{\frac{-\Omega}{10}}} = \frac{1}{1 + 10^{\frac{-11.7302}{10}}} = 0.9371$$

$$Upper\ limit = \frac{1}{1 + 10^{\frac{-\Omega}{10}}} = \frac{1}{1 + 10^{\frac{-35.9196}{10}}} = 0.9997$$

18.   **If μ is 0 or 1**
If the estimated μ is itself 0 or 1, Taguchi[9] suggests:

$$\mu = \sqrt{\left(F_{\alpha,v1,v2} \times V_e \times \frac{1}{2n_{eff}} \times \left(1 - \frac{1}{2n_{eff}}\right) \times \frac{1}{n_{eff}}\right)} \qquad (5{:}6.21)$$

The likely improvement in the optimum process is shown in Figure 5:6.5 with the data for the current process taken from experiment 1.

5.6.5 Critique of the accumulation analysis

A critical appraisal of ordered categorical data analysis showed that Taguchi advocates accumulation analysis as a superior alternative to the chi-squared method traditionally used in fixed marginal enumerative data. Taguchi's reasons are that the chi-squared method is unsuitable for evaluation of the magnitude of the factorial effect and evaluation of the error variance is poor in the chi-squared method.

---

9   Genichi Taguchi, *System of Experimental Design: Engineering Methods to Optimize Quality and Minimize Costs*, UNIPUB/Kraus, 1924.

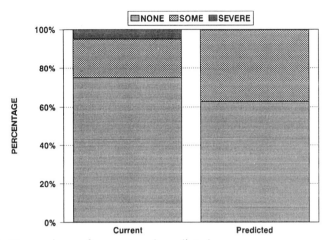

Figure 5:6.5 Comparison of current and predicted processes.

However, the method has been criticized[10] because for ordered categorical data the orthogonality of a design does not ensure independence of test statistics for different factors. They add that the accumulation analysis method proposed by Taguchi is inefficient, particularly for the multi-factor setting in industrial experiments and that the deficiency can and frequently leads to serious consequences such as: (i) Detection of spurious effects, (ii) order reversal of factor importance, and (iii) poor predictions at new settings.

Indeed, these seem to be the same reasons why Taguchi advocated accumulation analysis over the chi-squared method! Others[11] have suggested that the problems above are due to the use of the orthogonal array rather than the method of analysis itself. As such these problems are shared by all methods of analyzing ordered categorical data. They also caution engineers (and statisticians) to be careful when using the accumulation analysis and suggest that any error can be detected in the confirmation run, particularly since typical Taguchi users are looking for evidence of the importance of control factors. The consequences of this debate for the design engineer is that he/she should endeavour to measure the quality characteristic in terms of a continuous measurement. If that is not possible and an accumulation analysis has to be performed then use an unsaturated orthogonal array, take at least three classes and more than one observation per experiment. It is also recommended to use several inspectors with repeated evaluations in order to overcome the uncertainty that is inherent in subjective measurements.

10   Hamada and Wu, *A Critical Look at Accumulation Analysis and Related Methods*, Technometrics, May 1990, Volume 32, Number 2, pages 119 – 130.

11   Yanagisawa, Disney and Bendell, *Discussion on: A Critical Look at Accumulation Analysis and Related Methods*, Technometrics, May 1990, Volume 32, Number 2, pages 153 – 157.

5.6.6 Self-assessment questions

1.   Using the experimental data below, conduct an analysis to maximize the
Good category.

Exp	A	B	C	D	E	F	G	Good	Fair	Bad	Total
1	1	1	1	1	1	1	1	20	5	5	30
2	1	1	1	2	2	2	2	13	10	7	30
3	1	2	2	1	1	2	2	24	4	2	30
4	1	2	2	2	2	1	1	28	2	0	30
5	2	1	2	1	2	1	2	22	6	2	30
6	2	1	2	2	1	2	1	27	2	1	30
7	2	2	1	1	2	2	1	20	4	6	30
8	2	2	1	2	1	1	2	22	8	0	30

5.6.7 Answers to self-assessment questions

1.   Using the experimental data below, conduct an analysis to maximize the
     Good category.

Exp	A	B	C	D	E	F	G	Good	Fair	Bad	Total
1	1	1	1	1	1	1	1	20	5	5	30
2	1	1	1	2	2	2	2	13	10	7	30
3	1	2	2	1	1	2	2	24	4	2	30
4	1	2	2	2	2	1	1	28	2	0	30
5	2	1	2	1	2	1	2	22	6	2	30
6	2	1	2	2	1	2	1	27	2	1	30
7	2	2	1	1	2	2	1	20	4	6	30
8	2	2	1	2	1	1	2	22	8	0	30

Answer
The method of conducting the analysis of variance is outlined as follows:

1.   **Create a table of data with the results**
     See Figure above.

2.   **Establish the cumulative frequencies**
     For reasons of space, let Good = I, Fair = II and Bad = III. The
     cumulative frequencies (I) = I, (II) = I + II and (III) = I + II + III.

3.   **Calculate the total cumulative frequency in each class**

$$f_{(I)} = 176$$

$$f_{(II)} = 217$$

$$f_{(III)} = 240$$

Exp	Control Factors							Frequencies			Cumulative Frequencies		
	A	B	C	D	E	F	G	I	II	III	(I)	(II)	(III)
1	1	1	1	1	1	1	1	20	5	5	20	25	30
2	1	1	1	2	2	2	2	13	10	7	13	23	30
3	1	2	2	1	1	2	2	24	4	2	24	28	30
4	1	2	2	2	2	1	1	28	2	0	28	30	30
5	2	1	2	1	2	1	2	22	6	2	22	28	30
6	2	1	2	2	1	2	1	27	2	1	27	29	30
7	2	2	1	1	2	2	1	20	4	6	20	24	30
8	2	2	1	2	1	1	2	22	8	0	22	30	30
Column totals								176	41	23	176	217	240

4.   **Calculate the fraction defective of each class**

$$p_I = \frac{f_I}{f_{(III)}}$$

$$= \frac{176}{240}$$

$$p_I = 0.7333$$

$$p_{II} = 0.1708$$

$$p_{III} = 0.0958$$

5.   **Calculate the weight of each class**
For ease of calculation we use:

$$\omega_I = \frac{f_{(III)}^2}{f_I \times (f_{(III)} - f_I)}$$

$$= \frac{240^2}{176 \times (240 - 176)}$$

$$= 5.11$$

$$\omega_{II} = \frac{240^2}{41 \times (240 - 41)}$$

$$= 11.54$$

6. **Calculate the total sum of squares due to class I**

$$S_I = \text{total sum of squares of class I}$$

$$= f_{(III)}$$

Similarly, the total sum of squares of class II is also $f_{(III)}$.

7. **Calculate the overall total sum of squares due to class I and class II**

$$ST = \text{(total number of measurements)} \times \text{(number of classes - 1)}$$

$$= 240 \times (3 - 1)$$

$$= 480$$

8. **Calculate the degrees of freedom due to class I**
The degrees of freedom of class I is νI.

$$\nu_I = f_{(III)} - 1$$

Similarly, the degrees of freedom of class II is also $f_{(III)} - 1$.

9. **Calculate the total degrees of freedom**
The total degrees of freedom due to class I and class II are:

νT = *(total number of measurements* – 1) × *(number of classes* – 1)

$\quad$ = (240 – 1) × (3 – 1)

$\quad$ = 478

10. **Calculate the sum of squares due to the mean of each class**

$$Sm_I = \frac{f_I^2}{f_{(III)}} \; \omega_I$$

$$= \frac{176^2}{240} \times 5.11$$

$$= 660.00$$

$$Sm_{II} = \frac{217^2}{240} \times 11.54$$

$$= 2264.35$$

11. **Calculate the total sum of squares due to the mean**

$$Sm = Sm_I + Sm_{II}$$

$$= 2924.35$$

12. **Calculate the sum of squares due to a factor**
This is best done by first calculating the response table for the factors (and interactions) using the frequencies as shown below. Note that totals are used here.

		A	B	C	D	E	F	G
(I)	Level 1	85	82	75	86	93	92	95
	Level 2	91	94	101	90	83	84	81
(II)	Level 1	21	23	27	19	19	21	13
	Level 2	20	18	14	22	22	20	28

Subsequently,

$$SA = \frac{(f^2_{I:A1} + f^2_{I:A2}) \times \omega_I + (f^2_{II:A1} + f^2_{II:A2}) \times \omega_{II}}{120} - Sm$$

$$= \frac{(85^2 + 91^2) \times 5.11 + (21^2 + 20^2) \times 11.54}{120} - 2924.35$$

$$= 1.97$$

Similarly, the sums of squares due to the remaining factors (or interactions) are calculated.

13.   Calculate the degrees of freedom for a factor
The degrees of freedom for a factor, say factor A, is:

vA = (*number of classes* – 1) × (*number of levels* – 1)

$$= (3 - 1) \times (2 - 1)$$

$$= 2$$

The degrees of freedom of the remaining factors (or interactions) are similarly calculated.

14.   Calculate the error sum of squares
Since more than one measurement is made in each experiment, it is necessary to calculate the error sum of squares, Se.

$$Se = ST - (SA + SB + SC + SD + SE + SF + SG)$$

The degrees of freedom for ve:

$$ve = vT - (vA + vB + vC + vD + vE + vF + vG)$$

15.   Draw the analysis of variance
Notice that we have not used the F-ratio in these calculations. Pooling has been done based on the smallest variance (Mq).

Source	Pool	Sq	ν	Mq	Sq'	rho %
A	Y	1.97	2	0.98	-	-
B		5.42	2	2.71	3.53	0.74
C		22.53	2	11.27	20.64	4.30
D	Y	2.70	2	1.35	-	-
E	Y	4.49	2	2.24	-	-
F		5.26	2	2.63	3.37	0.70
G	Y	4.22	2	2.11	-	-
e	Y	433.41	464	0.93	-	-
Pooled e		446.79	472	0.95	452.47	94.26
ST		480.00	478	-	480.00	100.00

16. **Calculate the predicted optimum process**

From the analysis of variance, factors C, B and F are important. Since we want to maximize the **Good** category we have to maximise category I. We therefore select factor levels C2, B2 and F1 and characterize the predicted process means for categories I and II using the method of Omega transformation as shown in Section 4.3.5.

For μ_I,

$$\Omega_I \doteq \bar{\Omega}_I + (\Omega_{C2} - \bar{\Omega}_I) + (\Omega_{B2} - \bar{\Omega}_I) + (\Omega_{F1} - \bar{\Omega}_I)$$

$$= \Omega_{C2} + \Omega_{B2} + \Omega_{F1} - 2\,\bar{\Omega}_I$$

$$= 7.26 + 5.58 + 5.17 - 2 \times 4.39$$

$$= 9.22 \text{ dB}$$

$$P_I = \frac{1}{1 + 10^{\frac{\Omega}{-10}}}$$

$$= 0.8930$$

For μ_{II}:

$$\Omega_{II} = \bar{\Omega}_I + (\Omega_{C2} - \bar{\Omega}_I) + (\Omega_{B2} - \bar{\Omega}_I) + (\Omega_{FI} - \bar{\Omega}_I)$$

$$= \Omega_{C2} + \Omega_{B2} + \Omega_{FI} - 2\,\bar{\Omega}_{II}$$

$$= 13.62 + 11.46 + 12.08 - 2 \times 9.75$$

$$= 17.66 \text{ dB}$$

$$P_{II} = \frac{1}{1 + 10^{\frac{\Omega}{-10}}}$$

$$= 0.9832$$

17.   Calculate the confidence interval of a predicted mean

$$CI = \sqrt{\left(F_{\alpha,v1,v2} \times V_e \times \mu \times (1 - \mu) \times \frac{1}{n_{eff}} \right)}$$

where:

$$n_{eff} = \frac{total\ number\ of\ degrees\ of\ freedom}{sum\ of\ degrees\ of\ freedom\ used\ in\ estimate\ of\ mean}$$

$$= \frac{v_T}{v_\mu + v_C + v_B + v_F}$$

$$= \frac{478}{2 + 2 + 2 + 2}$$

$$= 59.75$$

Therefore,

$$CI_I = \sqrt{\left(F_{0.05,1,472} \times V_e \times \mu \times (1 - \mu) \times \frac{1}{n_{\text{eff}}}\right)}$$

$$= \sqrt{\left(3.84 \times 0.95 \times 0.8930 \times (1 - 0.8930) \times \frac{1}{59.75}\right)}$$

$$= \pm\ 0.0762$$

$$CI_{II} = \sqrt{\left(3.84 \times 0.95 \times 0.9832 \times (1 - 0.9832) \times \frac{1}{59.75}\right)}$$

$$= \pm\ 0.0317$$

Although the confidence limit for CI_{II} slightly exceeds 1.00 we shall ignore this. The following figure compares the likely improvement in the process with the data for the current process taken from experiment 1. In this example, clearly the Good category has been increased significantly.

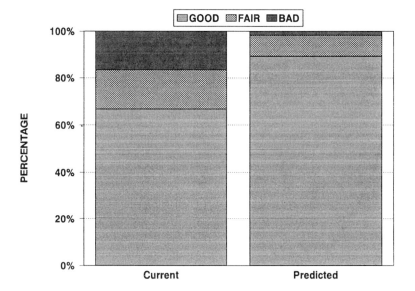

CHAPTER 6

MODIFYING ORTHOGONAL ARRAYS

AIMS:
To provide methods of modifying orthogonal arrays to suit industrial experimentation.

OBJECTIVES:
When you have completed studying this chapter you should be able to:
- use a multi-level design,
- use a dummy level design,
- use a combination factor design,
- use a pseudo-factor (nested) design,
- use a pseudo-factor (idle-column) design,
- use a distributed interaction design.

OVERVIEW:
Industrial experimentation does not always lend itself to direct applications of standard orthogonal arrays. Sometimes it is necessary to tailor an orthogonal array for a more specific purpose. This chapter introduces five ways of modifying orthogonal array designs for specific purposes. These are the multi-level design, dummy level design, combination factor design, pseudo-factor (nested) design and the pseudo-factor (idle-column) design. This chapter also introduces the use of distributed interaction designs, which although does not require any modification, nevertheless, provides a special method of factor assignment. Each of these techniques has its own uniqueness and merit for application.

6.1 Multi-level Design

6.1.1 Introduction
Production situations do not always permit the direct use of standard 2-level and 3-level orthogonal arrays. Often, it is necessary to use factors with different numbers of levels. This section shows how to accommodate mixed level factors.

6.1.2 Multi-level design
Some orthogonal arrays such as $L_4(2^3)$, $L_8(2^7)$, $L_{12}(2^{11})$, $L_{16}(2^{15})$ $L_{32}(2^{31})$ and $L_{64}(2^{63})$ have only 2-level columns. These orthogonal arrays are called *2^n series orthogonal arrays*. A 2^n series orthogonal array can only be used to assign n factors at two levels.

Some orthogonal arrays such as $L_9(3^4)$, $L_{27}(3^{13})$ and $L_{81}(3^{40})$ have only 3-level columns. These orthogonal arrays are called *3^n series orthogonal arrays*. A 3^n series orthogonal array can only be used to assign n factors at three levels.

Some orthogonal arrays such as $L_{18}(2^1 \times 3^7)$, $L_{36}(2^3 \times 3^{13})$, $L_{36}(2^{11} \times 3^{12})$ and $L_{54}(2^1 \times 3^{25})$ have both 2-level and 3-level columns. These orthogonal arrays are called *mixed series orthogonal arrays*. A mixed series orthogonal array can be used to assign factors at mixed levels.

Where possible, we may always use a standard orthogonal array. Sometimes, however, this is not possible and we may have to modify an existing orthogonal array to suit certain factor level combinations. As a specific example, consider the standard $L_8(2^7)$ orthogonal array. This array has seven degrees of freedom and can accommodate seven 2-level factors. By inspection of the degrees of freedom, it should be possible to modify the $L_8(2^7)$ orthogonal array into the $L_8(2^4 \times 4^1)$ orthogonal array. This is because the $L_8(2^4 \times 4^1)$ orthogonal array has four 2-level factors (four degrees of freedom) and one 4-level factor (three degrees of freedom) with a total of seven degrees of freedom. The technique that allows us to modify an orthogonal array in this way is called a *multi-level design*. Multi-level designs can be used to:

1.   fit a 4-level factor in a 2^n series orthogonal array,
2.   fit an 8-level factor in 2^n series orthogonal array,
3.   fit a 9-level factor in a 3^n series orthogonal array.

A multi-level design for fitting a 4-level factor in a 2^n series orthogonal array is shown in the next section. Fitting an 8-level factors in 2^n series orthogonal array is shown in the Answer to self-assessment question 1.

6.1.3 Incorporating a 4-level factor in a 2-level series
Suppose we wish to investigate the main effects of factors A, B, C, D and E where; A is a 4-level factor and B, C, D and E are 2-level factors. The degrees of freedom v, required for this experiment is:

$$vA = 1 \times (4 - 1) = 3$$

$$vB, \ vC, \ vD, \ vE = 4 \times (2 - 1) = 4$$

$$v = vA + vB + vC + vD + vE = 7$$

Hence, an $L_8(2^7)$ orthogonal array should be sufficient. However, an $L_8(2^7)$ orthogonal array does not have any column with three degrees of freedom to accommodate a 4-level factor. This problem can be overcome by using the multi-level column technique. The procedure for conducting the multi-level column technique is shown below:
- establish the degrees of freedom and the orthogonal array,
- choose a suitable linear graph,
- modify the linear graph,
- represent the new factor level,
- replace columns in the orthogonal array,
- create the corresponding linear graph.

1.   Establish the degrees of freedom and the orthogonal array
Establish the number of degrees of freedom and the orthogonal array required. For the experiment above, we require seven degrees of freedom and hence an $L_8(2^7)$ orthogonal array.

2.   Choose a suitable linear graph
Since we decided on seven degrees of freedom and an $L_8(2^7)$ orthogonal array we choose the linear graph[1] for an $L_8(2^7)$ orthogonal array. This is shown in Figure 6:1.1 (1).

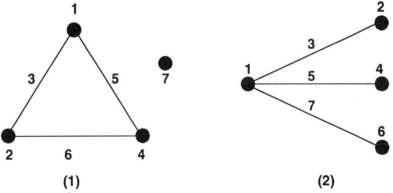

(1)   (2)

Figure 6:1.1 $L_8(2^7)$ Linear graphs.

1   Taguchi and Konishi, *Orthogonal Arrays and Linear Graphs: Tools for Quality Engineering*, 1987, ASI Press.

3. Modify the linear graph
Since factor A requires three degrees of freedom, we must allocate three degrees of freedom in the linear graph. We do this by combining two points and the connecting line corresponding to two factors and their interaction. For example, we choose two points, 1 and 2, together with the connecting line 3 in the linear graph shown in Figure 6:1.2.

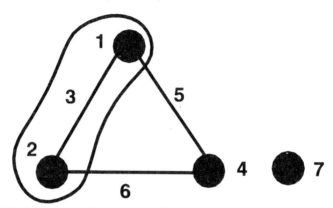

Figure 6:1.2 Selecting a line segment from the linear graph.

4. Represent the new factor level
This corresponds to the combination of two columns together with the associated interaction. Thus we may represent;

Column 1	Column 2	Line 3		Factor A
1	1	-		1
1	2	-		2
2	1	-		3
2	2	-		4

and hence create a 4-level column.

5. Replace columns in orthogonal array
In the $L_8(2^7)$ orthogonal array, we replace columns 1, 2 and 3 with the new 4-level column. The new array is represented as an $L_8(2^4 \times 4^1)$ orthogonal array since there are four columns at two levels and one column at four levels. Figure 6:1.3 (Layout 1) shows the standard $L_8(2^7)$ orthogonal array compared to the $L_8(2^4 \times 4^1)$ orthogonal array (Layout 2).

6. Create the corresponding linear graph
The corresponding linear graph is shown in Figure 6:1.4; notice that the interaction of a 4-level factor (dot 123) with a 2-level factor (dot 4) takes up three degrees of freedom. Thus factor A is assigned to column 1 and factors B, C, D and E are then allocated to the remaining columns.

Layout 1								Layout 2				
$(L_8(2^7))$								$L_8(2^4 \times 4^1)$				
	1	2	3	4	5	6	7		1	4	5	6	7
1	1	1	1	1	1	1	1	1	1	1	1	1	1
2	1	1	1	2	2	2	2	2	1	2	2	2	2
3	1	2	2	1	1	2	2	3	2	1	1	2	2
4	1	2	2	2	2	1	1	4	2	2	2	1	1
5	2	1	2	1	2	1	2	5	3	1	2	1	2
6	2	1	2	2	1	2	1	6	3	2	1	2	1
7	2	2	1	1	2	2	1	7	4	1	2	2	1
8	2	2	1	2	1	1	2	8	4	2	1	1	2

Figure 6:1.3 Creating a 4-level column.

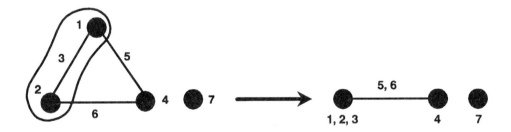

Figure 6:1.4 Linear graph for $L_8(2^4 \times 4^1)$ orthogonal array.

6.1.4 Self-assessment questions

1.   Using the $L_{16}(2^{15})$ orthogonal array, linear graph and interaction table incorporate an 8-level factor into the $L_{16}(2^{15})$ orthogonal array.

	1	2	3	4	5	6	7	8	9	10	11	12	13	14	15
1	1	1	1	1	1	1	1	1	1	1	1	1	1	1	1
2	1	1	1	1	1	1	1	2	2	2	2	2	2	2	2
3	1	1	1	2	2	2	2	1	1	1	1	2	2	2	2
4	1	1	1	2	2	2	2	2	2	2	2	1	1	1	1
5	1	2	2	1	1	2	2	1	1	2	2	1	1	2	2
6	1	2	2	1	1	2	2	2	2	1	1	2	2	1	1
7	1	2	2	2	2	1	1	1	1	2	2	2	2	1	1
8	1	2	2	2	2	1	1	2	2	1	1	1	1	2	2
9	2	1	2	1	2	1	2	1	2	1	2	1	2	1	2
10	2	1	2	1	2	1	2	2	1	2	1	2	1	2	1
11	2	1	2	2	1	2	1	1	2	1	2	2	1	2	1
12	2	1	2	2	1	2	1	2	1	2	1	1	2	1	2
13	2	2	1	1	2	2	1	1	2	2	1	1	2	2	1
14	2	2	1	1	2	2	1	2	1	1	2	2	1	1	2
15	2	2	1	2	1	1	2	1	2	2	1	2	1	1	2
16	2	2	1	2	1	1	2	2	1	1	2	1	2	2	1

Figure 6:1.5 The $L_{16}(2^{15})$ orthogonal array.

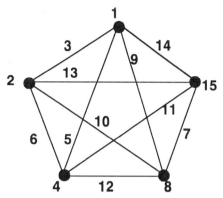

Figure 6:1.6 A linear graph of an $L_{16}(2^{15})$ orthogonal array.

1	2	3	4	5	6	7	8	9	10	11	12	13	14	15
1	3	2	5	4	7	6	9	8	11	10	13	12	15	14
	2	1	6	7	4	5	10	11	8	9	14	15	12	13
		3	7	6	5	4	11	10	9	8	15	14	13	12
			4	1	2	3	12	13	14	15	8	9	10	11
				5	3	2	13	12	15	14	9	8	11	10
					6	1	14	15	12	13	10	11	8	9
						7	15	14	13	12	11	10	9	8
							8	1	2	3	4	5	6	7
								9	3	2	5	4	7	6
									10	1	6	7	4	5
										11	7	6	5	4
											12	1	2	3
												13	3	2
													14	1
														15

Figure 6:1.7 The $L_{16}(2^{15})$ interaction table.

6.1.5 Answers to self-assessment questions

1.   Using the $L_{16}(2^{15})$ orthogonal array, linear graph and interaction table incorporate an 8-level factor into the $L_{16}(2^{15})$ orthogonal array.

Answer

1.   Establish the degrees of freedom and orthogonal array
We require seven degrees of freedom for an 8-level factor to be incorporated into an $L_{16}(2^{15})$ orthogonal array.

2.   Choose a suitable linear graph
We choose a suitable linear graph[2] for an $L_{16}(2^{15})$ orthogonal array. In this case, we use the linear graph provided as shown in Figure 6:1.8.

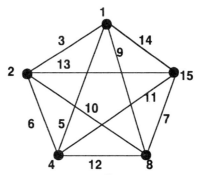

Figure 6:1.8 A linear graph of an $L_{16}(2^{15})$ orthogonal array.

3.   Modify the linear graph
Since an 8-level factor requires seven degrees of freedom, we must allocate seven degrees of freedom to a closed set of linear graph. We do this by taking a closed triangle of dots and connecting lines comprising seven degrees of freedom. This is shown in Figure 6:1.9. Note that the three vertices and the three sides of the triangle only take up six degrees of freedom and hence we have to add another line (13) between the vertex and base of the triangle for the seventh degree of freedom.

4.   Represent the new factor level
We now take the three columns of the vertices (1, 4 and 8) and let their combinations correspond to the new levels. This corresponds to the combination of three columns together with the associated interactions.

---

2   Taguchi and Konishi, *Orthogonal Arrays and Linear Graphs: Tools for Quality Engineering*, 1987, ASI Press.

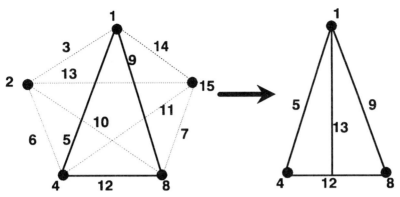

Figure 6:1.9 A closed of triangle of seven degrees of freedom.

Thus, we may represent;

Column 1	Column 4	Column 8	Factor A
1	1	1	1
1	1	2	2
1	2	1	3
1	2	2	4
2	1	1	5
2	1	2	6
2	2	1	7
2	2	2	8

and hence create an 8-level column.

5.   Replace columns in the orthogonal array

In the $L_{16}(2^{15})$ orthogonal array, we replace columns 1, 4, 5, 8, 9, 12 and 13 with the new 8-level column. The new array is represented as an $L_{16}(2^8 \times 8^1)$ orthogonal array since there are eight 2-level columns and one 8-level column, as shown in Figure 6:1.10.

6.   Create the corresponding linear graph

The interaction of an 8-level factor with a 2-level factor requires seven degrees of freedom. Thus, the interaction of columns 1 and 2 is given in columns 3, 6, 7, 10, 11, 14 and 15 of the array orthogonal shown in Figure 6:1.10.

	1	2	3	6	7	10	11	14	15
1	1	1	1	1	1	1	1	1	1
2	2	1	1	1	1	2	2	2	2
3	3	1	1	2	2	1	1	2	2
4	4	1	1	2	2	2	2	1	1
5	1	2	2	2	2	2	2	2	2
6	2	2	2	2	2	1	1	1	1
7	3	2	2	1	1	2	2	1	1
8	4	2	2	1	1	1	1	2	2
9	5	1	2	1	2	1	2	1	2
10	6	1	2	1	2	2	1	2	1
11	7	1	2	2	1	1	2	2	1
12	8	1	2	2	1	2	1	1	2
13	5	2	1	2	1	2	1	2	1
14	6	2	1	2	1	1	2	1	2
15	7	2	1	1	2	2	1	1	2
16	8	2	1	1	2	1	2	2	1

Figure 6:1.10   The $L_{16}(2^8 \times 8^1)$ orthogonal array.

6.2 Dummy Level Design

6.2.1 Introduction
The dummy level design is useful when the number of factor levels is less than the number of levels in a column of an orthogonal array to which the factor is assigned. Thus, if a 2-level factor is to be assigned to a 3-level column, the third level can be assigned one of the two levels and so form the *dummy level*. The dummy level design can only be used with a column that has three or more levels. This is because, to use a dummy level design in a column with only two levels would result with a meaningless 1-level factor.

6.2.2 Dummy level design
Sometimes it may be necessary to assign a 2-level factor to a 3-level column of an orthogonal array. This can happen when there is a 2-level factor but a 3-level column during the assignment of factors to an orthogonal array. In such a case, the factor is formally treated as a 3-level factor but with one of the two levels substituted for the third level. It does not matter which level is substituted although it is usual to substitute the preferred level. The preferred level here is simply the level that is more convenient, less expensive or simply more sensible. In any case, this allows more data points at the preferred level. The substituted level is then called a dummy level.

   The dummy treatment can be used when an m-level factor is assigned to an n-level column where $m < n$ and 1 of the m levels is repeated $(n - m + 1)$ times.

   Example 1
   A dummy treatment can be used when a 2-level factor is assigned to a 3-level column where $2 < 3$ and 1 of the 2 levels is repeated 2 (i.e. $3 - 2 + 1$) times.

   Example 2
   A dummy treatment can be used when a 3-level factor is assigned to a 4-level column where $3 < 4$ and 1 of the 3 levels is repeated 2 (i.e. $4 - 3 + 1$) times.

   Example 3
   A dummy treatment can be used when a 2-level factor is assigned to a 4-level column where $2 < 4$ and 1 of the 2 levels is repeated 3 (i.e. $4 - 2 + 1$) times.

6.2.3 Procedure for dummy level design
The procedure for conducting a dummy level design is as follows:
*   select a column for the 2-level factor from a 3-level series array,
*   select the preferred level of the 2-level factor,
*   assign the dummy level,
*   amend the corresponding calculations.

1.   Select a column for the 2-level factor from a 3-level series array
We select an $L_9(3^4)$ as shown in Figure 6:2.1. Next we select a column for the 2-level factor from the orthogonal array. Column 1 is selected from the $L_9(3^4)$ orthogonal array.

2.   Select the preferred level of the 2-level factor
Select the level of the 2-level factor which may be more important than the other. Level 1 is selected.

3.   Assign the dummy level
Rewrite the levels in the 2-level column, labelling such that the important level is replicated. In the present case, the 3's in column 1 are replaced with 1''s.

4.   Amend the corresponding calculations
Amend the corresponding calculations for mean effects using the appropriate numbers for the averages, etc.

	$L_9(3^4)$					$L_9(3^4)$			
	1	2	3	4		1	2	3	4
1	1	1	1	1	1	1	1	1	1
2	1	2	2	2	2	1	2	2	2
3	1	3	3	3	3	1	3	3	3
4	2	1	2	3	4	2	1	2	3
5	2	2	3	1	5	2	2	3	1
6	2	3	1	2	6	2	3	1	2
7	3	1	3	2	7	1'	1	3	2
8	3	2	1	3	8	1'	2	1	3
9	3	3	2	1	9	1'	3	2	1

Figure 6:2.1 The $L_9(3^4)$ orthogonal array and dummy level design.

6.2.4 Example of dummy level design analysis
Consider an experiment in which we need to study one 2-level factor and three 3-level factors. The associated degrees of freedom is eight. Since there are three 3-level factors we choose the $L_9(3^4)$ orthogonal array. However, this orthogonal array has four columns that can accommodate four 3-level factors. The dummy level design allows us

to fit a 2-level factor into a 3-level factor as shown Figure 6:2.2. The results of an experiment are also shown in Figure 6:2.2. The response table is shown in Figure 6:2.3. Notice that factor A level 3 corresponds to the dummy level. Although the mean effect of A3 is strictly incorrect, it is necessary for calculating the analysis of variance. The mean effect of A1 and the dummy level A1', is also given in the response table. Additionally, the ranking for 3-level factors, based on factor level differences cannot be established easily as was done for 2-level factors. However, we can use percent contribution of a factor in an analysis of variance to establish factor importance. The response graph including the confidence intervals is shown in Figure 6:2.4. The method of calculating the analysis of variance and confidence interval is also shown.

Exp	A	B	C	D	Results	
1	1	1	1	1	15	13
2	1	2	2	2	4	3
3	1	3	3	3	0	1
4	2	1	2	3	6	7
5	2	2	3	1	6	5
6	2	3	1	2	12	14
7	1'	1	3	2	6	5
8	1'	2	1	3	9	10
9	1'	3	2	1	0	2

Figure 6:2.2 Dummy level design.

	A	B	C	D
Level 1	6.00	8.67	12.17	6.83
Level 2	8.33	6.17	3.67	7.33
Level 3	5.33	4.83	3.83	5.50
Level 1&1'	5.67	-	-	-

Figure 6:2.3 Response table of dummy level design where A is formally treated as a 3-level factor with level 3 corresponding to 1'.

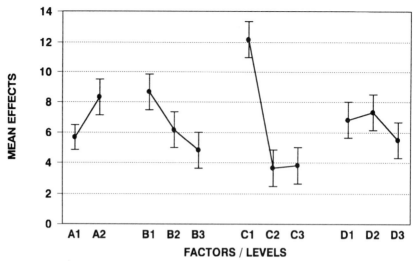

Figure 6:2.4 Response graph of dummy level design.

6.2.5 Analysis of variance
Calculations of the analysis of variance and the contribution ratio are shown as follows:
- calculation of ST,
- calculation of the overall mean,
- calculation of Sm,
- calculation of SA,
- calculation of SB, SC and SD,
- calculation of SeA,
- calculation of error,
- result of analysis of variance,
- pooling of insignificant factors,
- calculation of the pure sum of squares,
- calculation of the contribution ratio.

    1.   **Calculation of ST**

$$ST = 15^2 + 13^2 + ... + 0^2 + 2^2$$

$$= 1152.00$$

2.   **Calculation of the overall mean**

$$\bar{y} = \frac{\Sigma y}{n}$$

$$= \frac{15 + 13 + \dots + 0 + 2}{18}$$

$$= 6.56$$

3.   **Calculation of *Sm***

$$Sm = n \times \bar{y}^2$$

$$= 18 \times 6.56^2$$

$$= 773.56$$

4.   **Calculation of *SA***

Note that $n_{A1, A1'} = 12$ and $n_{A2} = 6$.

$$SA = n_{A1, A1'} \times \bar{y}_{A1, A1'}^2 + n_{A2} \times \bar{y}_{A2}^2 - Sm$$

$$= 12 \times 5.67^2 + 6 \times 8.33^2 - 773.56$$

$$= 28.44$$

5.   **Calculation of *SB*, *SC* and *SD***

Note that for factors B, C and D, $n_{\Psi1} = 6$ and $n_{\Psi2} = 6$ and $n_{\Psi3} = 6$ where Ψ is factor B, C or D. Therefore *SB*, *SC* and *SD* are calculated as follows.

$$SB = n_{B1} \times \bar{y}_{B1}^2 + n_{B2} \times \bar{y}_{B2}^2 + n_{B3} \times \bar{y}_{B3}^2 - Sm = 45.44$$

$$SC = n_{C1} \times \bar{y}_{C1}^2 + n_{C2} \times \bar{y}_{C2}^2 + n_{C3} \times \bar{y}_{C3}^2 - Sm = 283.44$$

$$SD = n_{D1} \times \bar{y}_{D1}^2 + n_{D2} \times \bar{y}_{D2}^2 + n_{D3} \times \bar{y}_{D3}^2 - Sm = 10.78$$

6.   **Calculation of *SeA***

The calculation of *SA* involved the mean effects of levels A1 and A2. But the effect of level A1 includes the effects of levels A1 and A3 (i.e. A1′) as shown in the response table. Thus, there is one degree of freedom between A1 and A3 which must be assigned to error since there is no distinguishable factor level between these. Since A1 and A1' mean effects are known, we can calculate the error sum of squares due to the dummy level SeA by formally treating factor A as a 3-level factor $S_{A1, A2, A1'}$ and then subtracting SA as calculated previously. The reader should note that:

$$SeA = S_{A1, A2, A1'} - S_A$$

$$= \left[n_{A1}\, \bar{y}_{A1}^2 + n_{A2}\, \bar{y}_{A2}^2 + n_{A1'}\, \bar{y}_{A1'}^2 - Sm \right] - \left[n_{A1, A1'}\, \bar{y}_{A1, A1'}^2 + n_{A2}\, \bar{y}_{A2}^2 - Sm \right]$$

$$= n_{A1}\, \bar{y}_{A1}^2 + n_{A1'}\, \bar{y}_{A1'}^2 - n_{A1, A1'}\, \bar{y}_{A1, A1'}^2$$

$$= 6 \times 6.00^2 + 6 \times 5.33^2 - 12 \times 5.67^2$$

$$= 1.33$$

7.   Calculation of the error
Se is calculated as follows:

$$Se = St - SA - SB - SC - SD - SA'$$

$$= 378.44 - 28.44 - 45.44 - 283.44 - 10.78 - 1.33$$

$$= 9.00$$

8.   Result of analysis of variance
The result of the analysis of variance is shown in Figure 6:2.5.

9.   Pooling of insignificant factors
From the analysis of variance, we pool insignificant factors D, eA and e as shown in Figure 6:2.5.

10.   Calculation of the pure sum of squares
The pure sum of squares due to factors A, B and C are calculated as follows:

$$S'A = SA - vA \times V_e$$

$$= 28.44 - 1 \times 1.76 = 26.69$$

$$S'B = SB - vB \times V_e$$

$$= 45.44 - 2 \times 1.76 = 41.93$$

Source	Pool	Sq	ν	Mq	F-ratio	Sq'	rho %
A		28.44	1	28.44	16.17	26.69	7.05
B		45.44	2	22.72	12.92	41.93	11.08
C		283.44	2	141.72	80.56	279.93	73.97
D	Y	10.78	2	5.39	-	-	-
eA	Y	1.33	1	1.33	-	-	-
e	Y	9.00	9	1.00	-	-	-
Pooled e		21.11	12	1.76	1.00	29.91	7.90
St		378.44	17	22.26	-	378.44	100.00
Mean		773.56	1	-	-	-	-
ST		1152.00	18	-	-	-	-

Figure 6:2.5 Analysis of variance for dummy level design.

$$S'C = SC - \nu C \times V_e$$

$$= 283.44 - 2 \times 1.76 = 279.93$$

$$SPooled\ e = St - S'A - S'B - S'C$$

$$= 378.44 - 26.69 - 41.93 - 279.93 = 29.91$$

11. **Calculation of the contribution ratio**

$$\rho\ A = \frac{S'A}{St} \times 100\ \% = \frac{26.69}{378.44} \times 100\ \% = 7.05\ \%$$

$$\rho\ B = \frac{S'B}{St} \times 100\ \% = \frac{41.93}{378.44} \times 100\ \% = 11.08\ \%$$

$$\rho\ C = \frac{S'C}{St} \times 100\ \% = \frac{279.93}{378.44} \times 100\ \% = 73.97\ \%$$

$$\rho\ Pooled\ e = \frac{SPooled\ e}{St} \times 100\ \% = \frac{29.91}{378.44} \times 100\ \% = 7.90\ \%$$

6.2.6 Calculation of the confidence interval
In calculating the confidence intervals, the F-ratio referred has $\alpha = 0.05$, $\nu 1 = 1$ always and $\nu 2 = $ degrees of freedom of the pooled error. Since factor A is essentially a 2-level factor with effectively 12 observations in level 1 and six observations in level 2, the confidence interval will be different for A1 and A2. Factors B, C and D are 3-level factors with six observations at each level and hence, the confidence interval for these factor levels will be same for all levels. Additionally, it may be noted that factor level A2 and factor levels B, C and D will all have the same confidence intervals since the effective number of observations at each level is the same. We now calculate the:

- confidence interval for level A1,
- confidence interval for level A2,
- confidence interval for factors B, C and D,
- confidence interval for the predicted mean,
- confidence interval for the confirmation experiment.

1.   Confidence interval for level A1
For the calculation of a confidence interval we use $\nu 1 = 1$ degree of freedom. Since pooled error was calculated on 12 degrees of freedom $\nu 2 = 12$. Therefore, $F_{\alpha,\nu 1,\nu 2} = F_{0.05,1,12} = 4.75$. Also the effect of A1 was averaged on 12 observations, i.e. $n = 12$. Therefore,

$$CI_{A1} = \sqrt{F_{0.05,1,12} \times V_e \times \left[\frac{1}{n}\right]}$$

$$= \sqrt{4.75 \times 1.76 \times \left[\frac{1}{12}\right]}$$

$$= \pm\ 0.83$$

2.   Confidence interval for level A2
For the calculation of a confidence interval we use $\nu 1 = 1$ degree of freedom. Since the pooled error was calculated on 12 degrees of freedom $\nu 2 = 12$. Therefore, $F_{\alpha,\nu 1,\nu 2} = F_{0.05,1,12} = 4.75$. Also the effect of A1 was averaged on six observations, i.e. $n = 6$. Therefore,

$$CI_{A2} = \sqrt{F_{0.05,1,12} \times V_e \times \left[\frac{1}{n}\right]}$$

$$= \sqrt{4.75 \times 1.76 \times \left[\frac{1}{6}\right]}$$

$$= \pm\ 1.18$$

3.   Confidence interval for factors B, C and D

Factors B and C are 3-level factors with six experiments at each level and hence the confidence intervals are the same for all levels of B and C. For the calculation of a confidence interval we use $\nu1 = 1$ degree of freedom. Since the pooled error was calculated on 12 degrees of freedom, $\nu2 = 12$. The number of observations in each factor average is six, i.e. $n = 6$. Therefore, $F_{\alpha,\nu1,\nu2} = F_{0.05,1,12} = 4.75$.

$$CI_{B,\ C,\ D} = \sqrt{F_{0.05,2,12} \times V_e \times \left[\frac{1}{n}\right]}$$

$$= \sqrt{4.75 \times 1.76 \times \left[\frac{1}{6}\right]}$$

$$= \pm\ 1.18$$

4.   Confidence interval of the predicted mean

Assuming a larger-the-better characteristic, the predicted mean is:

$$\mu_{Predicted} = \bar{y} + (\overline{A2} - \bar{y}) + (\overline{B1} - \bar{y}) + (\overline{C1} - \bar{y})$$

$$= \overline{A2} + \overline{B1} + \overline{C1} - 2\ \bar{y}$$

$$= 8.33 + 8.67 + 12.17 - 2 \times 6.56$$

$$= 16.06$$

If an *m*-level dummy treated factor is used in an *n*-level column, the number of effective degrees of freedom for that factor to be used in the estimation of the predicted mean is:

$$\nu_{eff} = \left[\frac{number\ of\ levels\ before\ dummy}{number\ of\ duplicated\ levels}\right] - 1$$

$$= \frac{n}{n - m + 1} - 1$$

Notice that if $m = n$, i.e. there is no duplication, then the equation reduces to $n - 1$ which is the usual degrees of freedom for an *n*-level factor.

Therefore, for the dummy treated factor A,

$$v_{eff} = \left[\frac{number\ of\ levels\ before\ dummy}{number\ of\ duplicated\ levels} \right] - 1$$

$$= \frac{3}{2} - 1$$

$$= 0.5$$

and the associated confidence interval is:

$$CI_{Predicted} = \sqrt{F_{\alpha,v1,v2} \times V_e \times \left[\frac{1}{n_{eff}} \right]}$$

Here, n_{eff} is the effective number of degrees of freedom used in calculating the predicted mean, where,

$$n_{eff} = \frac{total\ number\ of\ experiments}{sum\ of\ degrees\ of\ freedom\ used\ in\ estimate\ of\ mean}$$

$$= \frac{9 \times 2}{v_{\mu} + v_A + v_B + v_C}$$

$$= \frac{18}{1 + 0.5 + 2 + 2}$$

$$= 3.27$$

Substituting for $n_{eff} = 3.27$,

$$CI_{Predicted} = \sqrt{F_{0.05,1,12} \times V_e \times \left[\frac{1}{n_{eff}} \right]}$$

$$= \sqrt{4.75 \times 1.76 \times \left[\frac{1}{3.27} \right]}$$

$$= \pm\ 1.60$$

5.   Confidence interval of the confirmation experiment

Similarly, the confidence interval for the confirmation experiment with r trials is:

$$CI_{Confirmation} = \sqrt{F_{\alpha,v1,v2} \times V_e \times \left[\frac{1}{n_{eff}} + \frac{1}{r}\right]}$$

For five trials, $r = 5$ and the confidence interval is:

$$CI_{Confirmation} = \sqrt{F_{0.05,1,12} \times V_e \times \left[\frac{1}{n_{eff}} + \frac{1}{r}\right]}$$

$$= \sqrt{4.75 \times 1.76 \times \left[\frac{1}{3.27} + \frac{1}{5}\right]}$$

$$= \pm\ 2.06$$

Once the confidence interval calculations are completed, the response graph may be redrawn with the appropriate confidence intervals for the factors and levels. This gives a graphical representation of the differences between the factor levels. For the comparison of the predicted mean and the confirmation experiment, the experimenter should compare the range of the predicted mean $[\mu_{Predicted} \pm CI_{Predicted}]$ and the range of the confirmation experiment $[\mu_{Confirmation} \pm CI_{Confirmation}]$. If the ranges overlap, then the experiment can be regarded as reproducible.

6.2.7 Self-assessment questions

1. An engineer wishes to study a 3-level factor along with four 2-level factors. Outline how you could use an $L_8(2^7)$ orthogonal array to accommodate this design.

2. Given the following experimental data complete an analysis of variance and confidence intervals.

Exp	1	4	5	6	7	Results	
1	1	1	1	1	1	10	8
2	1	2	2	2	2	5	4
3	2	1	1	2	2	8	9
4	2	2	2	1	1	4	5
5	3	1	2	1	2	1	2
6	3	2	1	2	1	7	6
7	1'	1	2	2	1	6	5
8	1'	2	1	1	2	11	10

Figure 6:2.6 Introducing a dummy level into a 4-level column.

3. An experiment was conducted with one 2-level factor A, two 3-level factors B and C, and two 4-level factors D and E using an $L_{16}(4^5)$ orthogonal array. Assuming that there are 2 replications of each trial in the experiment and that factors A, B, and E are significant, what is the effective degrees of freedom used in the estimation of the predicted mean?

6.2.8 Answers to self-assessment questions

1.   An engineer wishes to study a 3-level factor along with four 2-level factors. Outline how you could use and $L_8(2^7)$ orthogonal array to accommodate this design.

Answer

One 4-level factor requires three degrees of freedom. Four 2-level factors require four degrees of freedom. Therefore we require seven degrees of freedom. The $L_8(2^7)$ orthogonal array should be sufficient. Using the multi-level design discussed in the last section, we modify an $L_8(2^7)$ orthogonal array into an $L_8(2^4 \times 4^1)$ orthogonal array. Subsequently, we treat one level of a factor with the dummy treatment.

Layout 1								Layout 2					
$L_8(2^7)$								$L_8(2^4 \times 4^1)$					
	1	2	3	4	5	6	7		1	4	5	6	7
1	1	1	1	1	1	1	1	1	1	1	1	1	1
2	1	1	1	2	2	2	2	2	1	2	2	2	2
3	1	2	2	1	1	2	2	3	2	1	1	2	2
4	1	2	2	2	2	1	1	4	2	2	2	1	1
5	2	1	2	1	2	1	2	5	3	1	2	1	2
6	2	1	2	2	1	2	1	6	3	2	1	2	1
7	2	2	1	1	2	2	1	7	4	1	2	2	1
8	2	2	1	2	1	1	2	8	4	2	1	1	2

Figure 6:2.7 Creating a 4-level column.

	Multi-level design						Multi-level design with Dummy level design				
	$L_8(2^4 \times 4^1)$						$L_8(2^4 \times 4^1)$				
	1	4	5	6	7		1	4	5	6	7
1	1	1	1	1	1	1	1	1	1	1	1
2	1	2	2	2	2	2	1	2	2	2	2
3	2	1	1	2	2	3	2	1	1	2	2
4	2	2	2	1	1	4	2	2	2	1	1
5	3	1	2	1	2	5	3	1	2	1	2
6	3	2	1	2	1	6	3	2	1	2	1
7	4	1	2	2	1	7	1'	1	2	2	1
8	4	2	1	1	2	8	1'	2	1	1	2

Figure 6:2.8 Introducing a dummy level into a 4-level column.

2.   Given the following experimental data complete an analysis of variance and confidence intervals.

Exp	1	4	5	6	7	Results	
1	1	1	1	1	1	10	8
2	1	2	2	2	2	5	4
3	2	1	1	2	2	8	9
4	2	2	2	1	1	4	5
5	3	1	2	1	2	1	2
6	3	2	1	2	1	7	6
7	1'	1	2	2	1	6	5
8	1'	2	1	1	2	11	10

Figure 6:2.9 Introducing a dummy level into a 4-level column.

Answer

The response table is as follows:

	A	B	C	D	E
Level 1	6.75	6.13	8.63	6.38	6.38
Level 2	6.50	6.50	4.00	6.25	6.25
Level 3	4.00	-	-	-	-
Level 1'	8.00	-	-	-	-
Level 1&1'	7.38	-	-	-	-

The analysis of variance is as follows:

Source	Pool	Sq	ν	Msq	F-ratio	Sq'	rho%
A		30.56	2	15.28	19.69	29.01	23.13
B	Y	0.56	1	0.56	-	-	-
C		85.56	1	85.56	110.26	84.79	67.59
D	Y	0.06	1	0.06	-	-	-
E	Y	0.06	1	0.06	-	-	-
eA	Y	3.13	1	3.13	-	-	-
e	Y	5.50	8	0.69	-	-	-
Pooled e		9.31	12	0.78	1.00	11.64	9.28
St		125.44	15	8.36	-	125.44	100.00
Mean		637.56	1	-	-	-	-
ST		763.00	16	-	-	-	-

1.   Confidence interval for level A1

For the calculation of a confidence interval we use $\nu 1 = 1$ degree of freedom. Since the pooled error was calculated on 12 degrees of freedom, $\nu 2 = 12$. Therefore, $F_{\alpha, \nu 1, \nu 2} = F_{0.05, 1, 12} = 4.75$. Also, the effect of A1 was averaged on eight observations, i.e. $n = 8$. Therefore,

$$CI_{A1} = \sqrt{F_{0.05,1,12} \times V_e \times \left[\frac{1}{n}\right]}$$

$$= \sqrt{4.75 \times 0.78 \times \left[\frac{1}{8}\right]}$$

$$= \pm\ 0.68$$

2.   Confidence interval for levels A2 and A3

For the calculation of a confidence interval we use $\nu1 = 1$ degree of freedom. Since the pooled error was calculated on 12 degrees of freedom, $\nu2 = 12$. Therefore, $F_{\alpha,\nu1,\nu2} = F_{0.05,1,12} = 4.75$. Also, the effect of A2 (and A3) were averaged on four observations, i.e. $n = 4$. Therefore,

$$CI_{A2,\ A3} = \sqrt{F_{0.05,1,12} \times V_e \times \left[\frac{1}{n}\right]}$$

$$= \sqrt{4.75 \times 0.78 \times \left[\frac{1}{4}\right]}$$

$$= \pm\ 0.96$$

3.   Confidence interval for factors B, C, D and E

Factors B, C, D and E are 2-level factors with four experiments at each level and hence the confidence intervals are the same for all levels of B, C, D and E. For the calculation of a confidence interval we use $\nu1 = 1$ degree of freedom. Since the pooled error was calculated on 12 degrees of freedom, $\nu2 = 12$. Therefore, $F_{\alpha,\nu1,\nu2} = F_{0.05,1,12} = 4.75$. The number of observations in each factor level average is four, i.e. $n = 8$.

$$CI_{B,\ C,\ D,\ E} = \sqrt{F_{0.05,1,12} \times V_e \times \left[\frac{1}{n}\right]}$$

$$= \sqrt{4.75 \times 0.78 \times \left[\frac{1}{8}\right]}$$

$$= \pm\ 0.68$$

4.   Confidence interval of the predicted mean

Assuming a larger-the-better characteristic, the predicted mean at the optimum condition is:

$$\mu_{Predicted} = \bar{y} + (\overline{A1} - \bar{y}) + (\overline{C1} - \bar{y})$$

$$= \overline{A1} + \overline{C1} - \bar{y}$$

$$= 7.38 + 8.63 - 6.31$$

$$= 9.69$$

If a dummy treated factor is used in the estimation of the predicted mean then the effective degrees of freedom for that factor is:

$$v_{eff} = \left[\frac{number\ of\ levels\ before\ dummy}{number\ of\ duplicated\ levels} \right] - 1$$

Therefore, for the dummy treated factor A,

$$v_{eff} = \left[\frac{4}{2} \right] - 1 = 1$$

Hence, the degrees of freedom used to calculate the predicted mean is 1 for A and 1 for C. Noting that $\alpha = 0.05$, $v1 = 1$ and $v2 = 12$, therefore, $F_{0.05,1,12} = 4.75$ and the confidence interval is:

$$CI_{Predicted} = \sqrt{F_{\alpha,v1,v2} \times V_e \times \left[\frac{1}{n_{eff}} \right]}$$

where, n_{eff} is the effective number of degrees of freedom used in calculating the confidence interval, and

$$n_{eff} = \frac{total\ number\ of\ experiments}{sum\ of\ degrees\ of\ freedom\ used\ in\ estimate\ of\ mean}$$

$$= \frac{8 \times 2}{v_\mu + v_A + v_C}$$

$$= \frac{16}{1 + 1 + 1}$$

$$= 5.33$$

Substituting for $n_{eff} = 5.33$,

$$CI_{Predicted} = \sqrt{F_{0.05,1,12} \times V_e \times \left[\frac{1}{n_{eff}}\right]}$$

$$= \sqrt{4.75 \times 0.78 \times \left[\frac{1}{5.33}\right]}$$

$$= \pm \; 0.83$$

5.   Confidence interval of the confirmation experiment

Similarly, the confidence interval for the confirmation experiment with r trials is:

$$CI_{Confirmation} = \sqrt{F_{\alpha,v1,v2} \times V_e \times \left[\frac{1}{n_{eff}} + \frac{1}{r}\right]}$$

For five trials, $r = 5$ and the confidence interval is:

$$CI_{Confirmation} = \sqrt{F_{0.05,1,12} \times V_e \times \left[\frac{1}{n_{eff}} + \frac{1}{r}\right]}$$

$$= \sqrt{4.75 \times 0.78 \times \left[\frac{1}{5.33} + \frac{1}{5}\right]}$$

$$= \pm \; 1.20$$

The corresponding response graph is as follows:

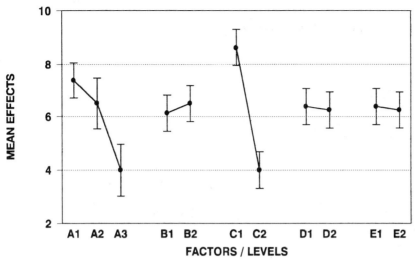

Figure 6:2.12   Response graph of factor effects with confidence intervals.

3.   An experiment was conducted with one 2-level factor A, two 3-level factors B and C, and two 4-level factors D and E using an $L_{16}(4^5)$ orthogonal array. Assuming that there are 2 replications of each trial in the experiment and that factors A, B, and E are significant, what is the effective degrees of freedom used in the estimation of the predicted mean?

Answer

$$n_{eff} = \frac{total\ number\ of\ experiments}{sum\ of\ degrees\ of\ freedom\ used\ in\ estimate\ of\ mean}$$

$$= \frac{16 \times r}{v_\mu + vA + vB + vE}$$

$$= \frac{16 \times 2}{1 + \left(\frac{4}{3} - 1\right) + \left(\frac{4}{2} - 1\right) + (4 - 1)}$$

$$= \frac{32}{1 + \frac{1}{3} + 1 + 3}$$

$$= 6.00$$

6.3 Combination Factor Design

6.3.1 Introduction
Combination factor design is a design to assign two 2-level factors to a 3-level column in a 3^n series orthogonal array.

6.3.2 Combination factor design
Two 2-level factors may be compared with two degrees of freedom. Such a comparison does not take account of the interaction between the two factors. To make a better comparison we would require three degrees of freedom. With three degrees of freedom we can assign one degree of freedom for factor A, one for factor B and one for the interaction $A \times B$. A combination factor design uses only two degrees of freedom to study two 2-level factors and ignores the interaction effect. This design is only possible if the interaction effect between the two factors is not important. Doing so, however, is not equivalent to assuming that there is no interaction because there are simply no degrees of freedom associated with the interaction. Therefore, *orthogonality is lost between two combination factors* and the interaction information of factors involved in the combination design is not available.

6.3.3 Procedure
Suppose we wish to study an experiment involving two 2-level and three 3-level factors. Assume that factors A and B are the 2-level factors and factors C, D and E are the 3-level factors. Two 2-level factors and three 3-level factors require eight degrees of freedom. The $L_9(3^4)$ orthogonal array has eight degrees of freedom. But the $L_9(3^4)$ orthogonal array does not have any 2-level column. The following procedure will now show how to incorporate two 2-level factors into a 3-level factor:
●   define the two 2-level factors and the orthogonal array,
●   assign combinations.

    1.   Define the two 2-level factors and the orthogonal array
    Suppose we need to assign two 2-level factors A and B; and three 3-level factors C, D and E to an $L_9(3^4)$ orthogonal array as shown in Figure 6:3.1.

    2.   Assign combinations
    To assign the two 2-level factors, we need to use the following coding:
    AB1   is assigned to A1B1
    AB2   is assigned to A2B1
    AB3   is assigned to A2B2
    If column 1 of the $L_9(3^4)$ orthogonal array is assigned to the combined factor AB then the experiment can be represented as Figure 6:3.2. The three 3-level factors C, D and E can then be assigned directly to columns 2, 3 and 4.

	1	2	3	4
1	1	1	1	1
2	1	2	2	2
3	1	3	3	3
4	2	1	2	3
5	2	2	3	1
6	2	3	1	2
7	3	1	3	2
8	3	2	1	3
9	3	3	2	1

Figure 6:3.1 The $L_9(3^4)$ orthogonal array.

	1	2	3	4			1		2	3	4
	AB	C	D	E			A	B	C	D	E
1	1	1	1	1		1	1	1	1	1	1
2	1	2	2	2		2	1	1	2	2	2
3	1	3	3	3		3	1	1	3	3	3
4	2	1	2	3		4	2	1	1	2	3
5	2	2	3	1		5	2	1	2	3	1
6	2	3	1	2		6	2	1	3	1	2
7	3	1	3	2		7	2	2	1	3	2
8	3	2	1	3		8	2	2	2	1	3
9	3	3	2	1		9	2	2	3	2	1

Figure 6:3.2 The $L_9(3^4)$ orthogonal array with combined factors.

6.3.4 Example of combination design

Suppose we need to assign the two 2-level factors A and B, and three 3-level factors C, D and E to an $L_9(3^4)$ orthogonal array. The $L_9(3^4)$ orthogonal array and the results of an experiment are shown in Figure 6:3.3. We now conduct the data analysis[3] for the results.

Exp	1		2	3	4	Results
	A	B	C	D	E	
1	1	1	1	1	1	3
2	1	1	2	2	2	8
3	1	1	3	3	3	12
4	2	1	1	2	3	6
5	2	1	2	3	1	11
6	2	1	3	1	2	1
7	2	2	1	3	2	16
8	2	2	2	1	3	6
9	2	2	3	2	1	12

Figure 6:3.3 The $L_9(3^4)$ orthogonal array with combined factors.

We proceed with the following calculations:
- calculation of the experimental average,
- calculation of the sum of squares due to mean,
- calculation of the total sum of squares,
- calculation of the sum of squares due to combination factors,
- calculation of the sum of squares due to other factors,
- calculation of the sum of squares due to error.

3   The reader should note that in this chapter, most of the experiments only have one or two observations for the results. This is simply to minimize mathematical complexity while illustrating the various orthogonal array modification techniques.

1.   **Calculation of the experimental average**

$$\bar{y} = \frac{\sum y}{n}$$

$$= \frac{3 + 8 + ... + 6 + 12}{9}$$

$$= 8.33$$

2.   **Calculation of the sum of squares due to mean**

$$Sm = n\,\bar{y}^2$$

$$= 9 \times 8.33^2$$

$$= 625.00$$

3.   **Calculation of the total sum of squares**

$$ST = 3^2 + 8^2 + ... + 6^2 + 12^2$$

$$= 811.00$$

4.   **Calculation of the sum of squares due to combination factors**
For ease of reference the response table for the results above is shown in Figure 6:3.4. The reader should note the level averages for factor A and B.

	AB	A	B	C	D	E
Level 1	7.67	7.67	6.00	8.33	3.33	8.67
Level 2	6.00	6.00	11.33	8.33	8.67	8.33
Level 3	11.33	-	-	8.33	13.00	8.00

Figure 6:3.4 Response table of combination design.

$$SAB = n_{AB1}\,\bar{y}_{AB1}^2 + n_{AB2}\,\bar{y}_{AB2}^2 + n_{AB3}\,\bar{y}_{AB3}^2 - Sm$$

$$= 3 \times 7.67^2 + 3 \times 6.00^2 + 3 \times 11.33^2 - 625.00$$

$$= 44.67$$

5.   **Calculation of the sum of squares due to other factors**

$$SC = n_{C1} \, \bar{y}_{C1}^2 + n_{C2} \, \bar{y}_{C2}^2 + n_{C3} \, \bar{y}_{C3}^2 - Sm$$

$$= 3 \times 8.33^2 + 3 \times 8.33^2 + 3 \times 8.333^2 - 625.00$$

$$= 0.00$$

$$SD = n_{D1} \, \bar{y}_{D1}^2 + n_{D2} \, \bar{y}_{D2}^2 + n_{D3} \, \bar{y}_{D3}^2 - Sm$$

$$= 3 \times 3.33^2 + 3 \times 8.67^2 + 3 \times 13.00^2 - 625.00$$

$$= 140.67$$

$$SE = n_{E1} \, \bar{y}_{E1}^2 + n_{E2} \, \bar{y}_{E2}^2 + n_{E3} \, \bar{y}_{E3}^2 - Sm$$

$$= 3 \times 8.67^2 + 3 \times 8.33^2 + 3 \times 8.00^2 - 625.00$$

$$= 0.67$$

6.   **Calculation of the sum of squares due to error**

$$Se = St - SAB - SC - SD - SE$$

$$= 186.00 - 44.67 - 0.00 - 140.67 - 0.67$$

$$= 0.00$$

Of course it may be unnecessary to calculate *Se* here. That is because, taken together, the AB combination maintains orthogonality and since there are no degrees of freedom associated with the error, *Se* must be zero. We say that there are no degrees of freedom associated with the error because, we only conducted nine experiments with one observation each. That gives us eight degrees of freedom which is taken up by the four factors AB, C, D and E, each with two degrees of freedom. If, however, we conducted two observations in each experiment, then, we would have some degrees of freedom (nine in this case) to estimate error.

The rest of the calculations follow the standard procedure given earlier. From the analysis of variance shown in Figure 6:3.5, it is obvious that the AB combination has 23.84 % contribution to the total sum of squares. However, the AB combination is the effect of two factors, A and B. We now break down the effects of the AB combination so that we can analyze the individual effects of factors A and B. This done as follows:

- calculate the sum of squares for A,
- calculate the sum of squares for B,
- breakdown of the orthogonality between factors A and B,
- difference of the total sum of squares between A and B.

Source	Pool	Sq	ν	Mq	F-ratio	Sq'	rho %
AB		44.67	2	22.33	134.00	44.33	23.84
C	Y	0.00	2	-	-	-	-
D		140.67	2	70.33	422.00	140.33	75.45
E	Y	0.67	2	0.33	-	-	-
e		-	-	-	-	-	-
Pooled e		0.66	4	0.17	1.00	1.33	0.72
St		186.00	8	23.25	-	186.00	100.00
Mean		625.00	1	-	-	-	-
ST		811.00	9	-	-	-	-

Figure 6:3.5 Analysis of variance for the combination design.

1.   Calculate the sum of squares for A
The sum of squares for factor A is calculated as follows:

$$SA = n_{AB1} \, \bar{y}^{2}_{AB1} + n_{AB2} \, \bar{y}^{2}_{AB2} - n_{AB1,\,AB2} \, \bar{y}^{2}_{AB1,\,AB2}$$

$$= 3 \times 7.67^2 + 3 \times 6.00^2 - 6 \times 6.83^2$$

$$= 4.17$$

This is because, we assigned AB1 to A1B1 and AB2 to A2B1. By comparing AB1 to AB2 we are comparing A1B1 and A2B1 and hence the effect of A over a constant B1. Of course this is not an orthogonal treatment.

2.   Calculate the sum of squares for B
The sum of squares for factor B is calculated as follows:

$$SB = n_{AB2} \, \bar{y}_{AB2}^2 + n_{AB3} \, \bar{y}_{AB3}^2 - n_{AB2,\,AB3} \, \bar{y}_{AB2,\,AB3}^2$$

$$= 3 \times 6.00^2 + 3 \times 11.33^2 - 6 \times 8.67^2$$

$$= 42.67$$

This is because, we assigned AB2 to A2B1 and AB3 to A2B2. By comparing AB2 to AB3 we are comparing A2B1 and A2B2 and hence the effect of B over a constant A2. Of course this is also not an orthogonal treatment.

3.   Breakdown of the orthogonality between factors A and B
Earlier, we calculated that

$$SAB = 44.67$$

Notice that:

$$SAB \neq SA + SB$$

$$44.67 \neq 4.17 + 42.67$$

because A and B are not orthogonal in a combination design and the sum of squares of AB is also consequently not equal to the sum of squares of A and B.

4.   Difference of the total sum of squares between A and B
For conservation of the sums of squares in the analysis of variance, we introduce the quantity δAB;

$$\delta AB = SAB - SA - SB$$

$$= 44.67 - 4.17 - 42.67$$

$$= -2.17$$

It is very important to recognize that this quantity is not the interaction $A \times B$ since there is no degree of freedom associated with δAB. It must also be noted that since $\delta AB \neq 0$, the sums of squares of A and B are shown shaded in Figure 6:3.6 to indicate that the total sum of squares will not be conserved. Consequently, any differences in Sq' are pooled into error (*Pooled e*). The modified analysis of variance is shown in Figure 6:3.6.

Source	Pool	Sq	ν	Mq	F-ratio	Sq'	rho %
AB		44.67	2	22.33	134.00	44.33	23.84
A	Y	4.17	1	-	-	-	-
B		42.67	1	42.67	80.00	42.13	22.65
δAB	Y	-2.17	0	-	-	-	-
C	Y	0.00	2	-	-	-	-
D		140.67	2	70.33	131.88	139.60	75.05
E	Y	0.66	2	0.33	-	-	-
Pooled e		2.67	5	0.53	1.00	4.27	2.29
St		186.00	8	23.25	-	186.00	100.00
Mean		625.00	1	-	-	-	-
ST		811.00	9	-	-	-	-

Figure 6:3.6 Analysis of variance for combination design.

6.3.5 Case when one combination factor is insignificant

From the analysis of variance in Figure 6:3.6, we note that factor A has a very small effect while factor B has a very large effect. For practical purposes, we may regard that factor A was not included in the experiment. In that case, we may regard factor B as dummy design treated as shown in Figure 6:3.7. Therefore, we may recalculate the sum of squares for factor B as dummy level design:

$$SB = n_{B1} \bar{y}_{B1}^2 + n_{B2} \bar{y}_{B2}^2 - Sm$$

$$= 6 \times 6.83^2 + 3 \times 11.33^2 - 625.00$$

$$= 40.50$$

In the dummy level design, the 2-level factor B takes up only 1 of the 2 degrees of freedom available in its column. The remaining degree of freedom is regarded as error. The sum of squares due to this error degree of freedom is:

	Starting design						Effective design			
	1	2	3	4			1	2	3	4
	A	B	C	D	E		A	B	C	D	E
1	1	1	1	1	1	1	-	1	1	1	1
2	1	1	2	2	2	2	-	1	2	2	2
3	1	1	3	3	3	3	-	1	3	3	3
4	2	1	1	2	3	4	-	1'	1	2	3
5	2	1	2	3	1	5	-	1'	2	3	1
6	2	1	3	1	2	6	-	1'	3	1	2
7	2	2	1	3	1	7	-	2	1	3	2
8	2	2	2	1	3	8	-	2	2	1	3
9	2	2	3	2	1	9	-	2	3	2	1

Figure 6:3.7 $L_9(3^4)$ orthogonal array combination design with insignificant factor A.

$$SeB = n_{B1} \, \bar{y}_{B1}^2 + n_{B1'} \, \bar{y}_{B1'}^2 - n_{B1, \, B1'} \, \bar{y}_{B1, \, B1'}^2$$

$$= 3 \times 7.67^2 + 3 \times 6.00^2 - 6 \times 6.83^2$$

$$= 4.17$$

The analysis of variance can now be redrawn as shown in Figure 6:3.8.

6.3.6 Calculation of the confidence intervals
The confidence intervals are calculated as follows:
- confidence interval for factor level B1,
- confidence interval for factor level B2,
- confidence intervals for factors C, D and E,
- confidence interval of the predicted mean,
- confidence interval of the confirmation experiment.

Source	Pool	Sq	ν	Mq	F-ratio	Sq'	rho %
B		40.50	1	40.50	41.90	39.53	21.25
C	Y	0.00	2	0.00	-	-	-
D		140.67	2	70.33	72.76	138.73	74.59
E	Y	0.67	2	0.33	-	-	-
eB	Y	4.17	1	4.17	-	-	-
e		-	-	-	-	-	-
Pooled e		4.83	5	0.97	1.00	7.73	4.16
St		186.00	8	23.25	-	186.00	100.00
Mean		625.00	1	-	-	-	-
ST		811.00	9	-	-	-	-

Figure 6:3.8 Analysis of variance for combination design.

1.   Confidence interval for factor level B1

The confidence interval for factor level B1 has six observations. For the calculation of a confidence interval we use $\nu1 = 1$ degree of freedom. Since the error was calculated on five degrees of freedom, $\nu2 = 5$. Therefore, $F_{\alpha,\nu1,\nu2} = F_{0.05,1,5} = 6.61$. Also the mean effect of B1 was averaged on six observations, i.e. $n = 6$. Therefore,

$$CI_{B1} = \sqrt{F_{0.05,1,5} \times V_e \times \left[\frac{1}{n}\right]}$$

$$= \sqrt{6.61 \times 0.97 \times \left[\frac{1}{6}\right]}$$

$$= \pm\ 1.03$$

2.   Confidence interval for factor level B2

The confidence interval for factor level B2 is calculated on three observations, i.e. $n = 3$. Therefore,

$$CI_{B2} = \sqrt{F_{0.05,1,5} \times V_e \times \left[\frac{1}{n}\right]}$$

$$= \sqrt{6.61 \times 0.97 \times \left[\frac{1}{3}\right]}$$

$$= \pm\ 1.46$$

3.   Confidence intervals for factors C, D and E

The confidence intervals for factor levels C, D and E are calculated on three observations, i.e. $n = 3$. Therefore,

$$CI_{C,\ D,\ E} = \sqrt{F_{0.05,1,5} \times V_e \times \left[\frac{1}{n}\right]}$$

$$= \sqrt{6.61 \times 0.97 \times \left[\frac{1}{3}\right]}$$

$$= \pm\ 1.46$$

4.   Confidence interval of the predicted mean

Assuming a larger-the-better characteristic, the predicted mean is:

$$\mu_{Predicted} = \bar{y} + (\overline{B2} - \bar{y}) + (\overline{D3} - \bar{y})$$

$$= \overline{B2} + \overline{D3} - \bar{y}$$

$$= 11.33 + 13.00 - 8.33$$

$$= 16.00$$

If a dummy treated factor is used in the estimation of the predicted mean then the effective degrees of freedom for that factor is:

$$v_{eff} = \left[\frac{number\ of\ levels\ before\ dummy}{number\ of\ duplicated\ levels}\right] - 1$$

Therefore, for the dummy treated factor B,

$$v_{eff} = \begin{bmatrix} 3 \\ 2 \end{bmatrix} - 1 = 0.50$$

The confidence interval for the predicted process mean is calculated based on the degrees of freedom of B and D. Hence, the degrees of freedom used to calculate $\mu_{Predicted}$ is 0.50 from B and 1 from D. Noting that $\alpha = 0.05$, $v1 = 1$ and $v2 = 5$, therefore, $F_{\alpha,v1,v2} = F_{0.05,1,5} = 6.61$, and the confidence interval is:

$$CI_{Predicted} = \sqrt{F_{\alpha,v1,v2} \times V_e \times \left[\frac{1}{n_{eff}}\right]}$$

where, n_{eff} is the effective number of degrees of freedom and,

$$n_{eff} = \frac{total\ number\ of\ experiments}{sum\ of\ degrees\ of\ freedom\ used\ in\ estimate\ of\ mean}$$

$$= \frac{9 \times 1}{v_\mu + v_B + v_D}$$

$$= \frac{9}{1 + 0.50 + 2}$$

$$= 2.57$$

Substituting for $n_{eff} = 2.57$,

$$CI_{Predicted} = \sqrt{F_{0.05,1,5} \times V_e \times \left[\frac{1}{n_{eff}}\right]}$$

$$= \sqrt{6.61 \times 0.97 \times \left[\frac{1}{2.57}\right]}$$

$$= \pm\ 1.58$$

5.   Confidence interval of the confirmation experiment
The confidence interval for the confirmation experiment with r trials is:

$$CI_{Confirmation} = \sqrt{F_{\alpha,v1,v2} \times V_e \times \left[\frac{1}{n_{eff}} + \frac{1}{r}\right]}$$

For five trials, $r = 5$ and the confidence interval is:

$$CI_{Confirmation} = \sqrt{F_{0.05,1,5} \times V_e \times \left[\frac{1}{n_{eff}} + \frac{1}{r}\right]}$$

$$= \sqrt{6.61 \times 0.97 \times \left[\frac{1}{2.57} + \frac{1}{5}\right]}$$

$$= \pm 1.94$$

The corresponding response graph is as follows.

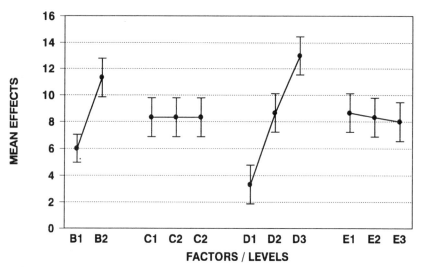

Figure 6:3.9 Response graph of factor effects.

6.3.7 Self-assessment questions

1.   The experiment below incorporates a combination factor design for factors A and B and a dummy treated factor design for factor E. Conduct a suitable analysis.

Exp	1		2	3	4	Results	
	A	B	C	D	E	P	Q
1	1	1	1	1	1	10	12
2	1	1	2	2	2	4	8
3	1	1	3	3	1'	4	5
4	2	1	1	2	1'	21	20
5	2	1	2	3	1	12	17
6	2	1	3	1	2	12	10
7	1	2	1	3	2	7	9
8	1	2	2	1	1'	6	7
9	1	2	3	2	1	6	9

Figure 6:3.10   The $L_9(3^4)$ orthogonal array with combination and dummy level design.

6.3.8 Answers to self-assessment questions

1. The experiment below incorporates a combination factor design for factors A and B and a dummy treated factor design for factor E. Conduct a suitable analysis.

Exp	1		2	3	4	Results	
	A	B	C	D	E	P	Q
1	1	1	1	1	1	10	12
2	1	1	2	2	2	4	8
3	1	1	3	3	1'	4	5
4	2	1	1	2	1'	21	20
5	2	1	2	3	1	12	17
6	2	1	3	1	2	12	10
7	1	2	1	3	2	7	9
8	1	2	2	1	1'	6	7
9	1	2	3	2	1	6	9

Figure 6:3.11   The $L_9(3^4)$ orthogonal array with combination and dummy level design.

Answer
The response table is shown as follows:

	AB	A	B	C	D	E
Level 1	7.17	7.17	15.33	13.17	9.50	10.75
Level 2	15.33	15.33	7.33	9.00	11.33	8.33
Level 3	7.33	-	-	7.67	9.00	-

An analysis of variance is shown as follows:

Source	Pool	Sq	ν	Ms	F-ratio	Sq'	rho %
AB		261.44	2	130.72	30.99	253.01	58.17
C		98.78	2	49.39	11.71	90.34	20.77
D	Y	18.11	2	9.06	-	-	-
E		24.11	1	24.11	5.72	19.89	4.57
eE	Y	0.75	1	0.75	-	-	-
e	Y	31.75	9	3.53	-	-	-
Pooled e		50.61	12	4.22	1.00	71.70	16.48
St		434.94	17	25.58	-	434.94	100.00
Mean		1780.06	1	-	-	-	-
ST		2215.00	18	-	-	-	-

Since the combination factor is significant, we need to break down the combination factor effects as shown earlier in the text. Only the results are shown in the following analysis of variance.

Source	Pool	Sq	ν	Ms	F-ratio	Sq'	rho %
AB		261.44	2	130.72	30.99	253.01	58.17
A		200.08	1	200.08	23.08	191.41	44.01
B	Y	0.08	1	0.08	-	-	-
delta	Y	61.28	0	-	-	-	-
C		98.78	2	49.39	5.70	81.44	18.72
D	Y	18.11	2	9.06	-	-	-
E		23.36	1	23.36	2.69	14.69	3.38
eE	Y	0.75	1	0.75	-	-	-
e	Y	32.50	9	3.61	-	-	-
Pooled e		112.72	13	8.67	1.00	147.41	33.89
St		434.94	17	25.58	-	434.94	100.00
Mean		1780.06	1	-	-	-	-
ST		2215.00	18	-	-	-	-

Treating factor A as a dummy level design and taking factors C and E as significant, the recalculated analysis of variance is shown below:

Source	Pool	Sq	ν	Ms	F-ratio	Sq'	rho %
A		261.36	1	261.36	66.05	257.40	59.18
C		98.78	2	49.39	12.48	90.86	20.89
E		23.36	1	23.36	5.90	19.40	4.46
Pooled e		51.44	13	3.96	1.00	67.27	15.47
St		434.94	17	25.58	-	434.94	100.00
Mean		1780.06	1	-	-	-	-
ST		2215.00	18	-	-	-	-

The confidence intervals are calculated as follows:
- confidence intervals for factor levels A1 and E1,
- confidence intervals for factor levels A2 and E2,
- confidence intervals for factors C and D,
- confidence interval of the predicted mean,
- confidence interval of the confirmation experiment.

1.   Confidence intervals for factor levels A1 and E1
The confidence interval for factor levels A1 and E1 are the same as they both have 12 observations in their levels. For the calculation of a confidence interval we use $\nu1 = 1$ degree of freedom. Since error was calculated on 13 degrees of freedom $\nu2 = 13$. Therefore, $F_{\alpha,\nu1,\nu2} = F_{0.05,1,13} = 4.67$. Also the mean effect of A1 and E1 was averaged on 12 observations, i.e. $n = 12$. Therefore,

$$CI_{A1,\ E1} = \sqrt{F_{0.05,1,13} \times V_e \times \left[\frac{1}{n}\right]}$$

$$= \sqrt{4.67 \times 3.96 \times \left[\frac{1}{12}\right]}$$

$$= \pm\ 1.24$$

2.   Confidence intervals for factor levels A2 and E2
For the calculation of a confidence interval we use $\nu1 = 1$ degree of freedom. The confidence intervals for A2 and E2 are calculated on six observations each, i.e. $n = 6$. Therefore,

$$CI_{A2,\ E2} = \sqrt{F_{0.05,1,13} \times V_e \times \left[\frac{1}{n}\right]}$$

$$= \sqrt{4.67 \times 3.96 \times \left[\frac{1}{6}\right]}$$

$$= \pm\ 1.76$$

3.   Confidence intervals for factors C and D
For the calculation of a confidence interval we use $\nu1 = 1$ degree of freedom. Also, the factor level averages was calculated on six observations each, i.e. $n = 6$. Therefore,

$$CI_{C,D} = \sqrt{F_{0.05,1,13} \times V_e \times \left[\frac{1}{n}\right]}$$

$$= \sqrt{4.67 \times 3.96 \times \left[\frac{1}{6}\right]}$$

$$= \pm\ 1.76$$

4.   Confidence interval of the predicted mean

Assuming a larger-the-better characteristic, the predicted mean is:

$$\mu_{Predicted} = \bar{y} + (\overline{A2} - \bar{y}) + (\overline{C1} - \bar{y}) + (\overline{E1} - \bar{y})$$

$$= \overline{A2} + \overline{C1} + \overline{E1} - 2 \times \bar{y}$$

$$= 15.33 + 13.17 + 10.75 - 2 \times 9.94$$

$$= 19.36$$

The confidence interval for the predicted process mean is calculated based on the degrees of freedom of A, C and E. For a dummy level factor, the number of effective degrees of freedom for calculating the confidence interval of the predicted mean is:

$$v_{eff} = \left[\frac{number\ of\ levels\ before\ dummy}{number\ of\ duplicated\ levels}\right] - 1$$

Therefore, for the dummy treated factors A and E,

$$v_{eff} = \left[\frac{3}{2}\right] - 1 = 0.50$$

Hence, the degrees of freedom used to calculate the predicted mean is 0.5 from A, 2 from C and 0.5 from E. Noting that $\alpha = 0.05$, $v1 = 1$ and $v2 = 13$, therefore, $F_{\alpha,v1,v2} = F_{0.05,1,13} = 4.67$ and the confidence interval is:

$$CI_{Predicted} = \sqrt{F_{\alpha,v1,v2} \times V_e \times \left[\frac{1}{n_{eff}}\right]}$$

where, n_{eff} is the effective number of degrees of freedom used in calculating the confidence interval, where

$$n_{eff} = \frac{total\ number\ of\ experiments}{sum\ of\ degrees\ of\ freedom\ used\ in\ estimate\ of\ mean}$$

$$= \frac{9 \times 2}{v_\mu + v_A + v_C + v_E}$$

$$= \frac{18}{1 + 0.5 + 2 + 0.5}$$

$$= 4.50$$

Substituting for $n_{eff} = 4.50$,

$$CI_{Predicted} = \sqrt{F_{0.05,1,13} \times V_e \times \left[\frac{1}{n_{eff}}\right]}$$

$$= \sqrt{4.67 \times 3.96 \times \left[\frac{1}{4.50}\right]}$$

$$= \pm 2.03$$

5. Confidence interval of the confirmation experiment

The confidence interval for the confirmation experiment with r trials is:

$$CI_{Confirmation} = \sqrt{F_{\alpha,v1,v2} \times V_e \times \left[\frac{1}{n_{eff}} + \frac{1}{r}\right]}$$

For five trials, $r = 5$ and the confidence interval is:

$$CI_{Confirmation} = \sqrt{F_{0.05,1,13} \times V_e \times \left[\frac{1}{n_{eff}} + \frac{1}{r}\right]}$$

$$= \sqrt{4.67 \times 3.96 \times \left[\frac{1}{4.50} + \frac{1}{5}\right]}$$

$$= \pm\ 2.79$$

The associated response graph of factor effects is as follows:

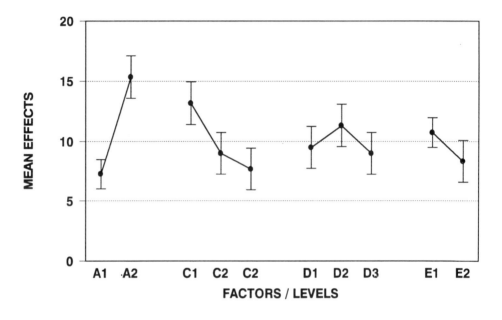

6.4 Pseudo-factor Design – Nested Factors

6.4.1 Introduction
Pseudo-factor design is a very useful technique for assigning complex factors in an orthogonal array. There are two such designs namely, nested factor method (discussed in this section) and idle-column method (discussed in Section 6.5).

6.4.2 Nested factor method
Suppose a baking experiment involves mixing a material at two mixing ratios (factor A) and two mixing methods (factor B). The resulting material can then be baked in two types of ovens (factor C) which we shall call Electric (C1) and Gas (C2). Suppose further that the electric oven requires two controls, current and feed rate, while the gas oven also requires two controls, gas rate and cure time. In this experiment, when factor C is in level 1 (Electric oven), the two controls required are current (D') and feed rate (E'). Notice that current (D') and feed rate (E') are not relevant to the Gas oven (C2). Similarly, when factor C is in level 2 (Gas oven), the two controls required are gas rate (D") and cure time (E"). Notice also that gas rate (D") and cure time (E") are not relevant to the Electric oven (C1). Factors such as D and E which take specific values depending upon another factor are called *nested factors*. A factor such as C which incorporates nested factors is called an *axial factor*. The relationship between axial and nested factors is shown in Figure 6:4.1.

From a design point of view, we must note that in this experiment, factors A and B are common throughout the experiment. But there are two baking methods (of which only one will be selected), namely, electric oven and gas oven . The electric oven baking has two factors namely current and feed rate. Similarly, the gas oven baking has two factors, namely, gas rate and cure time. Material baked by one method is not baked by the other method. Hence, there is a dissimilar treatment which must be performed while still maintaining orthogonality. The better baking method can only be determined after each baking method is optimized.

6.4.3 Procedure
We now give an outline of how to conduct a nested factor design together with the appropriate data analysis:
- draw a linear graph of the axial and nested factors,
- match this linear graph to an appropriate linear graph,
- leave the interaction columns unassigned,
- assign the factors to the orthogonal array.

A Material mix ratio	A1		A2	
B Mixing method	B1		B2	
C Baking method **Axial factor**	C1 Electric oven		C2 Gas oven	
D (Current / Gas rate) **Nested factor**	D′ Current		D″ Gas rate	
	D′1	D′2	D″1	D″2
E (Feed rate / Cure time) **Nested factor**	E′ Feed rate		E″ Cure time	
	E′1	E′2	E″1	E″2

Figure 6:4.1 Nested factor design.

1.   Draw a linear graph of the axial and nested factors
This is done by connecting the dot of the axial factor to the nested factors. One
nested factor from each level of the axial factor is combined at one dot. For the
baking experiment above:
A and B are two 2-level factors,
C is an axial factor with two nested factors, D and E,
D′ and D″ are placed at the same dot,
E′ and E″ are placed at the same dot, as shown in Figure 6:4.2.

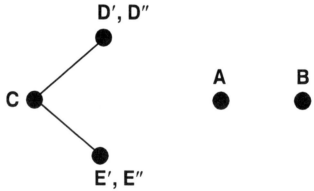

Figure 6:4.2 Required linear graph for nested factor experiment.

The total degrees of freedom required for experimentation is:

A and B $2 \times (2 - 1) = 2$
C $1 \times (2 - 1) = 1$
D' and D" $2 \times (2 - 1) = 2$
E' and E" $2 \times (2 - 1) = 2$

or seven degrees of freedom. Thus an $L_8(2^7)$ orthogonal array may be sufficient.

2.   Match this linear graph to an appropriate linear graph
Matching of the required linear graph to standard linear graph must be done for the appropriate orthogonal array that is to be used for the experiment. Since an $L_8(2^7)$ orthogonal array is to be used, we use an $L_8(2^7)$ linear graph.

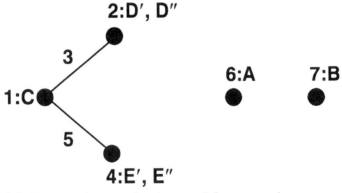

Figure 6:4.3 Factor assignment for the nested factor experiment.

3.   Leave the interaction columns unassigned
The interaction between the axial factor and nested factors must always be left unassigned to avoid confounding of interaction. In this case, line segments 3 and 5 (see Figure 6:4.3) are left unassigned in order to avoid confounding of interaction. Notice also that doing so allows two degrees of freedom between columns 2 and 3 as well as columns 4 and 5. This is the necessary degrees of freedom to separate the effects of D' and D" as well as E' and E".

4.   Assign the factors to the orthogonal array
The factors are assigned to the $L_8(2^7)$ orthogonal array as shown in Figure 6:4.4. The results for this experiment are also given in Figure 6:4.4. A response table of factor effects is given in Figure 6:4.5 to facilitate an analysis of variance.

	1	2	3	4	5	6	7	
	C1	D'		E'		A	B	
Exp	C2	D"		E"				Results
1	1	1	1	1	1	1	1	13
2	1	1	1	2	2	2	2	9
3	1	2	2	1	1	2	2	12
4	1	2	2	2	2	1	1	11
5	2	1	2	1	2	1	2	14
6	2	1	2	2	1	2	1	12
7	2	2	1	1	2	2	1	10
8	2	2	1	2	1	1	2	12

Figure 6:4.4 The $L_8(2^7)$ orthogonal array with a nested factor design.

	C	D'	D"	E'	E"	A	B
Level 1	11.25	11.00	13.00	12.50	12.00	12.50	11.50
Level 2	12.00	11.50	11.00	10.00	12.00	10.75	11.75

Figure 6:4.5 Response table of nested factor effects.

We proceed to the:
- calculation of the overall mean,
- calculation of the sum of squares due to the mean,
- calculation of the sum of squares due to an axial factor,
- calculation of the sum of squares due to nested factors,
- estimation of the predicted mean,
- confidence interval of the axial factor,
- confidence interval of the nested factors,
- confidence interval of the predicted mean,
- confidence interval of the confirmation experiment.

1. **Calculation of the overall mean**

$$\bar{y} = \frac{\sum y}{n}$$

$$= \frac{13 + 9 + ... + 10 + 12}{8}$$

$$= 11.63$$

2. **Calculation of the sum of squares due to the mean**

$$Sm = n\,\bar{y}^2$$

$$= 8 \times 11.63^2$$

$$= 1081.13$$

3. **Calculation of the sum of squares due to an axial factor**

$$SC = n_C\,\bar{y}_{C1}^2 + n_{C2}\,\bar{y}_{C2}^2 - n_C\,\bar{y}_C^2$$

$$= 4 \times 11.25^2 + 4 \times 12.00^2 - Sm$$

$$= 1.13$$

4. **Calculation of the sum of squares due to nested factors**
Remembering that D', D" and E', E" are different factors, the sum of squares are calculated as follows:

$$SD' = n_{D'1}\,\bar{y}_{D'1}^2 + n_{D'2}\,\bar{y}_{D'2}^2 - n_{D'}\,\bar{y}_{D'}^2$$

$$= 2 \times 11.00^2 + 2 \times 11.50^2 - 4 \times 11.25^2$$

$$= 0.25$$

$$SD'' = n_{D''1}\,\bar{y}_{D''1}^2 + n_{D''2}\,\bar{y}_{D''2}^2 - n_{D''}\,\bar{y}_{D''}^2$$

$$= 2 \times 13.00^2 + 2 \times 11.00^2 - 4 \times 12.00^2$$

$$= 4.00$$

$$SE' = n_{E'1}\,\bar{y}^2_{E'1} + n_{E'2}\,\bar{y}^2_{E'2} - n_{E'}\,\bar{y}^2_{E'}$$

$$= 2 \times 12.50^2 + 2 \times 10.00^2 - 4 \times 11.25^2$$

$$= 6.25$$

$$SE'' = n_{E''1}\,\bar{y}^2_{E''1} + n_{E''2}\,\bar{y}^2_{E''2} - n_{E''}\,\bar{y}^2_{E''}$$

$$= 2 \times 12.00^2 + 2 \times 12.00^2 - 4 \times 12.00^2$$

$$= 0.00$$

From these results the analysis of variance is drawn as shown in Figure 6:4.6. From this figure, it is obvious that D', E'' and B are insignificant, and these are pooled using the method discussed earlier.

Source	Pool	Sq	v	Mq	F-ratio	Sq'	rho %
C		1.13	1	1.13	9.00	1.00	5.59
D'	Y	0.25	1	0.25	-	-	-
D''		4.00	1	4.00	32.00	3.88	21.68
E'		6.25	1	6.25	50.00	6.13	34.27
E''	Y	0.00	1	0.00	-	-	-
A		6.13	1	6.13	49.00	6.00	35.57
B	Y	0.13	1	0.13	-	-	-
Pooled e		0.38	3	0.13	1.00	0.88	4.90
St		17.88	7	2.55	-	17.88	100.00
Mean		1081.13	1	-	-	-	-
ST		1099.00	8	-	-	-	-

Figure 6:4.6 Analysis of variance for the nested factor experiment.

5.   Estimation of the predicted mean

Assuming that the quality characteristic is larger-the-better we proceed to select the optimum process. First, it is necessary to determine which of the processes C1 (electric oven) or C2 (gas oven) is better. To do this, we need to analyze the nested factors. Considering only the axial and nested factors, the possible optimum conditions could be C1 or C2.

For C1 (Electric oven)
The optimum condition at C1 is D'2 and E'1. Since D'2 is insignificant, we consider only C1 and E'1 for the estimation. In this case:

$$\mu_{\overline{C1,\ E'1}} = \overline{C1} + (\overline{E'1} - \overline{C1})$$

$$= \overline{E'1}$$

$$= 12.50$$

For C2 (Gas oven)
The optimum condition at C2 is D''1 and E''1 or E''2. Since E'' is insignificant, we consider only C2 and D''1 for the estimation:

$$\mu_{\overline{C2,\ D''1}} = \overline{C2} + (\overline{D''1} - \overline{C2})$$

$$= \overline{D''1}$$

$$= 13.00$$

A comparison of $\mu_{C1,\ E'1}$ and $\mu_{C2,\ D''1}$ shows that the C2, D''1 combination is better because $\mu_{C1,\ E'1} < \mu_{C2,\ D''1}$. The remaining factors can now be considered for overall optimization. Since factor B is also insignificant, we need only include factor A. Noting that:

$$\bar{y} = \frac{\sum y}{n}$$

$$= \frac{93.00}{8}$$

$$= 11.63$$

The optimum process average at C2, D"1, A1 is:

$$\mu_{Predicted} = \mu_{\overline{A1,\,C2,\,D''1}}$$

$$= \bar{y} + (\overline{A1} - \bar{y}) + (\overline{C2D''1} - \bar{y})$$

$$= \overline{A1} + \overline{C2D''1} - \bar{y}$$

$$= 12.50 + 13.00 - 11.63$$

$$= 13.88$$

6.   Confidence interval of the axial factor

For the calculation of a confidence interval we use $\nu1 = 1$ degree of freedom. The number of experiments involved in an axial factor level is four, i.e. $n = 4$. Therefore,

$$CI_{Axial} = \sqrt{F_{0.05,1,3} \times V_e \times \left[\frac{1}{n}\right]}$$

$$= \sqrt{10.13 \times 0.13 \times \left[\frac{1}{4}\right]}$$

$$= \pm\ 0.56$$

7.   Confidence interval of the nested factors

For the calculation of a confidence interval we use $\nu1 = 1$ degree of freedom. The number of experiments involved in a nested factor level is two, i.e. $n = 2$. Therefore,

$$CI_{Nested} = \sqrt{F_{0.05,1,3} \times V_e \times \left[\frac{1}{n}\right]}$$

$$= \sqrt{10.13 \times 0.13 \times \left[\frac{1}{2}\right]}$$

$$= \pm\ 0.80$$

8.   Confidence interval of the predicted mean

The confidence interval for the predicted process mean is calculated based on the degrees of freedom of A, C and D. Additionally, since we have used both C and D in estimating the mean, we need to include the degrees of freedom associated with $C \times D$ as well. Hence, the degrees of freedom used to calculate the predicted mean is one from A, one from C, one from D and one from $C \times D$. Noting that $\alpha = 0.05$, $\nu 1 = 1$ and $\nu 2 = 3$, therefore, $F_{\alpha, \nu 1, \nu 2} = F_{0.05, 1, 3} = 10.13$ and the confidence interval is:

$$CI_{Predicted} = \sqrt{F_{\alpha, \nu 1, \nu 2} \times V_e \times \left[\frac{1}{n_{eff}} \right]}$$

where n_{eff} is the effective number of degrees of freedom used in estimating the predicted mean, and

$$n_{eff} = \frac{total\ number\ of\ experiments}{sum\ of\ degrees\ of\ freedom\ used\ in\ estimate\ of\ mean}$$

$$= \frac{8 \times 1}{\nu_\mu + \nu_C + \nu_D + \nu_A + \nu_{C \times D}}$$

$$= \frac{8}{1 + 1 + 1 + 1 + 1}$$

$$= 1.60$$

Substituting for $n_{eff} = 1.60$,

$$CI_{Predicted} = \sqrt{F_{0.05, 1, 3} \times V_e \times \left[\frac{1}{n_{eff}} \right]}$$

$$= \sqrt{10.13 \times 0.13 \times \left[\frac{1}{1.60} \right]}$$

$$= \pm\ 0.89$$

If n_{eff} becomes difficult in complicated assignments Taguchi[4] suggests:

$$CI_{Predicted} = \pm\ 3\ \sqrt{Ve}$$

---

4   Genichi Taguchi, *System of Experimental Design: Engineering Methods to Optimize Quality and Minimize Costs*, UNIPUB/Kraus, 1924.

9.   Confidence interval of the confirmation experiment

The confidence interval for the confirmation experiment with r trials is:

$$CI_{Confirmation} = \sqrt{F_{\alpha,v1,v2} \times V_e \times \left[\frac{1}{n_{eff}} + \frac{1}{r}\right]}$$

For five trials, $r = 5$ and the confidence interval is:

$$CI_{Confirmation} = \sqrt{F_{0.05,1,3} \times V_e \times \left[\frac{1}{n_{eff}} + \frac{1}{r}\right]}$$

$$= \sqrt{10.13 \times 0.13 \times \left[\frac{1}{1.6} + \frac{1}{5}\right]}$$

$$= \pm\ 1.02$$

The response graph of factor effects is as follows:

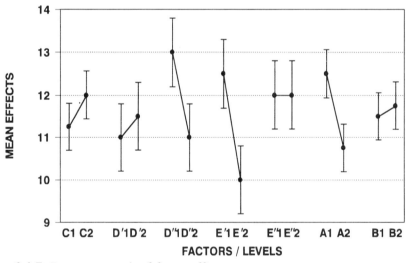

Figure 6:4.7 Response graph of factor effects.

6.4.4 Self-assessment questions

1. In the following experiment, there are two processes namely, acid wash (C1) and water wash (C2), to be considered before optimizing the overall process. Acid wash has two 2-levels factors (D′ and D″). Water wash has only one 2-level factor (E′) to be studied. Conduct an appropriate analysis of variance. Two other 2-level factors (A and B) are also studied. The results of the experiment are as follows:

Exp	1 C1 C2	2 D′ D″	3	4 E′ E″	5	6 A	7 B	Results
1	1	1	1	1	1	1	1	9
2	1	1	1	2	2	2	2	10
3	1	2	2	1	1	2	2	12
4	1	2	2	2	2	1	1	8
5	2	1	2	1	2	1	2	9
6	2	1	2	2	1	2	1	12
7	2	2	1	1	2	2	1	6
8	2	2	1	2	1	1	2	4

6.4.5 Answers to self-assessment questions

1.   In the following experiment, there are two processes namely, acid wash (C1) and water wash (C2), to be considered before optimizing the overall process. Acid wash has two 2-levels factors (D′ and D″). Water wash has only one 2-level factor (E′) to be studied. Conduct an appropriate analysis of variance. Two other 2-level factors (A and B) are also studied. The results of the experiment are as follows:

	1	2	3	4	5	6	7	
	C1	D′		E′		A	B	
Exp	C2	D″		E″				Results
1	1	1	1	1	1	1	1	9
2	1	1	1	2	2	2	2	10
3	1	2	2	1	1	2	2	12
4	1	2	2	2	2	1	1	8
5	2	1	2	1	2	1	2	9
6	2	1	2	2	1	2	1	12
7	2	2	1	1	2	2	1	6
8	2	2	1	2	1	1	2	4

The response table is as follows:

	C	D′	D″	E′	E″	A	B
Level 1	9.75	9.50	10.50	10.50	7.50	7.50	8.75
Level 2	7.75	10.00	5.00	9.00	8.00	10.00	8.75

1.   **Calculate the overall mean**

$$\bar{y} = \frac{\sum y}{n}$$

$$= \frac{9 + 10 + ... + 6 + 4}{8}$$

$$= 8.75$$

2.   **Sum of squares due to the mean**

$$Sm = n \, \bar{y}^2$$

$$= 8 \times 8.75^2$$

$$= 612.50$$

3.   **Sum of squares for the axial factor**

$$SC = n_{C1} \, \bar{y}_{C1}^2 + n_{C2} \, \bar{y}_{C2}^2 - n_C \, \bar{y}_C^2$$

$$= 4 \times 9.75^2 + 4 \times 7.75^2 - Sm$$

$$= 8.00$$

4.   **Sum of squares due to nested factors**
Remembering that D′, D″ and E′, E″ are different factors, the associated sum of squares for an analysis of variance is calculated as follows:

$$SD' = n_{D'1} \, \bar{y}_{D'1}^2 + n_{D'2} \, \bar{y}_{D'2}^2 - n_{D'} \, \bar{y}_{D'}^2$$

$$= 2 \times 9.50^2 + 2 \times 10.00^2 - 4 \times 9.75^2$$

$$= 0.25$$

$$SD'' = n_{D''1} \, \bar{y}_{D''1}^2 + n_{D''2} \, \bar{y}_{D''2}^2 - n_{D''} \, \bar{y}_{D''}^2$$

$$= 2 \times 10.50^2 + 2 \times 5.00^2 - 4 \times 7.75^2$$

$$= 30.25$$

$$SE' = n_{E'1}\, \bar{y}^2_{E'1} + n_{E'2}\, \bar{y}^2_{E'2} - n_{E'}\, \bar{y}^2_{E'}$$

$$= 2 \times 10.50^2 + 2 \times 9.00^2 - 4 \times 9.75^2$$

$$= 2.25$$

$$SE'' = n_{E''1}\, \bar{y}^2_{E''1} + n_{E''2}\, \bar{y}^2_{E''2} - n_{E''}\, \bar{y}^2_{E''}$$

$$= 2 \times 7.50^2 + 2 \times 8.00^2 - 4 \times 7.75^2$$

$$= 0.25$$

For convenience of calculation, although E″ does not exist as a control factor we treat it as if it existed as such and then simply pool it into the error. From these results, the analysis of variance is drawn as shown in the figure below. From this figure, it is obvious that D′, E′, E″ and B are insignificant and these are pooled using the method discussed earlier.

Source	Pool	Sq	ν	Ms	F-ratio	Sq'	rho %
C		8.00	1	8.00	11.64	7.31	13.67
D′	Y	0.25	1	0.25	-	-	-
D″		30.25	1	30.25	44.00	29.56	55.26
E′	Y	2.25	1	2.25	-	-	-
E″	Y	0.25	1	0.25	-	-	-
A		12.50	1	12.50	18.18	11.81	22.08
B	Y	0.00	1	0.00	-	-	-
Pooled e		2.75	4	0.69	1.00	4.81	9.00
St		53.50	7	7.64	-	53.50	100.00
Mean		612.50	1	-	-	-	-
ST		666.00	8	-	-	-	-

5. Estimation of the predicted mean

Assuming that the quality characteristic is larger-the-better we proceed to select the optimum process. First it is necessary to determine which of the processes C1 (acid wash) or C2 (water wash) is better. To do this, we need to analyze the nested factors. Considering only the axial and nested factors, the possible optimum conditions could be C1 or C2.

For C1 (Acid wash):
The optimum condition at C1 is D'2 and E'1. Since D'2 and E'1 are both insignificant, we do not consider C1 for the estimation.

For C2 (Water wash):
The optimum condition at C2 is D"1 and E"2. Since E"2 is insignificant and indeed only an error, we consider only C2 and D"1 for the estimation:

$$\mu_{\overline{C2,\ D''1}} = \overline{C2} + (\overline{D''1} - \overline{C2})$$

$$= \overline{D''1}$$

$$= 10.50$$

Comparing μ_{C1} and μ_{C2}, we choose $\mu_{C2,\ D''1}$ because μ_{C1} is an insignificant process. The remaining factors can now be considered for overall optimization. However, since factor B is insignificant, we need only include factor A. Noting that:

$$\bar{y} = \frac{\sum y}{n}$$

$$= \frac{70.00}{8}$$

$$= 8.75$$

The optimum process average at C2, D"1, A2 is:

$$\mu_{\overline{A2,\ C2,\ D''1}} = \bar{y} + (\overline{A2} - \bar{y}) + (\overline{C2D''1} - \bar{y})$$

$$= \overline{A2} + \overline{C2D''1} - \bar{y}$$

$$= 10.00 + 10.50 - 8.75$$

$$= 11.75$$

An interesting conclusion from this experiment is that the acid wash process has no significance and it may be suggested that this process be discontinued.

6.   Confidence interval of the axial factor

For the calculation of a confidence interval we use $\nu 1 = 1$ degree of freedom. The number of experiments involved at an axial factor level is four, i.e. $n = 4$. Therefore,

$$CI_{Axial} = \sqrt{F_{0.05,1,4} \times V_e \times \left[\frac{1}{n}\right]}$$

$$= \sqrt{7.71 \times 0.69 \times \left[\frac{1}{4}\right]}$$

$$= \pm\ 1.15$$

7.   Confidence interval of the nested factors

For the calculation of a confidence interval we use $\nu 1 = 1$ degree of freedom. The number of experiments involved at a nested factor level is two, i.e. $n = 2$. Therefore,

$$CI_{Nested} = \sqrt{F_{0.05,1,4} \times V_e \times \left[\frac{1}{n}\right]}$$

$$= \sqrt{7.71 \times 0.69 \times \left[\frac{1}{2}\right]}$$

$$= \pm\ 1.63$$

8.   Confidence interval of the predicted mean

The confidence interval for the predicted process mean is calculated based on the degrees of freedom of A, C and D. Additionally, since we have used both C and D in estimating the predicted mean, we need to include the degrees of freedom associated with C×D as well. Hence, the degrees of freedom used to calculate the predicted mean is one from A, one from C, one from D and one from C×D. Noting that $\alpha = 0.05$, $\nu 1 = 1$ and $\nu 2 = 4$, therefore, $F_{\alpha,1,\nu 2} = F_{0.05,1,4} = 7.71$ and the confidence interval is:

$$CI_{Predicted} = \sqrt{F_{\alpha,\nu 1,\nu 2} \times V_e \times \left[\frac{1}{n_{eff}}\right]}$$

where n_{eff} is the effective number of degrees of freedom used in estimating the predicted mean, and

$$n_{eff} = \frac{\textit{total number of experiments}}{\textit{sum of degrees of freedom used in estimate of mean}}$$

$$= \frac{8 \times 1}{v_\mu + v_A + v_C + v_D + v_{C \times D}}$$

$$= \frac{8}{1 + 1 + 1 + 1 + 1}$$

$$= 1.60$$

Substituting for $n_{eff} = 1.60$,

$$CI_{Predicted} = \sqrt{F_{0.05,1,4} \times V_e \times \left[\frac{1}{n_{eff}}\right]}$$

$$= \sqrt{7.71 \times 0.69 \times \left[\frac{1}{1.60}\right]}$$

$$= \pm\ 1.82$$

9.   Confidence interval of the confirmation experiment

The confidence interval for the confirmation experiment with r trials is:

$$CI_{Confirmation} = \sqrt{F_{\alpha,v1,v2} \times V_e \times \left[\frac{1}{n_{eff}} + \frac{1}{r}\right]}$$

For five trials, $r = 5$ and the confidence interval is:

$$CI_{Confirmation} = \sqrt{F_{0.05,1,4} \times V_e \times \left[\frac{1}{n_{eff}} + \frac{1}{r}\right]}$$

$$= \sqrt{7.71 \times 0.69 \times \left[\frac{1}{1.60} + \frac{1}{5}\right]}$$

$$= \pm 2.09$$

The associated response graph of factor effects is shown below.

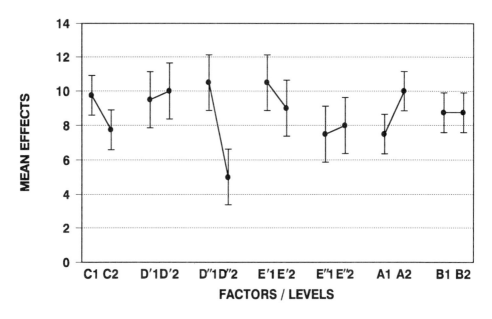

6.5 Pseudo-factor Design – Idle-column

6.5.1 Introduction
The idle-column design is another pseudo-factor design. This is a very efficient design by which, for example, 3-level factors can be assigned to a 2-level series orthogonal array. When using a 2-level series orthogonal array that has no allocation for a 3-level factor, we could create a 4-level factor by using a multi-level design and then assign a dummy level. Such an experimental design is not efficient, however, since one degree of freedom is wasted in the error of the dummy level. Additionally, although one degree of freedom is taken up by the idle-column, subsequent pairs of 2-level factors only need two columns. Hence, the idle-column method is an efficient method of experimentation.

6.5.2 Idle-column method
Suppose it is necessary to incorporate one 3-level factor into an $L_8(2^7)$ orthogonal array. This can be done by creating one 4-level factor and then assigning a 3-level factor with one dummy level. However, such a method wastes one degree of freedom. Using the idle-column method, we will now show how to assign two 3-level factors and two 2-level factors into an $L_8(2^7)$ orthogonal array. In this method, we need to leave one column, the *idle-column*, unassigned. Such a method uses all the available degrees of freedom and represents more efficient experimentation.

6.5.3 Procedure
We now give an outline of how to conduct an idle-column design together with the appropriate data analysis:
- factor assignment,
- draw the required linear graph,
- match the linear graph to an appropriate orthogonal array,
- assign the factors to the orthogonal array.

1.   Factor assignment
Consider an assignment of experiment where A and B are two 3-level factors and C and D are two 2-level factors. Two 3-level factors require four degrees of freedom. Two 2-level factors require two degrees of freedom. Hence we require a total of six degrees of freedom.

2.   Draw the required linear graph
We assign column 1 to be the idle column and then draw the required linear graph. We draw a dashed circle around the dot and connecting line segment to indicate that the factor uses both columns.

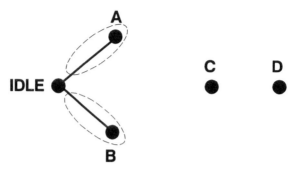

Figure 6:5.1 Required linear graph.

3. Match the linear graph to an appropriate orthogonal array

Match the linear graph to an appropriate 2^n series linear graph and assign the experiment. We match the linear graph above to an appropriate linear graph. In this case we use an $L_8(2^7)$ linear graph.

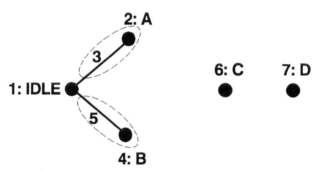

Figure 6:5.2 Matching the required linear graph to an appropriate orthogonal array.

4. Assign the factors to the orthogonal array

Using the linear graph and orthogonal array, we assign the idle-column to column 1, factor A to columns 2 and 3, factor B to columns 4 and 5, factor C to column 6 and factor D to column 7, as shown in Figure 6:5.3. Note the dummy treatments for factors A and B.

	I	A ⋀		B ⋀		C	D			I	A	B	C	D
	1	2	3	4	5	6	7			1	23	45	6	7
1	1	1	1	1	1	1	1		1	1	1	1	1	1
2	1	1	1	2	2	2	2		2	1	1	2	2	2
3	1	2	2	1	1	2	2		3	1	2	1	2	2
4	1	2	2	2	2	1	1		4	1	2	2	1	1
5	2	1	2	1	2	1	2		5	2	2'	2'	1	2
6	2	1	2	2	1	2	1		6	2	2'	3	2	1
7	2	2	1	1	2	2	1		7	2	3	2'	2	1
8	2	2	1	2	1	1	2		8	2	3	3	1	2

Figure 6:5.3 Orthogonal array for idle-column factor design.

6.5.4 Example of experimental analysis
We now outline the data analysis for the results of an idle-column design experiment shown in Figure 6:5.4. We proceed by calculating:
- the response table,
- the overall mean,
- the sum of squares due to the mean,
- the sum of squares due to an idle-column,
- the sums of squares due to other factors,
- the sums of squares due to idle-column factors,
- the confidence intervals of idle-column factors,
- the confidence intervals of other factors,
- the confidence interval of the predicted mean,
- the confidence interval of the confirmation experiment.

1.   The response table
The response table is calculated in the usual way except for factors A and B. Since these factors are treated with an idle-column, we need to calculate the mean effects of levels 1 and 2 with idle-column level 1, and levels 2' and 3 with idle-column level 2, as shown in Figure 6:5.5.

Exp	1 I	2, 3 A	4, 5 B	6 C	7 D	Results
1	1	1	1	1	1	4
2	1	1	2	2	2	10
3	1	2	1	2	2	14
4	1	2	2	1	1	8
5	2	2'	2'	1	2	3
6	2	2'	3	2	1	8
7	2	3	2'	2	1	7
8	2	3	3	1	2	2

Figure 6:5.4 Orthogonal array for idle-column factors.

	I	C	D			A	B
Level 1	9.00	4.25	6.75	Idle 1	Level 1	7.00	9.00
					Level 2	11.00	9.00
Level 2	5.00	9.75	7.25	Idle 2	Level 2'	5.50	5.00
					Level 3	4.50	5.00

Figure 6:5.5 Response table of idle-column factor effects.

2.   **The overall mean**

$$\bar{y} = \frac{\sum y}{n}$$

$$= \frac{4 + 10 + ... + 7 + 2}{8}$$

$$= 7.00$$

3.   **The sum of squares due to the mean**

$$Sm = n\,\bar{y}^2$$

$$= 8 \times 7.00^2$$

$$= 392.00$$

4.   **The sum of squares due to an idle-column**

The sum of squares due to an idle column is calculated as follows:

$$SI = n_{I1}\,\bar{y}_{I1}^2 + n_{I2}\,\bar{y}_{I2}^2 - Sm$$

$$= 4 \times 9.00^2 + 4 \times 7.00^2 - 392.00$$

$$= 32.00$$

5.   **The sums of squares due to other factors**

The sums of squares for factors C and D are calculated by the method given earlier.

6.   **The sums of squares due to idle-column factors**

For A1 A2:

$$S_{A1\,A2} = n_{A1}\,\bar{y}_{A1}^2 + n_{A2}\,\bar{y}_{A2}^2 - n_{A1\,A2}\,\bar{y}_{A1\,A2}^2$$

$$= 2 \times 7.00^2 + 2 \times 11.00^2 - 4 \times 9.00^2$$

$$= 16.00$$

For A2' A3:

$$S_{A2'\,A3} = n_{A2'}\,\bar{y}_{A2'}^2 + n_{A3}\,\bar{y}_{A3}^2 - n_{A2'\,A3}\,\bar{y}_{A2'\,A3}^2$$

$$= 2 \times 5.50^2 + 2 \times 4.50^2 - 4 \times 5.00^2$$

$$= 1.00$$

For B1 B2:

$$S_{B1\ B2} = n_{B1}\ \bar{y}_{B1}^2 + n_{B2}\ \bar{y}_{B2}^2 - n_{B1\ B2}\ \bar{y}_{B1\ B2}^2$$

$$= 2 \times 9.00^2 + 2 \times 9.00^2 - 4 \times 9.00^2$$

$$= 0.00$$

For B2' B3:

$$S_{B2'\ B3} = n_{B2'}\ \bar{y}_{B2'}^2 + n_{B3}\ \bar{y}_{B3}^2 - n_{B2'\ B3}\ \bar{y}_{B2'\ B3}^2$$

$$= 2 \times 5.00^2 + 2 \times 5.00^2 - 4 \times 5.00^2$$

$$= 0.00$$

Using these results we can draw the analysis of variance as shown in Figure 6:5.6.

Source	Pool	Sq	ν	Mq	F-ratio	Sq'	rho %
IDLE		32.00	1	32.00	85.33	31.63	28.75
A1 A2		16.00	1	16.00	42.67	15.63	14.20
A2' A3	Y	1.00	1	1.00	-	-	-
B1 B2	Y	0.00	1	0.00	-	-	-
B2' B3	Y	0.00	1	0.00	-	-	-
C		60.50	1	60.50	161.33	60.13	54.66
D	Y	0.50	1	0.50	-	-	-
Pooled e		1.50	4	0.38	1.00	2.63	2.39
St		110.00	7	15.71	-	110.00	100.00
Mean		392.00	1	-	-	-	-
ST		502.00	8	-	-	-	-

Figure 6:5.6 Analysis of variance for the idle-column experiment.

From the analysis of variance, we can say that factors A2' A3, B1 B2, B2' B3 and D are insignificant and we pool them as shown in earlier examples. To compare the effects of A1, A2 and A3 we need to make an adjustment between

A1 A2 and A2′ A3. Since A2 and A2′ are the same factor levels, we find the half-difference between their averages:
Half-difference = (11.0 − 5.5)/2 = 2.75,
by which A1 and A2 must be decreased and A2′ and A3 must be increased. Therefore, A1 and A2 will need to be decreased by 2.75 while A2′ A3 will need to be increased by 2.75 as shown below:

A1   = 7.00 → 7.00 − 2.75   = 4.25
A2   = 11.0 → 11.0 − 2.75   = 8.25
A2′   = 5.50 → 5.50 + 2.75   = 8.25
A3   = 4.50 → 4.50 + 2.75   = 7.25

Of course A2′ is only shown for comparison and may be omitted in subsequent calculations. The effects of B1, B2 and B3 are similarly adjusted by a half-difference of 2.00 as follows:
B1   = 9.00 → 9.00 − 2.00   = 7.00
B2   = 9.00 → 9.00 − 2.00   = 7.00
B2   = 5.00 → 5.00 + 2.00   = 7.00
B3   = 5.00 → 5.00 + 2.00   = 7.00

The matching of idle column factor A is shown in Figure 6:5.7 and the appropriate modified response table is shown in Figure 6:5.8.

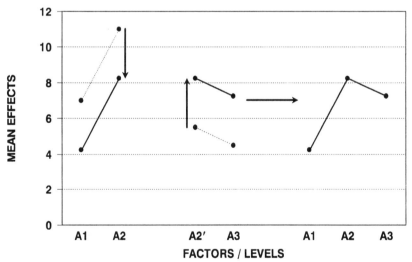

Figure 6:5.7 Matching the idle column levels A2 and A2′ by the half-difference.

	A	B	C	D
Level 1	4.25	7.00	4.25	6.75
Level 2	8.25	7.00	9.75	7.25
Level 3	7.25	7.00	-	-

Figure 6:5.8 Modified response table of idle-column factor effects.

7.   Confidence intervals of idle-column factors

The confidence intervals for idle-column factors A and B are calculated on the basis that two observations have been made at each level. For the calculation of a confidence interval we use $\nu 1 = 1$ degree of freedom. The confidence interval is therefore:

$$CI_{A,\,B} = \sqrt{F_{0.05,1,4} \times V_e \times \left[\frac{1}{n}\right]}$$

$$= \sqrt{7.71 \times 0.38 \times \left[\frac{1}{2}\right]}$$

$$= \pm\ 1.20$$

8.   Confidence intervals of other factors

The confidence intervals for 2-level factors C and D are calculated on the basis that four observations have been made at each level. For the calculation of a confidence interval we use $\nu 1 = 1$ degree of freedom. The confidence interval is therefore:

$$CI_{C,\,D} = \sqrt{F_{0.05,1,4} \times V_e \times \left[\frac{1}{n}\right]}$$

$$= \sqrt{7.71 \times 0.38 \times \left[\frac{1}{4}\right]}$$

$$= \pm\ 0.85$$

9.   Confidence interval of the predicted mean

Assuming a larger-the-better characteristic, we choose levels A2 and C2. The predicted process mean at this condition is:

$$\mu_{Predicted} = \bar{y} + (\overline{A2} - \bar{y}) + (\overline{C2} - \bar{y})$$

$$= \overline{A2} + \overline{C2} - \bar{y}$$

$$= 8.25 + 9.75 - 7.00$$

$$= 11.00$$

The confidence interval for the predicted mean is:

$$CI_{Predicted} = \sqrt{F_{\alpha,v1,v2} \times V_e \times \left[\frac{1}{n_{eff}}\right]}$$

where n_{eff} is the effective number of degrees of freedom used in estimating the predicted mean, and

$$n_{eff} = \frac{total\ number\ of\ experiments}{sum\ of\ degrees\ of\ freedom\ used\ in\ estimate\ of\ mean}$$

$$= \frac{8 \times 1}{v_\mu + v_A + v_C}$$

$$= \frac{8}{1 + 2 + 1}$$

$$= 2$$

Note that factor A has effectively two degrees of freedom since it occupies columns 2 and 3 of the $L_8(2^7)$ orthogonal array (see Figure 6:5.4). Substituting $n_{eff} = 2$,

$$CI_{Predicted} = \sqrt{F_{0.05,1,4} \times V_e \times \left[\frac{1}{n_{eff}}\right]}$$

$$= \sqrt{7.71 \times 0.38 \times \left[\frac{1}{2}\right]}$$

$$= \pm\ 1.20$$

If n_{eff} becomes difficult in complicated assignments Taguchi suggests:

$$CI_{Predicted} = \pm 3 \sqrt{Ve}$$

10.   Confidence interval of the confirmation experiment

The confidence interval for the confirmation experiment with *r* trials is:

$$CI_{Confirmation} = \sqrt{F_{\alpha,v1,v2} \times V_e \times \left[\frac{1}{n_{eff}} + \frac{1}{r}\right]}$$

For five trials, $r = 5$ and the confidence interval is:

$$CI_{Confirmation} = \sqrt{F_{0.05,1,4} \times V_e \times \left[\frac{1}{n_{eff}} + \frac{1}{r}\right]}$$

$$= \sqrt{7.71 \times 0.38 \times \left[\frac{1}{2} + \frac{1}{5}\right]}$$

$$= \pm 1.42$$

The response graph for this experiment is shown in Figure 6:5.9.

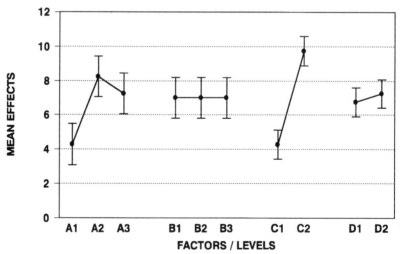

Figure 6:5.9 Response graph of idle-column design experiment.

6.5.5 Self-assessment questions

1.   Complete the analysis of variance for the idle-column technique
       experiment below with a suitable data analysis:

| Exp | 1 | 2, 3 | 4, 5 | 6 | 7 | Results |
	I	A	B	C	D	
1	1	1	1	1	1	12
2	1	1	2	2	2	14
3	1	2	1	2	2	6
4	1	2	2	1	1	5
5	2	2'	2'	1	2	1
6	2	2'	3	2	1	10
7	2	3	2'	2	1	4
8	2	3	3	1	2	9

Figure 6:5.10   Orthogonal array for idle-column factor design.

6.5.6 Answers to self-assessment questions

1.   Complete the analysis of variance for the idle-column technique experiment below with a suitable data analysis:

Exp	1	2, 3	4, 5	6	7	Results
	I	A	B	C	D	
1	1	1	1	1	1	12
2	1	1	2	2	2	14
3	1	2	1	2	2	6
4	1	2	2	1	1	5
5	2	2'	2'	1	2	1
6	2	2'	3	2	1	10
7	2	3	2'	2	1	4
8	2	3	3	1	2	9

Figure 6:5.11   Orthogonal array for idle-column factor design.

We now outline the data analysis for the results of the idle-column design experiment above.
- the response table,
- the overall mean,
- the sum of squares due to the mean,
- the sum of squares due to an idle-column,
- the sums of squares due to other factors,
- the sums of squares due to idle-column factors,
- the confidence intervals of idle-column factors,
- the confidence intervals of other factors,
- the confidence interval of the predicted mean,
- the confidence interval of the confirmation experiment.

1.   The response table
The response table is calculated in the usual way except for factors A and B. Since these factors are treated with an idle column, we need to calculate the mean effects of levels 1 and 2 with idle-column level 1, and levels 2' and 3 with idle-column level 2, as shown below.

	I	C	D			A	B
Level 1	9.25	6.75	7.75	Idle 1	Level 1	13.00	9.00
					Level 2	5.50	9.50
Level 2	6.00	8.50	7.50	Idle 2	Level 2'	5.50	2.50
					Level 3	6.50	9.50

Figure 6:5.12   Response table of idle-column factor effects.

2.   **The overall mean**

$$\bar{y} = \frac{\sum y}{n}$$

$$= \frac{12 + 14 + ... + 4 + 9}{8}$$

$$= 7.63$$

3.   **The sum of squares due to the mean**

$$Sm = n \, \bar{y}^2$$

$$= 8 \times 7.63^2$$

$$= 465.13$$

4.   **The sum of squares due to an idle column**
The sum of squares due to an idle-column is calculated as follows:

$$SI = n_{I1} \, \bar{y}_{I1}^2 + n_{I2} \, \bar{y}_{I2}^2 - Sm$$

$$= 4 \times 9.25^2 + 4 \times 6.00^2 - 465.13$$

$$= 21.13$$

5.   **The sum of squares due to other factors**
The sums of squares for factors C and D are calculated by the method given earlier.

6.   The sums of squares due to idle-column factors

For A1 A2:

$$S_{A1\ A2} = n_{A1}\ \bar{y}^2_{A1} + n_{A2}\ \bar{y}^2_{A2} - n_{A1\ A2}\ \bar{y}^2_{A1\ A2}$$

$$= 2 \times 13.00^2 + 2 \times 5.50^2 - 4 \times 9.25^2$$

$$= 56.25$$

For A2′ A3:

$$S_{A2'\ A3} = n_{A2'}\ \bar{y}^2_{A2'} + n_{A3}\ \bar{y}^2_{A3} - n_{A2'\ A3}\ \bar{y}^2_{A2'\ A3}$$

$$= 2 \times 5.50^2 + 2 \times 6.50^2 - 4 \times 6.00^2$$

$$= 1.00$$

For B1 B2:

$$S_{B1\ B2} = n_{B1}\ \bar{y}^2_{B1} + n_{B2}\ \bar{y}^2_{B2} - n_{B1\ B2}\ \bar{y}^2_{B1\ B2}$$

$$= 2 \times 9.00^2 + 2 \times 9.50^2 - 4 \times 9.25^2$$

$$= 0.25$$

For B2′ B3:

$$S_{B2'\ B3} = n_{B2'}\ \bar{y}^2_{B2'} + n_{B3}\ \bar{y}^2_{B3} - n_{B2'\ B3}\ \bar{y}^2_{B2'\ B3}$$

$$= 2 \times 2.50^2 + 2 \times 9.50^2 - 4 \times 6.00^2$$

$$= 49.00$$

Using these results we can draw the analysis of variance as shown in Figure 6:5.13. From the analysis of variance, we can say that factors A2′ A3, B1 B2, C and D are insignificant, and we pool them as shown in earlier examples. To compare the effects of A1, A2 and A3 we need to make an adjustment between A1 A2 and A2′ A3. Since A2 and A2′ is the same factor level, we find the half-difference between their averages:

Half-difference = $(5.50 - 5.50)/2 = 0.00$.

For completeness of calculation, we decrease A1 and A2 by 0.00 while A2′ and A3 will need to be increased by 0.00. Therefore,

A1	= 13.0 → 13.00 − 0.00	= 13.00
A2	= 5.50 → 5.50 − 0.00	= 5.50
A2′	= 5.50 → 5.50 + 0.00	= 5.50
A3	= 6.50 → 6.50 + 0.00	= 6.50

Source	Pool	Sq	v	Mq	F-ratio	Sq'	rho %
IDLE		21.13	1	21.13	11.27	19.25	14.38
A1A2		56.25	1	56.25	30.00	54.38	40.62
A2'A3	Y	1.00	1	1.00	-	-	-
B1B2	Y	0.25	1	0.25	-	-	-
B2'B3		49.00	1	49.00	26.13	47.13	35.20
C	Y	6.13	1	6.13	-	-	-
D	Y	0.13	1	0.13	-	-	-
Pooled e		7.50	4	1.88	1.00	13.13	9.80
St		133.88	7	19.13	-	133.88	100.00
Mean		465.13	1	-	-	-	-
ST		599.00	8	-	-	-	-

Figure 6:5.13   Analysis of variance for the idle-column experiment.

In this particular case, the half-difference is zero and no adjustment is necessary. Of course A2' is only shown for comparison and may be omitted in subsequent calculations. The effects of B1, B2 and B3 are similarly adjusted by a half-difference of 3.50 as follows:

B1 $= 9.00 \rightarrow 9.00 - 3.50$ $= 5.50$
B2 $= 9.50 \rightarrow 9.50 - 3.50$ $= 6.00$
B2 $= 2.50 \rightarrow 2.50 + 3.50$ $= 6.00$
B3 $= 9.50 \rightarrow 9.50 + 3.50$ $= 13.00$

A modified response table is shown in Figure 6:5.14.

	A	B	C	D
Level 1	13.00	5.50	6.75	7.75
Level 2	5.50	6.00	8.00	7.50
Level 3	6.50	13.00	-	-

Figure 6:5.14   Modified response table of idle-column factor effects.

The matching of idle-column factor B is shown in Figure 6:5.15.

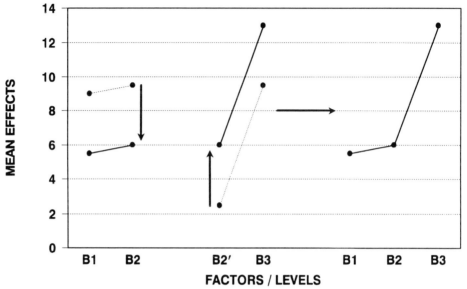

Figure 6:5.15   Matching the idle-column levels B2 and B2' by the half difference.

7.   Confidence interval of idle-column factors

The confidence intervals for idle-column factors A and B are calculated on the basis that two observations have been made at each level. For the calculation of a confidence interval we use $\nu 1 = 1$ degree of freedom. The confidence interval is therefore:

$$CI_{A,\,B} = \sqrt{F_{0.05,1,4} \times V_e \times \left[\frac{1}{n}\right]}$$

$$= \sqrt{7.71 \times 1.88 \times \left[\frac{1}{2}\right]}$$

$$= \pm\ 2.69$$

8.   Confidence intervals of other factors

The confidence intervals for 2-level factors C and D are calculated on the basis that four observations have been made at each level. For the calculation of a confidence interval we use $\nu 1 = 1$ degree of freedom. The confidence interval is therefore:

$$CI_{C,D} = \sqrt{F_{0.05,1,4} \times V_e \times \left[\frac{1}{n}\right]}$$

$$= \sqrt{7.71 \times 1.88 \times \left[\frac{1}{4}\right]}$$

$$= \pm \ 1.90$$

9.   Confidence interval of predicted mean

Assuming a larger-the-better characteristic, we choose levels A1 and B3. The predicted process average is:

$$\mu_{Predicted} = \bar{y} + (\overline{A1} - \bar{y}) + (\overline{B3} - \bar{y})$$

$$= \overline{A1} + \overline{B3} - \bar{y}$$

$$= 13.00 + 13.00 - 7.63$$

$$= 18.37$$

The confidence interval for the predicted mean is:

$$CI_{Predicted} = \sqrt{F_{\alpha,\nu1,\nu_2} \times V_e \times \left[\frac{1}{n_{eff}}\right]}$$

where

$$n_{\mathit{eff}} = \frac{\textit{total number of experiments}}{\textit{sum of degrees of freedom used in estimate of mean}}$$

$$= \frac{8}{v_{\mu} + v_{A} + v_{B}}$$

$$= \frac{8}{1 + 2.00 + 2.00}$$

$$= 1.60$$

Note that factor A has effectively two degrees of freedom since it occupies columns 2 and 3 of the $L_8(2^7)$ orthogonal array. Similarly, factor B has effectively two degrees of freedom since it occupies columns 4 and 5 of the $L_8(2^7)$ orthogonal array. Substituting $n_{\mathit{eff}} = 1.60$,

$$CI_{Predicted} = \sqrt{F_{0.05,1,4} \times V_e \times \left[\frac{1}{n_{\mathit{eff}}}\right]}$$

$$= \sqrt{7.71 \times 1.88 \times \left[\frac{1}{1.60}\right]}$$

$$= \pm \, 3.01$$

10.   Confidence interval of the confirmation experiment
The confidence interval for the confirmation experiment with r trials is:

$$CI_{Confirmation} = \sqrt{F_{\alpha,v1,v2} \times V_e \times \left[\frac{1}{n_{\mathit{eff}}} + \frac{1}{r}\right]}$$

For five trials, $r = 5$ and the confidence interval is:

$$CI_{Confirmation} = \sqrt{F_{0.05,1,4} \times V_e \times \left[\frac{1}{n_{\mathit{eff}}} + \frac{1}{r}\right]}$$

$$= \sqrt{7.71 \times 1.88 \times \left[\frac{1}{1.60} + \frac{1}{5}\right]}$$

$$= \pm \, 3.45$$

The response graph for this experiment is shown in Figure 6:5.16.

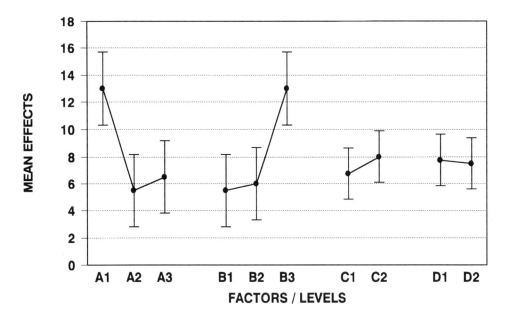

Figure 6:5.16   Response graph of idle-column design.

6.6   Distributed Interaction

6.6.1 Introduction
Another way of overcoming interactions in experiments is to use distributed interactions[5]. This is done by using orthogonal arrays in which the interaction effects are distributed more or less uniformly among all the columns. In this way interaction effects are averaged over many columns and hence many control factors can be studied simultaneously. Additionally, since it is encouraged that engineers should avoid studying interactions in the design stages, distributed interaction is a recommended way of studying many control factors. Consequently, of course, interaction effects cannot be determined from such orthogonal arrays.

6.6.2 Special designs
When conducting experiments to identify optimum levels for control factors, we avoid assigning interactions. However, we do not want the effects of interactions to disturb the experimental results. In such cases, we can use special designs in which the interaction effects are distributed more or less uniformly over the remaining columns to prevent erroneous conclusions. Some of these special designs are discussed below.

6.6.3 Distributed interaction design $L_{12}(2^{11})$
Interactions in this orthogonal array are distributed more or less uniformly among all columns. Hence, there is no linear graph for this array and it should not be used to analyze interactions. The advantage of this design is its ability to investigate 11 main effects and is thus a highly recommended orthogonal array for experiments.

6.6.4 Distributed interaction design $L_{18}(2^1 \times 3^7)$
The interaction between column 1 (2-level column) and column 2 (3-level column) can be obtained without sacrificing any other column.

one 2-level column	= 1 degree of freedom,
seven 3-level columns	= 14 degrees of freedom,
total	= 15 degrees of freedom

The remaining two degrees of freedom are taken up by the interaction of one 2-level column and one 3-level column, i.e. $(2 - 1) \times (3 - 1) = 2$ degrees of freedom. These two degrees of freedom are taken up by the *built-in* interaction between columns 1 and 2. This interaction information can be obtained without sacrificing any other column. Interactions between three-level columns are distributed more or less uniformly

---

5   Taguchi and Konishi, *Orthogonal Arrays and Linear Graphs: Tools for Quality Engineering*, 1987, ASI Press.

among all the other 3-level columns. This permits investigation of eight main effects and is thus a highly recommended orthogonal array for experiments.

6.6.5 Distributed interaction design $L_{32}(2^1 \times 4^9)$

The *built-in* interaction between columns 1 (2-level column) and 2 (4-level column) can be obtained without sacrificing any other column.

one 2-level column	= 1	degree of freedom,
nine 4-level columns	= 27	degrees of freedom,
total	= 28	degrees of freedom.

The remaining three degrees of freedom are taken up by the interaction of one 2-level column and one 4-level column, i.e. $(2 - 1) \times (4 - 1) = 3$ degrees of freedom. These three degrees of freedom are taken up by the built-in interaction between columns 1 and 2. Two-way interactions between any of the 4-level columns are distributed more or less uniformly among the 4-level columns. This permits the investigation of ten main effects and is thus a recommended orthogonal array for experiments. However, being largely a 4^n series it is less used compared to the $L_{36}(2^{11} \times 3^{12})$ and the $L_{36}(2^3 \times 3^{13})$ orthogonal arrays discussed next.

6.6.6 Distributed interaction design $L_{36}(2^{11} \times 3^{12})$

In the $L_{36}(2^{11} \times 3^{12})$ orthogonal array interactions are not orthogonal to other columns.

eleven 2-level columns	= 11 degrees of freedom,
twelve 3-level columns	= 24 degrees of freedom,
total	= 35 degrees of freedom.

However, the $L_{36}(2^{11} \times 3^{12})$ orthogonal array should not be used for obtaining interactions. In any case, there is no interaction table or linear graph for this orthogonal array. Two-way interactions between any of the columns is distributed more or less uniformly among the other columns. This permits the investigation of 23 main effects and is thus a recommended orthogonal array.

6.6.7 Distributed interaction design $L_{36}(2^3 \times 3^{13})$

We should not confuse the $L_{36}(2^{11} \times 3^{12})$ and the $L_{36}(2^3 \times 3^{13})$ orthogonal arrays. It must be noted however, that replacing eleven 2-level columns (1 to 11) from the $L_{36}(2^{11} \times 3^{12})$ with three 2-level (1′, 2′, 3′) and one 3-level column (4′) results in the $L_{36}(2^3 \times 3^{13})$ orthogonal array. In this array, there are:

three 2-level columns	= 6 degrees of freedom,
thirteen 3-level columns	= 26 degrees of freedom,
total	= 32 degrees of freedom.

The remaining three degrees of freedom are taken up by interactions between columns $1'$, $2'$, $3'$ and $4'$. There is no interaction table for this orthogonal array. Two-way interactions between any of the columns is distributed more or less uniformly among the other columns. This permits the investigation of 16 main effects and is thus a recommended orthogonal array.

The $L_{36}(2^3 \times 3^{13})$ orthogonal array is the most frequently used in parameter and tolerance designs for the following reasons:
- largely 3^n series array,
- accomodates many factors,
- distributed interaction.

1.   Largely 3^n series array
The $L_{36}(2^3 \times 3^{13})$ orthogonal array is a largely 3^n series orthogonal array making it ideal for investigating the plus and minus sides of a starting or nominal value. Mathematically, this means it is ideal for studying quadratic effects.

2.   Accomodates many factors
The $L_{36}(2^3 \times 3^{13})$ orthogonal array can accommodate thirteen 3-level factors. Since, it is very important to study many control factors together in robust design, this orthogonal array is ideal for studying many control factors simultaneously.

3.   Distributed interaction
Interactions between control factor columns are more or less uniformly distributed. This feature is particularly important when investigating large numbers of control factors in which interactions need to be treated as error.

6.6.8 Number of factor levels in distributed interaction designs
The purpose of using distributed interaction designs is to determine the best levels for design parameters or process factors without being bothered with interactions. In the design stages it is vital to study as many control factors as possible. Similarly, it is necessary to treat, as far as possible, interactions as noise so that the response can be predicted by control factors alone. If interactions are included in this stage then downstream reproducibility will become difficult.

If experimentation is relatively inexpensive, it may be possible to use 3-levels for control factors as well as noise factors. If experimentation is relatively expensive, then it may be necessary to use 2-level factors.

6.6.9 Self-assessment questions

1.   At what stage in the design process would you use an orthogonal array with a distributed interaction design?

6.6.10   Answers to self-assessment questions

1.   At what stage in the design process would you use an orthogonal array
         with a distributed interaction design?

Answer
         The distributed interaction design in particularly useful at the parameter
         design stage, where the primary objective is to efficiently select robust
         conditions which are reproducible downstream. Therefore, additivity is
         a requisite for efficient, reliable and reproducible experimentation. Thus,
         it is important to study the main effects of factors and to treat interaction
         effects as noise. The distributed interaction is ideal for this purpose.

CHAPTER 7

COMPUTER-AIDED PARAMETER DESIGN

AIMS:
To provide a method of performing parameter design using 2-level control factors and 2-level noise factors in a direct product design.

OBJECTIVES:
When you have completed studying this chapter you should be able to:
- perform a direct product parameter design using 2-level control and noise factors,
- calculate the associated signal-to-noise ratio,
- conduct an appropriate data analysis,
- determine the robust parameters.

OVERVIEW:
This chapter introduces computer-aided parameter design. A relatively simple idea of a cannon ball trajectory is used to illustrate the importance of parameter design. To demonstrate the power of parameter design, the problem is initially solved using a conventional design approach and later by using the parameter design approach. The results of both design techniques are compared in terms of the normal distributions of the trajectory distance for a controlled noise environment.

7.1   Computer-Aided Parameter Design

7.1.1 Introduction
Quality improvement is a key objective in reducing costs and increasing productivity. At the design stage of a product, it is most important that the setting of parameter levels are selected very carefully. Deviation of the objective characteristics from the intended ideal causes a financial loss. This loss can be due to variations in the system parameters (internal noise) as well as the environmental (external noise) conditions. The objective then is to identify parameter levels such that the objective characteristic is insensitive to noise effects.

      Thus, at the design stage of a new product development, it is important that the settings of nominal parameter levels (and not the tolerances of the parameters) are selected very carefully. The goal is to identify parameter levels such that the objective characteristic is insensitive to noise. Such a method allows the least expensive components to be used and hence to develop a process or product that is not only relatively inexpensive but also robust.

7.1.2 Conventional design for cannon ball distance
Consider the trajectory of a cannon ball as shown in Figure 7:1.1.

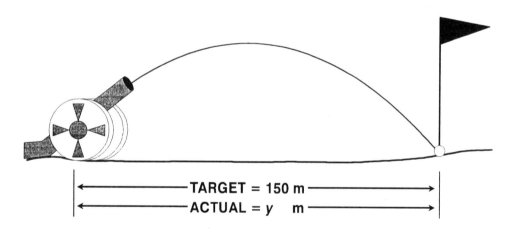

Figure 7:1.1 Trajectory of a cannon ball.

Suppose that a cannon ball of mass m is ejected with a force F at an angle α. The equation of this trajectory is:

$$y = \frac{F^2 \times \sin 2\alpha}{m^2 \times g} \qquad (7:1.1)$$

where g may be assumed to be 9.81 m s^{-2}. If the target distance is 150 metres, then by conventional design, we may arbitrarily set $F = 10$ N, $m = 0.20$ kg and proceed to calculate α using the formula:

$$\sin 2\alpha = \frac{y \times m^2 \times g}{F^2}$$

$$\alpha = 0.5 \times \sin^{-1}\left(\frac{y \times m^2 \times g}{F^2}\right) \qquad (7:1.2)$$

Using the formula above,

$$\alpha = 0.5 \times \sin^{-1}\left(\frac{y \times m^2 \times g}{F^2}\right)$$

$$= 0.5 \times \sin^{-1}\left(\frac{150 \times 0.20^2 \times 9.81}{10^2}\right)$$

$$= 0.5 \times \sin^{-1} 0.5886$$

$$= 18.0^\circ \ or \ 72^\circ$$

Hence, the conventional design would suggest that to reach a target distance of 150 m we may use a cannon ball of mass 0.20 kg, thrust with a force of 10 N at a minor angle of 18°. Of course there is also the major angle $(90^\circ - 18^\circ = 72^\circ)$ which can be used for the ejection. However, the principle of robustization is the same in both cases; here we consider only the case of the minor angle.

7.1.3 Introducing variation

There is nothing intrinsically wrong with the conventional method of calculating the angle α. If there is no variation in the force F, the angle of ejection α or the cannon ball mass m, there is little need for robustization. In reality, however, there will always be variation in the cannon ball mass m, the thrust F and the angle α of ejection which will cause the variations in the actual distance the cannon ball travels. Let us simulate this condition by building some variation about the force, angle and mass. We can do this simply by taking plus or minus a small value about the force, angle and mass, as shown in Figure 7:1.2. Using this variation, we can build a table of values as shown in Figure 7:1.3. This table simply creates eight combinations of low and high values for the force, angle and mass. Taking $F = 10$ N, $\alpha = 18°$ and $m = 0.20$ kg as the nominal values, we can calculate the distance travelled y, by the cannon ball, as shown in Figure 7:1.4.

Factor	Lower value	Nominal	Upper value
Force	$0.9 \times F$	F	$1.1 \times F$
Angle	$\alpha - 5°$	α	$\alpha + 5°$
Mass	$m - 0.01$	m	$m + 0.01$

Figure 7:1.2 Building variation.

Force	\multicolumn{4}{c}{$0.9 \times F$}				\multicolumn{4}{c}{$1.1 \times F$}				Summary	
Angle	\multicolumn{2}{c}{$\alpha-5°$}		\multicolumn{2}{c}{$\alpha+5°$}		\multicolumn{2}{c}{$\alpha-5°$}		\multicolumn{2}{c}{$\alpha+5°$}		results	
Mass	$m-$ 0.01	$m+$ 0.01	$m-$ 0.01	$m+$ 0.01	$m-$ 0.01	$m+$ 0.01	$m-$ 0.01	$m+$ 0.01	$\bar{y}$	σ
Conventional										
Parameter										

Figure 7:1.3 Building noise around a conventional design.

Force	\multicolumn{4}{c}{$0.9 \times F$}				\multicolumn{4}{c}{$1.1 \times F$}				Summary	
Angle	\multicolumn{2}{c}{$\alpha-5°$}		\multicolumn{2}{c}{$\alpha+5°$}		\multicolumn{2}{c}{$\alpha-5°$}		\multicolumn{2}{c}{$\alpha+5°$}		results	
Mass	$m-$ 0.01	$m+$ 0.01	$m-$ 0.01	$m+$ 0.01	$m-$ 0.01	$m+$ 0.01	$m-$ 0.01	$m+$ 0.01	$\bar{y}$	σ
Conventional	100.5	82.3	164.7	134.8	150.0	122.8	246.0	201.3	150	53.56
Parameter										

Figure 7:1.4 Distance travelled in a noise environment (conventional design).

Using the calculated values of the distance y, we determine the average distance ($\bar{y}$) and the standard deviation (σ). We also calculate the SN ratio (η) using the formula for nominal-the-best:

$$\eta = 10 \log_{10} \left(\frac{\bar{y}^2}{\sigma^2} \right) \tag{7:1.3}$$

Substituting, $\bar{y} = 150$ and $\sigma = 53.56$, $\eta = 8.96$ dB. At this stage $\bar{y}$, σ and η do not mean very much until we perform the parameter design, whereupon we shall compare similar results obtained by the parameter design.

7.1.4 Computer-aided parameter design

The parameter design method focuses on the formula:

$$y = \frac{F^2 \times \sin 2\alpha}{m^2 \times g}$$

Although this is a relatively simple formula[1] it is not possible to establish which of the factors F, α and m has the greatest influence on the distance travelled for the given variation. However, we can investigate this by using a parameter design experiment arranged as a direct product design, as shown in Figure 7:1.5. In this direct product design, there are two $L_8(2^7)$ orthogonal arrays; one on the left and one on the top. The array on the left is the *control factor array* or *inner array*. The array on the top is the *noise factor array* or *outer array*.

7.1.5 Creating control and noise factor combinations

In this experiment, essentially, we create a set of values for the control factors we intend to study. This is shown in Figure 7:1.6. Note that for convenience we have renamed F, α and m into factors A, B and C for consistency of use. The control factor levels are set as shown in Figure 7:1.6.

Having set the factors this way, we can substitute the factor values for level 1 and level 2 in the control factor array and determine the mean distance travelled. In reality, however, when we set factors A, B and C, there will always be some variation due to noise. To introduce noise about these factor levels, we create noise about the control factor levels as shown in Figure 7:1.7.

---

1   A reader with a calculus background may guess that the best value of α is 45°. However, it would still be difficult to guess the effects of F and m. We proceed with a formal procedure.

								G	1	2	2	1	2	1	1	2			
								F	1	2	2	1	1	2	2	1			
								E	1	2	1	2	2	1	2	1			
								D	1	2	1	2	1	2	1	2			
								C	1	1	2	2	2	2	1	1			
								B	1	1	2	2	1	1	2	2			
								A	1	1	1	1	2	2	2	2	$\bar{y}$	σ	η
CF	A	B	C	D	E	F	G		1	2	3	4	5	6	7	8			
1	1	1	1	1	1	1	1	1											
2	1	1	1	2	2	2	2	2											
3	1	2	2	1	1	2	2	3											
4	1	2	2	2	2	1	1	4											
5	2	1	2	1	2	1	2	5											
6	2	1	2	2	1	2	1	6											
7	2	2	1	1	2	2	1	7											
8	2	2	1	2	1	1	2	8											

Figure 7:1.5 The $L_8(2^7) \times L_8(2^7)$ direct product design.

Factor		Level 1	nominal	Level 2
A	Force, F	5.0	10 ± 5 N	15.0
B	Angle, α	10.0	$25 \pm 15°$	40
C	Mass, m	0.19	0.20 ± 0.01	0.21

Figure 7:1.6 Control factors for experiment.

Factor		Level 1	nominal value of control factor	Level 2
A	Force, F	$-10\ \%$	-	$+10\ \%$
B	Angle, α	$-5°$	-	$+5°$
C	Mass, m	$-10\ \%$	-	$+10\ \%$

Figure 7:1.7 Introducing noise about nominal values of the control factors.

Since we have a 2-level control factor and a 2-level noise factor there are four combinations of factor level values, as shown in Figure 7:1.8. In general, the combinations (1,1), (1,2), (2,1) and (2,2) may be denoted $\Psi_{1,1}$, $\Psi_{1,2}$, $\Psi_{2,1}$ and $\Psi_{2,2}$ respectively. Similarly, there will be four combinations for factors B and C. For convenience these are tabulated in Figure 7:1.9 where Ψ simply represents factors A, B or C.

Factor A		Noise factor levels (N)	
		Level 1	Level 2
Control factor levels	Level 1	$A_{1,1} = 4.5$	$A_{1,2} = 5.5$
	Level 2	$A_{2,1} = 13.5$	$A_{2,2} = 16.5$

Figure 7:1.8 Two-level control factor and 2-level noise factor combinations.

Factor	$\Psi_{1,1}$	$\Psi_{1,2}$	$\Psi_{2,1}$	$\Psi_{2,2}$
A (N)	4.5	5.5	13.5	16.5
B (°)	5	15	35	45
C (kg)	0.17	0.21	0.19	0.23

Figure 7:1.9 Control and noise factor combinations for the experiment.

Next we assign the control factors to the direct product design. Since we have seven degrees of freedom and only three 2-level control factors we can choose to assign factor A to column 1 and factor B to column 2. In that case, column 3 represents the interaction of columns 1 and 2. Therefore column 3 is left unassigned. Factor C is then assigned to column 4.

At this stage, namely the parameter design stage, it is important to identify control factors that are robust to noise. Therefore, we should not be unduly concerned about interactions. If there are many control factors to be studied, it is a good practice to use orthogonal arrays in which interaction effects are more or less uniformly distributed. Thus, we may choose to use an $L_{36}(2^{11} \times 3^{12})$ orthogonal array. In such a case there will be 36 experiments for the control factor array and 36 experiments for the noise factor array. Additionally, two 3-level factors will have nine combinations of factor levels. Thus the computational effort will be formidable. For this reason, the $L_8(2^7)$ orthogonal array will suffice for our present 'introductory' experiment. The $L_8(2^7)$ orthogonal array with the appropriately assigned columns is shown Figure 7:1.10.

									1	2	3	4	5	6	7	8			
								e	1	2	2	1	2	1	1	2			
								e	1	2	2	1	1	2	2	1			
								e	1	2	1	2	2	1	2	1			
								C	1	2	1	2	1	2	1	2			
								e	1	1	2	2	2	2	1	1			
								B	1	1	2	2	1	1	2	2			
								A	1	1	1	1	2	2	2	2	$\bar{y}$	σ	η
CF	A	B	e	C	e	e	e		1	2	3	4	5	6	7	8			
1	1	1	1	1	1	1	1	1											
2	1	1	1	2	2	2	2	2											
3	1	2	2	1	1	2	2	3											
4	1	2	2	2	2	1	1	4											
5	2	1	2	1	2	1	2	5											
6	2	1	2	2	1	2	1	6											
7	2	2	1	1	2	2	1	7											
8	2	2	1	2	1	1	2	8											

Figure 7:1.10   The $L_8(2^7) \times L_8(2^7)$ direct product design with factor assignment.

Having set the control factor and noise factor combinations, we can proceed to calculate the trajectory distance $y_{i,j}$ under the control factor level i and noise factor level j condition. We start with control factor array experiment $i = 1$ and noise factor array experiment $j = 1$. Reiterating, we note that for:
Experiment number $y_{1,1}$:

factor A has control factor level 1, noise factor level 1 → $A_{1,1}$
factor B has control factor level 1, noise factor level 1 → $B_{1,1}$
factor C has control factor level 1, noise factor level 1 → $C_{1,1}$
as shown in Figure 7:1.11.

Experiment number $y_{1,1}$					
Factor	Control factor level	Noise factor level	Level combination	Factor value	Calculated distance of y
A	1	1	$A_{1,1}$	4.5	
B	1	1	$B_{1,1}$	5.0	12.3
C	1	1	$C_{1,1}$	0.17	

Figure 7:1.11   Selection of factor values.

Indeed, we may denote the distance travelled y as a function of A, B and C:

$$y_{1,1} = f \{A_{1,1}, B_{1,1}, C_{1,1}\}$$

$$= f \{4.5, 5.0, 0.17\}$$

Substituting $F = A_{1,1} = 4.5$, $\alpha = B_{1,1} = 5.0$, $m = C_{1,1} = 0.17$ into:

$$y = \frac{F^2 \times \sin 2\alpha}{m^2 \times g}$$

$$= \frac{4.5^2 \times \sin (2 \times 5)}{0.17^2 \times 9.81}$$

$$= 12.3 \text{ m}$$

we obtain $y_{1,1} = 12.3$ m. Using a similar procedure, we note that:
$y_{1,2}$ $= f \{A_{1,1}, B_{1,1}, C_{1,2}\} = f \{4.5, 5.0, 0.21\} = 8.2$ m and so on until,
$y_{8,8}$ $= f \{A_{2,2}, B_{2,2}, C_{2,2}\} = f \{16.5, 45, 0.23\} = 520.1$ m.

    Thus there are a total of 64 calculations. These calculations are best done using a programmable calculator or a personal computer. For verification purposes, the results of these calculations are shown in Figure 7:1.12 together with three additional columns for the average ($\bar{y}$), standard deviation (σ) and the signal-to-noise ratio (η) for each row of data. The signal-to-noise ratio is calculated by the formula:

$$\eta = 10 \log_{10} \left(\frac{\bar{y}^2}{\sigma^2} \right)$$

representing a nominal-the-best type problem. This is because we have assumed a target of 150 m and are primarily interested in reducing variation about the mean. Using the results for $\bar{y}$ and η, we tabulate the response tables as shown in Figure 7:1.13.

		\multicolumn{8}{c}{Noise array}		\multicolumn{3}{c}{Results}
		1	2	3	4	5	6	7	8	$\bar{y}$	σ	η
C o n t r o l a r r a y	1	12.3	8.2	35.3	23.6	18.3	12.3	52.7	35.3	24.7	15.3	4.2
	2	10.0	6.7	28.9	19.3	15.0	10.0	43.2	28.9	20.3	12.5	4.2
	3	66.3	44.4	70.6	47.3	99.1	66.3	105.5	70.6	71.3	21.7	10.3
	4	54.3	36.3	57.8	38.7	81.1	54.3	86.3	57.8	58.3	17.7	10.3
	5	110.3	73.8	317.6	212.6	164.8	110.3	474.5	317.6	222.7	137.5	4.2
	6	90.3	60.4	260.0	174.0	134.9	90.3	388.4	260.0	182.3	112.5	4.2
	7	597.0	399.6	635.3	425.3	891.8	597.0	949.1	635.3	641.3	195.1	10.3
	8	488.7	327.1	520.1	348.2	730.0	488.7	776.9	520.1	525.0	159.7	10.3

Figure 7:1.12 Results of computer-aided parameter design.

Mean: $\bar{y}$	A	B	C
Level 1	43.65	112.50	240.00
Level 2	392.81	323.96	196.46
SN ratio: η	A	B	C
Level 1	7.26	4.19	7.26
Level 2	7.26	10.33	7.26

Figure 7:1.13 Response table of factor effects for $\bar{y}$ and η.

From the response tables we draw the response graphs for $\bar{y}$ and η. The graph of factor averages is shown in Figure 7:1.14. Graphically, we note that factors A and B both affect the average. Of course the student who wishes to perform an analysis of variance is urged to so.

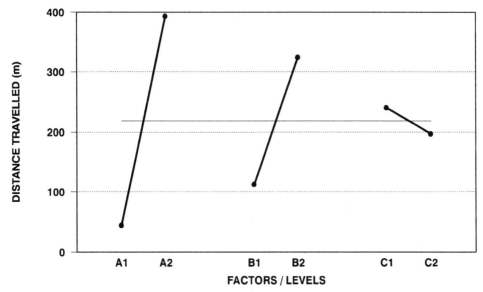

Figure 7:1.14   Response graph of averages.

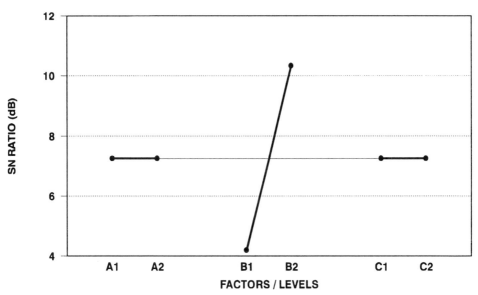

Figure 7:1.15   Response graph of SN ratios.

The response graph for SN ratio is shown in Figure 7:1.15. This graph shows a dramatic result. That is, only factor B has a significant effect on the signal-to-noise ratio. Since a high SN ratio implies a low variation, it is important to choose factor B level 2 to reduce sensitivity to noise.

Hence, we choose control factor level B2 as our value for the angle of ejection. This corresponds to $\alpha = 40°$. We then set the nominal value of $m = 0.20$ kg and calculate the force F required to attain the target distance $y = 150$ m using the formula:

$$F = + \sqrt{\frac{y^2 \times m^2 \times g}{\sin 2\alpha}}$$

The result of the substitution is $F = 7.73$ N. Thus, the parameter design would suggest that for a target distance $y = 150$ m and a cannon ball of mass $m = 0.20$ kg ejected at an angle of $\alpha = 40°$ we require a force $F = 7.73$ N.

7.1.6 Significance of result

For ease of comparison we tabulate the factor levels suggested by the conventional and parameter design experiments as follows:

Design method	Force (N)	Angle (°)	Mass (kg)
Conventional	10.00	18.0	0.20
Parameter	7.73	40.0	0.20

Figure 7:1.16   Comparison of factor settings.

Let us now compare the effect of noise on the conventional and the parameter designs. Figure 7:1.4 is repeated here (Figure 7:1.17) with the results of noise effects already calculated for the conventional design. We now calculate the trajectory distances for the parameter design under the same noise conditions using nominal values of $F = 7.73$ N, $\alpha = 40°$ and $m = 0.20$ kg. The results of the average and standard deviation of this noise effect are also tabulated in Figure 7:1.17.

Force	0.9 × F				1.1 × F				Summary	
Angle	$\alpha-5°$		$\alpha+5°$		$\alpha-5°$		$\alpha+5°$		results	
Mass	$m-$ 0.01	$m+$ 0.01	$m-$ 0.01	$m+$ 0.01	$m-$ 0.01	$m+$ 0.01	$m-$ 0.01	$m+$ 0.01	$\bar{y}$	σ
Conventional	100.5	82.3	164.7	134.8	150.0	122.8	246.0	201.3	150	53.56
Parameter	128.5	105.2	136.7	111.9	191.8	157	204.2	167.1	150	36.14

Figure 7:1.17   Comparison of conventional and parameter designs.

Figure 7:1.17 shows that while both methods achieve the same target distance, $\eta_{Conventional}$ = 8.95 db (calculated earlier) and $\eta_{Parameter}$ = 12.38 corresponding to a gain of 3.42 dB. Note that the standard deviation for the parameter design is 33 % less than the conventional design. It is also important to note that this reduction in variation is achieved without any additional new material, process or specification.

7.1.7 Normal curve representation
The results of Figure 7:1.17 are then used to draw the respective normal distributions using the probability distribution for the normal curve:

$$\Phi = \frac{1}{\sqrt{(2 \times \pi \times \sigma^2)}} e^{-\frac{1}{2}\left(\frac{y-\bar{y}}{\sigma}\right)^2} \qquad (7:1.4)$$

where Φ is the probability, σ is the standard deviation, π = 3.142 and e =2.718. For the conventional design, $\bar{y}$ = 150 and σ = 53.56. Substituting into Equation (7:1.4), the normal curve for the conventional design is:

$$\Phi = \frac{1}{134.26} 2.718^{-\frac{1}{2}\left(\frac{y-150}{53.56}\right)^2}$$

We may thus draw the curve for y over the range of 0 to 300 m. Similarly, the normal curve for the parameter design is:

$$\Phi = \frac{1}{90.58} 2.718^{-\frac{1}{2}\left(\frac{y-150}{36.14}\right)^2}$$

Both of these curves are shown in Figure 7:1.18. The importance of the parameter design is now clearly demonstrated.

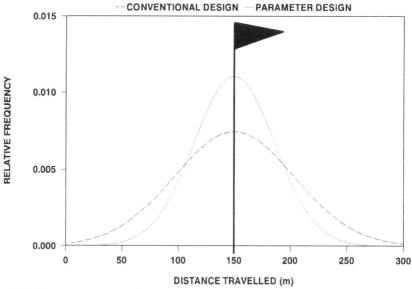

Figure 7:1.18   Result of comparison of conventional and parameter design.

7.1.8 Discussion

The cannon ball experiment clearly demonstrates the importance of parameter design. However, we only had three 2-level factors in the experiment. In most experiments, there may be several factors, many of which are not well understood. In such cases, we should use 3-level factors for the parameter design. The $L_{36}(2^3 \times 3^{13})$ is ideal for such experimentation as it can accommodate thirteen 3-level factors with distributed interaction. In particular, 3-level factors are most suited for parameter design experimentation since we can investigate a current or nominal value against a slightly smaller value as well as a slightly larger value and so simulate the effect of noise. Using a direct product design with an $L_{36}(2^3 \times 3^{13})$ control factor array and an $L_{36}(2^3 \times 3^{13})$ for the noise factor array, there are a total of 1296 experiments (36 × 36) each with 9 combinations of factor levels. Although the computational effort may be formidable it can be programmed relatively easily into a computer. For the next example in the Self-assessment question, however, we merely illustrate the principle of computer-aided parameter design by using an $L_8(2^7) \times L_8(2^7)$ direct product design for both the control factor and noise factor orthogonal arrays. Doing so, however, limits the scope of experiments since it is not possible to study both sides of a nominal value. Therefore, the author has contrived the Wheatstone bridge example in the Self-assessment question to illustrate the principle of computer-aided parameter design. Having understood the principle, the interested reader is greatly encouraged to extend the principle to larger 3-level series orthogonal arrays particularly the $L_{36}(2^3 \times 3^{13})$.

7.1.9 Self-assessment questions

1.   The problem with the Wheatstone bridge is how to set the nominal values of the resistors A, B, C, D and F so that the unknown resistance *y* can be measured accurately (See Figure 7:1.19). The resistances A, C, D and F are controllable. The resistance of B is not controllable since it needs to be adjusted during the experiment. Since we also want to make the Wheatstone bridge robust to voltage changes, we include the factor E, the electro-motive force of the cell and factor G, the residual current. The equation for the resistance *y* is then given by the equation:

$$y = \frac{BD}{C} - \frac{G\,(AD + AC + BD + CD)\,(BC + BD + BF + CF)}{C^2\,E}$$

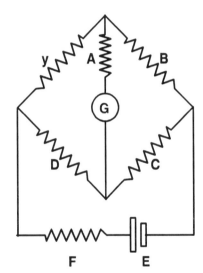

Figure 7:1.19   Wheatstone bridge arrangement.

Using the existing and suggested parameter values below, conduct a parameter design to establish the optimum nominal values for the Wheatstone bridge using tolerances of 0.3 % for factors A, B, C, D and F, 5 % for E and ± 0.0002 for G.

	A	B	C	D	E	F	G
Existing	100	2	10	10	6	10	0
Suggested	20	2	50	2	30	2	0

2.   What are the limitations in using 2-level control factors and 2-level noise factors in the direct product parameter design?

7.1.10   Answers to self-assessment questions

1.   The problem with the Wheatstone bridge is how to set the nominal
values of the resistors A, B, C, D and F so that the unknown resistance
y can be measured accurately (See Figure 7:1.20). The resistances A, C,
D and F are controllable. The resistance of B is not controllable since it
needs to be adjusted during the experiment. Since we also want to make
the Wheatstone bridge robust to voltage changes, we include the factor
E, the electro-motive force of the cell and factor G, the residual current.
The equation for the resistance y is then given by the equation:

$$y = \frac{BD}{C} - \frac{G\,(AD + AC + BD + CD)\,(BC + BD + BF + CF)}{C^2\,E}$$

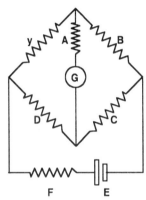

Figure 7:1.20   Wheatstone bridge arrangement.

Using the existing and suggested parameter values below, conduct a
parameter design to establish the optimum nominal values for the
Wheatstone bridge using tolerances of 0.3 % for factors A, B, C, D and
F, 5 % for E and ± 0.0002 for G.

	A	B	C	D	E	F	G
Existing	100	2	10	10	6	10	0
Suggested	20	2	50	2	30	2	0

Answers

The answer is given in a systematic way as follows:

- tabulate the control factor nominal values,
- create the noise factor levels,
- tabulate control and noise factor combinations,
- calculate the data array,
- results of the simulation,
- draw the response table,
- conduct an analysis of variance,
- confidence intervals for factors,
- select recommended factor levels,
- predict the SN ratio at the optimum condition,
- confirmation experiment,
- comparison of improvement.

1.   Tabulate the control factor nominal values

The initial nominal values for the existing and suggested levels are tabulated as follows:

	A	B	C	D	E	F	G
Existing	100	2	10	10	6	10	0
Suggested	20	2	50	2	30	2	0

Figure 7:1.21   Control factor nominal values.

2.   Create the noise factor levels

We create the noise factor levels by allowing some tolerance about each control factor level. Let us use the tolerances shown in Figure 7:1.22.

3.   Tabulate control and noise factor combinations

Since there are seven 2-level factors we introduce a table of combinations where Ψ1 and Ψ2 represent control factor levels while Ψ'1 and Ψ'2 represents noise factor levels 1 and 2, as shown in Figure 7:1.23

4.   Calculate the data array

We then calculate y_{ij} where i is the ith experiment in the control factor array and j is the jth experiment in the noise factor array, where

Factor	Level 1	Nominal value of control factor	Level 2
A	−0.3 %	-	+0.3 %
B	−0.3 %	-	+0.3 %
C	−0.3 %	-	+0.3 %
D	−0.3 %	-	+0.3 %
E	−5.0 %	-	+5.0 %
F	−0.3 %	-	+0.3 %
G	−0.0002	-	+0.0002

Figure 7:1.22   Introducing noise about nominal values of the control factors.

	A 1	A2		B1	B2		C1	C2
A′1	99.7	19.9	B′1	1.99	1.99	C′1	9.97	49.85
A′2	100.3	20.1	B′2	2.01	2.01	C′2	10.03	50.15
	D1	D2		E1	E2		F1	F2
D′1	9.97	1.99	E′1	5.7	28.50	F′1	9.97	1.99
D′2	10.03	2.01	E′2	6.3	31.50	F′2	10.03	2.01
	G1	G2						
G′1	−0.002	−0.002						
G′2	+0.002	+0.002						

Figure 7:1.23   Control-noise combinations for simulation experiment.

i $= \{1, 2, 3, 4, 5, 6, 7, 8\}$ and $j = \{1, 2, 3, 4, 5, 6, 7, 8\}$ and
y_{ij} $= f \{A, B, C, D, E, F, G\}$ according to the factor level combination for experiment y_{ij} as shown earlier in Figure 7:1.10.

For example,
$y_{1,1}$ $= f \{A_{1,1}, B_{1,1}, C_{1,1}, D_{1,1}, E_{1,1}, F_{1,1}, G_{1,1}\}$
      $= f \{99.7, 1.99, 9.97, 9.97, 5.70, 9.97, -0.0002\}$

Substituting the respective values of $A_{1,1}$, $B_{1,1}$, $C_{1,1}$, $D_{1,1}$, $E_{1,1}$, $F_{1,1}$ and $G_{1,1}$ into

$$y = \frac{BD}{C} - \frac{G\,(AD + AC + BD + CD)\,(BC + BD + BF + CF)}{C^2\,E}$$

we obtain $y_{1,1} = 2.1123$ ohms. Similarly,

$$
\begin{aligned}
y_{1,2} &= f\{A_{1,1}, B_{1,1}, C_{1,1}, D_{1,2}, E_{1,2}, F_{1,2}, G_{1,2}\} \\
&= f\{99.7, 1.99, 9.97, 10.03, 6.30, 10.03, 0.0002\}
\end{aligned}
$$

for which we obtain $y_{1,2} = 1.8981$ ohms.

5.   Results of the simulation

The results shown in Figure 7:1.24 were obtained using a suitable computer program to simulate the experiment. Although the results shown are rounded to two decimal places it is necessary to carry as many decimals as possible for the actual calculation. The SN ratio is calculated using:

$$\eta = 10 \log_{10}\left(\frac{\bar{y}^2}{\sigma^2}\right)$$

Exp	1	2	3	4	5	6	7	8	$\bar{y}$	η
1	2.11	1.90	1.88	2.11	1.87	2.11	2.11	1.90	2.00	24.33
2	0.40	0.40	0.39	0.40	0.39	0.40	0.40	0.40	0.40	38.61
3	0.42	0.38	0.38	0.42	0.38	0.42	0.42	0.38	0.40	25.62
4	0.09	0.07	0.07	0.09	0.07	0.09	0.09	0.07	0.08	18.54
5	0.40	0,40	0.40	0.40	0.39	0.40	0.40	0.40	0.40	40.45
6	0.08	0.08	0.08	0.08	0.08	0.08	0.08	0.08	0.08	27.38
7	2.00	2.00	1.99	2.01	1.98	2.00	2.01	2.02	2.00	44.97
8	0.41	0.39	0.39	0.41	0.38	0.41	0.41	0.39	0.40	29.30

Figure 7:1.24   Results of the computer simulation.

6.   Draw the response table

From Figure 7:1.25, the factor levels with the highest SN ratios are E2, A2, C1 and F2. Of course, factor B is an adjustable resistor and cannot be set. Similarly, G is the residual current and cannot be set.

	A	B	C	D	E	F	G
Level 1	26.78	32.70	34.30	33.84	26.66	28.16	28.81
Level 2	35.53	29.61	28.00	28.46	35.64	34.15	33.50

Figure 7:1.25   Response table of factor effects.

7.   Conduct an analysis of variance

The calculations for analysis of variance are conducted as described previously. Only the results of the analysis are given in Figure 7:1.26.

Source	Pool	Sq	ν	Mq	Sq'	rho %
A		153.16	1	153.16	112.80	19.21
B	Y	19.07	1	19.07	-	-
C		79.55	1	79.55	39.19	6.67
D	Y	58.00	1	58.00	-	-
E		161.48	1	161.48	121.11	20.63
F		71.80	1	71.80	31.43	5.35
G	Y	44.02	1	44.02	-	-
Pool		121.09	3	40.36	282.54	48.14
St		587.08	7	-	587.08	100.00
Mean		7763.25	1	-	-	-
ST		8350.32	8	-	-	-

Figure 7:1.26   Analysis of variance for the Wheatstone bridge experiment.

8.   Confidence interval for factors

The confidence interval for the factors is calculated from the total number of SN ratios for the whole experiment. Since there are only a total of eight experiments with four experiments at each factor level, $n = 4$:

$$CI = \sqrt{F_{0.05,1,3} \times V_e \times \left[\frac{1}{n}\right]}$$

$$= \sqrt{10.13 \times 40.36 \times \left[\frac{1}{4}\right]}$$

$$= \pm\ 10.11\ \text{dB}$$

9.   Select recommended factor levels

In computer-aided parameter design with the signal-to-noise ratio as the objective characteristic, it is important to see how the response varies with different levels of the controllable factors. Determining the significance of the control factor would be unimportant. The combination that gives the best signal-to-noise ratio should be selected *even when the difference is only slight.* In this experiment, therefore, we would select factor levels A2, C1, D1, E2 and F2. Of course, the factors B and G are not selected.

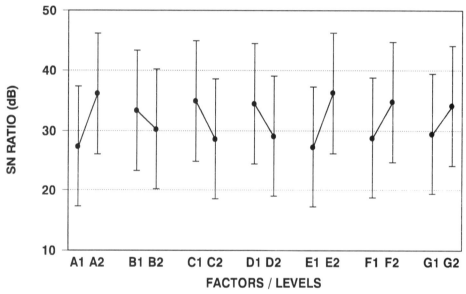

Figure 7:1.27   Response graph of factor effects.

10.   Predict the SN ratio at the optimum condition

$$\eta_{Predicted} = \bar{\eta} + (\overline{E2} - \bar{\eta}) + (\overline{A2} - \bar{\eta}) + (\overline{C1} - \bar{\eta}) + (\overline{F2} - \bar{\eta})$$

$$= \overline{E2} + \overline{A2} + \overline{C1} + \overline{F2} - 3 \times \bar{\eta}$$

$$= 35.64 + 35.53 + 34.30 + 34.15 - 3 \times 31.15$$

$$= 46.17 \text{ dB}$$

with confidence interval:

$$CI_{Predicted} = \sqrt{F_{0.05,1,3} \times V_e \times \left[\frac{1}{n_{eff}}\right]}$$

$$= \sqrt{10.13 \times 40.36 \times \left[\frac{1}{1.60}\right]}$$

$$= \pm \ 15.99 \text{ dB}$$

where:

$$n_{eff} = \frac{total \ number \ of \ experiments}{sum \ of \ degrees \ of \ freedom \ used \ in \ estimate \ of \ mean}$$

$$= \frac{8 \times 1}{v_\mu + v_E + v_A + v_C + v_F}$$

$$= \frac{8}{1 + 1 + 1 + 1 + 1}$$

$$= 1.60$$

The confidence interval for the confirmation experiment is calculated as follows with $r = 1$ as we can only simulate one set of results.

$$CI_{Confirmation} = \sqrt{F_{0.05,1,3} \times V_e \times \left[\frac{1}{n_{eff}} + \frac{1}{r}\right]}$$

$$= \sqrt{10.13 \times 40.36 \times \left[\frac{1}{1.60} + \frac{1}{1}\right]}$$

$$= \pm\ 25.78\ dB$$

10. Confirmation experiment

We perform a confirmation experiment by using the optimum control factor levels with the appropriate noise factor levels. The results of this confirmation simulation are:

Confirmation Experiment									
1	2	3	4	5	6	7	8	$\bar{y}$	η
2.00	2.00	1.99	2.01	1.98	2.00	2.01	2.02	2.00	44.97

with an SN ratio of 44.97 dB, which is well within the predicted confidence interval.

11. Comparison of improvement

The gain in signal-to-noise ratio is:

$$\eta_{Actual\ Gain} = \eta_{Optimum} - \eta_{Existing}$$

$$= 44.97 - 24.33$$

$$= 20.64\ dB$$

The SN ratio for the existing condition is taken from the control factor array experiment 1. Alternatively, the gain for this experiment can also be calculated by taking the difference between the optimum and the existing levels. Since the optimum condition is A2, C1, D1, E2 and F2, and the existing condition is A1, C1, D1, E1 and F1 we calculate the gain as follows:

$$\eta_{\text{Predicted Gain}} = (\overline{A2} - \overline{A1}) + (\overline{C1} - \overline{C1}) + (\overline{D1} - \overline{D1})$$

$$+ (\overline{E2} - \overline{E1}) + (\overline{F2} - \overline{F1})$$

$$= 8.75 + 0.00 + 0.00 + 8.99 + 5.99$$

$$= 23.73 \text{ dB}$$

Note that the *Actual Gain* is approximately the same as the *Predicted Gain*. A comparison of improvement is shown in Figure 7:1.28 for a resistance of 2.00 ohms across a range of noise conditions.

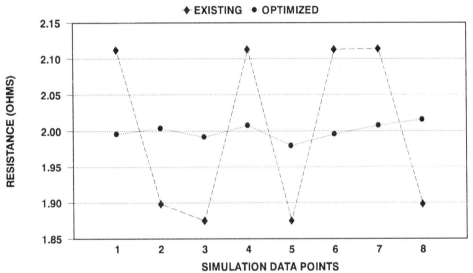

Figure 7:1.28   Comparison of improvement.

2.   What are the limitations in using 2-level control factors and 2-level noise factors in the direct product parameter design?

Answer

The main limitation in using 2-level control factors is that we can only study linear effects of the control factors. If we use 3-level control factors, we can begin with the mid-value as the starting point and investigate both sides (plus and minus) of the factor. The limitation in using 2-level noise factors reduces the number of data points available for each experiment. At the parameter design stage, it is important to investigate wide ranges of control factors over a full spectrum of noise factors.

CHAPTER 8

COMPUTER-AIDED TOLERANCE DESIGN

AIMS:
To provide a method of identifying the components of a system which contribute significantly to the functional variation in the system.

OBJECTIVES:
When you have completed studying this chapter you should be able to:
- identify which component(s) in a system affects the functional variation,
- perform an analysis of variance on the results of a tolerance design,
- perform a cost analysis on upgrading components.

OVERVIEW:
This chapter introduces a method of trade-off between reductions in quality loss due to performance variation and increases in manufacturing costs. Tolerance design identifies components (or subsystems) of a system that contribute significantly to the functional variation of the system. Only such significant components need to be replaced using higher grade components. Of course, the unselective replacement of components using higher grade components will lead to unnecessary manufacturing costs. Where the functional characteristic can be adequately modelled by a mathematical equation, the tolerance design can be fully performed by the computer-aided tolerance design.

8.1   Computer-Aided Tolerance Design

8.1.1 Introduction

Tolerance design should only be performed after the sensitivity to noise has been minimized by using parameter design. Many Western companies rely heavily on tolerance design to improve quality. When that fails, they revert to system design. Tolerance design itself makes products more expensive to manufacture since it requires higher grade materials and components with tighter tolerances. System design requires breakthrough technologies which are difficult to develop. Therefore, it is imperative to perform parameter design to make the process or product insensitive to noise as far as possible. When quality loss levels are still not satisfactory, only then should tolerance design be conducted.

8.1.2 Tolerance design

System design provides the prototype for a product or process. Parameter design is performed to identify nominal values of a process or product that maximize the SN ratio. If further quality improvement is necessary, we may address the tolerances of components by tolerance design so that only selected components are upgraded. Since tolerance design requires components to be upgraded, the unit cost of the product will increase.

8.1.3 Tolerance design of an electronic circuit

Consider an electronic circuit that controls a water heater as shown in Figure 8:1.1. The function of the electronic circuit is to maintain the temperature of the water at a set value, i.e. target value. A temperature sensor senses the water temperature and the temperature control circuit compares the water temperature against a nominal set temperature value and determines whether the heater should be turned ON or OFF. The temperature is sensed by a thermistor. Let us denote the value of the thermistor resistance of the set temperature as R_{TH-SET}. When the water temperature drops, the thermistor resistance increases to a value R_{TH-ON} at which the heater turns ON. When the water heats up, the thermistor resistance decreases to a value R_{TH-OFF} at which the heater turns OFF. The difference between R_{TH-ON} and R_{TH-OFF} from R_{TH-SET} is called *hysteresis*. Hysteresis allows the heater to remain ON for a period necessary to raise the water temperature above the set temperature, whereupon the heater turns OFF for a period necessary to lower the water temperature below the set temperature. To illustrate the principle of tolerance design, as well as for simplicity of calculations, we shall assume that R_{TH-SET} is midway between R_{TH-ON} and R_{TH-OFF}, as shown in Figure 8:1.2.

Figure 8:1.2 also shows that there will be some variation in R_{TH-ON}, which turns the heater ON. Similarly, there will also be some variation in R_{TH-OFF}, which turns the heater OFF. For our present purposes we shall assume that improving the performance of R_{TH-ON} will also improve the performance of R_{TH-OFF}. In this case, therefore, considering only the case of $R_{TH\text{-}ON}$, the objective of tolerance design is to reduce the variation in $R_{TH\text{-}ON}$.

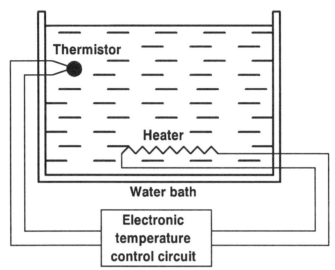

Figure 8:1.1 Schematic diagram of the electronic temperature control circuit.

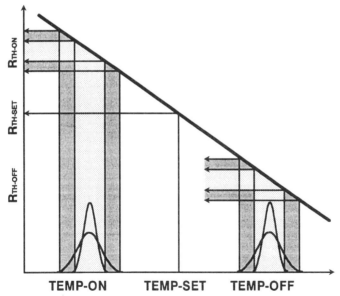

Figure 8:1.2 Functional variation in temperature control.

8.1.4 Factors affecting the functional variation

For the electronic temperature control circuit, R_{TH-ON} is calculated by the equation:

$$R_{TH-ON} = \frac{\left(\dfrac{R_3\,R_5}{R_3 + R_5}\right) R_2\,(E_Z\,R_4 + E_0\,R_1)}{R_1\,(E_Z\,R_2 + E_Z\,R_4 - E_0\,R_2)} \qquad (8:1.1)$$

where R_1, R_2, R_3, R_4 and R_5 are resistors, E_0 is the circuit voltage and E_Z is the Zener voltage on the thermistor, R_{TH}. The circuit diagram is shown in Figure 8:1.3. The component specifications are given in Figure 8:1.4. Notice that the lower and upper specifications are regarded as the factor levels for experimentation. In order to keep the calculations manageable, we shall use only 2-level factors. For ease of reference, the components or voltages have been coded as factors A to G. Since we have seven 2-levels factors we require seven degrees of freedom. An $L_8(2^7)$ orthogonal array (Figure 8:1.5) may be sufficient.

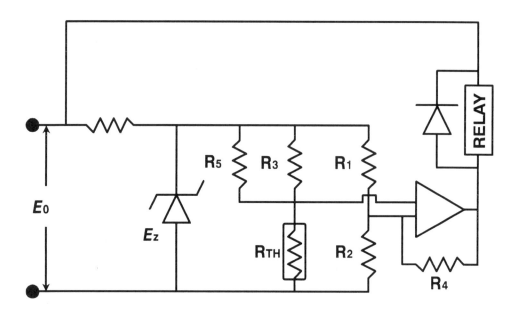

Figure 8:1.3 Electronic control circuit.

Component	Specification	Factor
R_1	3.9 kΩ $\pm$ 2 %	A
R_2	7.5 kΩ $\pm$ 2 %	B
R_3	1.0 kΩ $\pm$ 2 %	C
R_4	360.0 kΩ $\pm$ 2 %	D
R_5	3.3 kΩ $\pm$ 2 %	E
E_Z	5.3 $\pm$ 0.3 V	F
E_0	10.1 $\pm$ 0.3 V	G

Figure 8:1.4 Specifications for component tolerances.

Exp	A	B	C	D	E	F	G
1	1	1	1	1	1	1	1
2	1	1	1	2	2	2	2
3	1	2	2	1	1	2	2
4	1	2	2	2	2	1	1
5	2	1	2	1	2	1	2
6	2	1	2	2	1	2	1
7	2	2	1	1	2	2	1
8	2	2	1	2	1	1	2

Figure 8:1.5 The $L_8(2^7)$ orthogonal array.

8.1.5 Assignment of factor levels
There are many ways in which we can assign the values for factor levels in a tolerance design. We shall discuss two ways:
- percentage method,
- standard deviation method.

1.   Percentage method

The percentage method uses convenient tolerances such as ± 5 %, ± 2 % or ± 1 %. (See Figure 8:1.6). This method is simple and easy to use. Often tolerance levels such as 5 % or 1 % can be estimated from engineering experience.

	2-level factor		3-level factor
Level 1	X − 2 % = 0.98 X	Level 1	X − 2 % = 0.98 X
No level (nominal)	X	Level 2 (nominal)	X
Level 2	X + 2 % = 1.02 X	Level 3	X + 2 % = 1.02 X

Figure 8:1.6 Range of factor levels (Percentage).

2.   Standard deviation method

The standard deviation method uses the standard deviation of a characteristic as shown in Figure 8:1.7. Here, σ may be regarded as the standard deviation of the characteristic X due to deterioration and degradation over the expected life of the product. Again, σ can only be obtained from engineering experience or previous data.

	2-level factor		3-level factor
Level 1	X − σ	Level 1	X − $\sqrt{1.5} \times \sigma$
No level (nominal)	X	Level 2 (nominal)	X
Level 2	X + σ	Level 3	X + $\sqrt{1.5} \times \sigma$

Figure 8:1.7 Range of factor levels (Standard deviation).

For our experiment we shall be content to use a 2-level factors with a 2 % tolerance since we are merely interested in illustrating the principle involved in computer-aided tolerance design. Having learned the principle, the interested reader may use 3-level factors in an $L_{36}(2^3 \times 3^{13})$ orthogonal array. Thus, we use (X − 2 %) for level 1 and (X + 2 %) for level 2. This can also be regarded as 0.98 X for level 1 and 1.02 X for level 2. The factor levels are shown in Figure 8:1.8.

Factor	Component	Specification	Factor level 1	Factor level 2
			Lower specification	Upper specification
A	R_1	3.9 kΩ $\pm$ 2 %	3.82	3.98
B	R_2	7.5 kΩ $\pm$ 2 %	7.35	7.65
C	R_3	1.0 kΩ $\pm$ 2 %	0.98	1.02
D	R_4	360.0 kΩ $\pm$ 2 %	352.80	367.20
E	R_5	3.3 kΩ $\pm$ 2 %	3.23	3.37
F	E_Z	5.3 $\pm$ 0.3 V	5.0	5.6
G	E_0	10.1 $\pm$ 0.3 V	9.8	10.4

Figure 8:1.8 Specifications for component tolerances.

8.1.6 Conducting the experiment

The experiment may be conducted by making eight control circuits with the appropriate components according to the $L_8(2^7)$ orthogonal array. In experiment 1, for example, we would use (set or allow) A1 = 3.82 kΩ, B1 = 7.35 kΩ, C1 = 0.98 kΩ, D1 = 352.80 kΩ, E1 = 3.23 kΩ, F1 = 5.0 V and G1 = 9.8 V. We would then measure the thermistor resistance R_{TH1} for this combination. Similarly, we could measure the thermistor resistances for experiments 2 (R_{TH2}) through 8 (R_{TH8}). Doing so would be impractical and time consuming.

However, since the quality characteristic for the electronic control circuit can be calculated by Equation (8:1.1), we can use the factor levels for component specifications to calculate the thermistor resistances R_{TH1} to R_{TH8}. Although R_{TH} appears to be a complex formula, it is nevertheless possible to use a calculator or a personal computer to calculate R_{TH1} through R_{TH8}.

The results of such a calculation are given in Figure 8:1.9. The response table is shown in Figure 8:1.10. The analysis of variance is shown in Figure 8:1.11. From the analysis of variance it is obvious that components A, B and C are the significant components.

Exp	A	B	C	D	E	F	G	Results
1	1	1	1	1	1	1	1	1.507
2	1	1	1	2	2	2	2	1.514
3	1	2	2	1	1	2	2	1.613
4	1	2	2	2	2	1	1	1.631
5	2	1	2	1	2	1	2	1.514
6	2	1	2	2	1	2	1	1.482
7	2	2	1	1	2	2	1	1.513
8	2	2	1	2	1	1	2	1.513

Figure 8:1.9 The $L_8(2^7)$ orthogonal array with simulation results.

	A	B	C	D	E	F	G
Level 1	1.566	1.504	1.512	1.537	1.529	1.541	1.533
Level 2	1.506	1.568	1.560	1.535	1.543	1.531	1.539
Difference	0.061	0.063	0.049	0.002	0.014	0.011	0.005
Rank	2	1	3	7	4	5	6

Figure 8:1.10   Response table of simulation results.

Source	Pool	Sq	ν	Mq	F-ratio	Sq'	rho %
A		0.0074	1	0.0074	42.02	0.0072	34.66
B		0.0080	1	0.0080	45.54	0.0078	37.64
C		0.0047	1	0.0047	26.77	0.0045	21.78
D	Y	0.0000	1	0.0000	-	-	-
E	Y	0.0004	1	0.0004	-	-	-
F	Y	0.0002	1	0.0002	-	-	-
G	Y	0.0001	1	0.0001	-	-	-
Pooled e		0.0007	4	0.0002	1.00	0.0012	5.92
St		0.0208	7	0.0030	-	0.0208	100.00
Mean		18.8750	1	-	-	-	-
ST		18.8958	8	-	-	-	-

Figure 8:1.11   Analysis of variance for the simulation experiment.

8.1.7 Confidence interval for factors

The confidence interval for the factors is calculated from the total number of SN ratio data for the whole experiment. Since there are a total of eight experiments with four experiments at each factor level, $n = 4$. For the calculation of a confidence interval we use $\nu 1 = 1$ degree of freedom. The confidence interval for a factor is:

$$CI = \sqrt{F_{0.05,1,4} \times V_e \times \left[\frac{1}{n}\right]}$$

$$= \sqrt{7.71 \times 0.0002 \times \left[\frac{1}{4}\right]}$$

$$= \pm\ 0.02$$

The response graph of factor effects is shown in Figure 8:1.12. Of course, it is unnecessary to estimate a predicted process average since we are not interested in any factor level combination. We are merely interested in the magnitude of variation due to a factor.

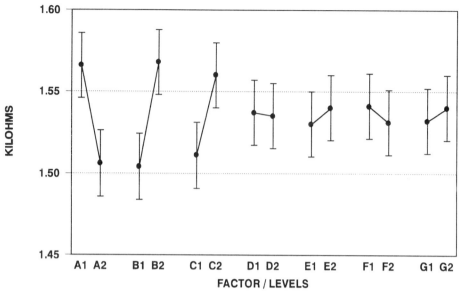

Figure 8:1.12   Response graph of factor effects.

8.1.8 Tolerance calculations

Suppose the specification for this circuit is $R_{\text{TH-ON}} = 1.54 \pm 0.10$ kΩ. If $R_{\text{TH-ON}}$ exceeds this tolerance, the circuit must be discarded by the customer with a loss of $ 20.00 When the characteristic is out of specification in the assembly process, adjustments can be made at a cost of $ 2.00. The cost of upgrading a resistor from $\pm$ 2 % to $\pm$ 1 % is $ 0.50 per resistor. The cost of upgrading a voltage regulator from $\pm$ 0.3 V to $\pm$ 0.1 V is $ 2.00. We now wish to calculate the total loss of each step, starting from upgrading the resistor with the highest percent contribution down to the lowest. We also wish to determine where to stop for minimum total loss.

8.1.9 Establishing the quality loss function

Given that the specification for the control circuit is:

$$R_{TH-ON} = 1.54 \pm 0.10 \text{ k}\Omega$$

This is clearly a case of nominal-the-best with a target value of 1.54 kΩ. The quality loss function is:

$$L(y) = k (y - m)^2 \qquad (8:1.2)$$

The cost coefficient k can be calculated as follows:

$$k = \frac{L(y)}{(y - m)^2}$$

$$k = \frac{A_0}{\Delta_0^2} \qquad (8:1.3)$$

$L(y) = A_0$ is the average customer loss at the customer tolerance $(y - m) = \Delta_0$. Of course, we must always remember that the cost coefficient k, must be calculated on the customer loss and customer tolerance. Given that the cost of repair or replacement when this circuit fails is on average \$ 20.00, we substitute A_0 = \$ 20.00 and Δ_0 = 0.10 kΩ into the equation:

$$k = \frac{A_0}{\Delta_0^2}$$

$$= \frac{20 \ \$}{(0.10 \ \text{k}\Omega)^2}$$

$$= 2000 \ \$ \ \text{k}\Omega^{-2}$$

Substituting $k = 2000$, the quality loss function is rewritten:

$$L(y) = k (y - m)^2$$

$$= 2000 (y - m)^2 \qquad (8:1.4)$$

For many pieces, the quality loss function can be rewritten as:

$$L(y) = k (y - m)^2$$

$$= k [MSD]$$

$$= k [\sigma^2 + (\bar{y} - m)^2] \qquad (8:1.5)$$

Since, tolerance design is only attempted after parameter design has been used to centre the process or product on target, the deviation $(\bar{y} - m)$ must be approximately zero. The reader may verify this for the control circuit simulation experiment. Consequently, the quality loss function may be reduced to:

$$L(y) = k\left[\sigma^2 + (\bar{y} - m)^2\right]$$

$$= k\left[\sigma^2 + (0)^2\right]$$

$$= k\,\sigma^2$$

(8:1.6)

For the current example of the control circuit, $k = 2000$. Therefore,

$$L(y) = k\,\sigma^2$$

$$L(y) = 2000\,\sigma^2$$

(8:1.7)

where σ^2 is the variance of a sample of circuits. Referring to our experiment and the analysis of variance in Figure 8:1.11, σ^2 corresponds to the variance Vt of the source St. Hence,

$$\sigma^2 = Vt$$

$$= \frac{St}{vt}$$

$$= \frac{0.0208}{7}$$

$$= 0.0030$$

For the current condition, the tolerances of the components are as given in Figure 8:1.4. In particular, the tolerance for the resistors is $\pm\,2\,\%$ and the tolerance for the voltages is $\pm\,0.3$ volts. Under this condition, the variance $Vt = 0.0030$. Substituting into:

$$L(y) = 2000\,\sigma^2$$

$$= 2000 \times 0.0030$$

$$= \$\,6.00$$

Therefore, the average loss per piece is $ 6.00. It must be noted that this average corresponds to the total variance, which includes the variances of all the degrees of freedom $(n - 1)$ studied in the $L_8(2^7)$ experiment. The next section shows how we can determine the loss due to an individual factor (i.e. component).

8.1.10   Establishing the loss due to a factor (component)

From the last section, the mathematical equation for loss-per-piece averaged over many pieces is:

$$L(y) = k \, \sigma^2$$

However, σ^2 is the variance due to all the components and the error variance with contribution ratios ρ. Therefore;

$$L(y) = k \left[\frac{\rho_A}{100} + \frac{\rho_B}{100} + \frac{\rho_C}{100} + \frac{\rho_D}{100} + \frac{\rho_E}{100} + \frac{\rho_F}{100} + \frac{\rho_G}{100} \right] \sigma^2 \quad (8{:}1.8)$$

and the loss due to an individual component can be calculated from its percent contribution to the variation. Further, since factors D, E, F and G are insignificant:

$$L(y) = k \left[\frac{\rho_A}{100} + \frac{\rho_B}{100} + \frac{\rho_C}{100} + \frac{\rho_D + \rho_E + \rho_F + \rho_G}{100} \right] \sigma^2$$

$$= k \left[\frac{\rho_A}{100} + \frac{\rho_B}{100} + \frac{\rho_C}{100} + \frac{\rho_{Pool}}{100} \right] \sigma^2 \quad (8{:}1.9)$$

Thus, the loss-per-piece due to a particular factor such as factor A is:

$$L_{Factor\ A} = k \left[\frac{\rho_A}{100} \right] \sigma^2 \quad (8{:}1.10)$$

where ρ is the percent contribution of that factor. Substituting:

$$L_{Resistor\ A} = k \left[\frac{\rho_A}{100} \right] \times \sigma^2$$

$$= 2000 \times \left[\frac{34.66}{100} \right] \times 0.0030 \quad (8{:}1.11)$$

$$= \$ \, 2.08$$

The loss due to factor B is:

$$L_{Resistor\ B} = k\left[\frac{\rho_B}{100}\right] \times \sigma^2$$

$$= 2000 \times \left[\frac{37.64}{100}\right] \times 0.0030$$

$$= \$\ 2.26$$

The loss due to factor C is:

$$L_{Resistor\ C} = k\left[\frac{\rho_C}{100}\right] \times \sigma^2$$

$$= 2000 \times \left[\frac{21.78}{100}\right] \times 0.0030$$

$$= \$\ 1.31$$

In fact, since the total loss is \$ 6.00 and the losses due to significant factors A, B and C are \$ 2.08, \$ 2.26 and \$ 1.31, respectively, we can say that

$$L(y) = L_A + L_B + L_C + L_{All\ others}$$

$$\$\ 6.00 = \$\ 2.08 + \$\ 2.26 + \$\ 1.31 + L_{All\ others}$$

$$L_{All\ others} = \$\ 6.00 - \$\ 2.08 - \$\ 2.26 - \$\ 1.31$$

$$= \$\ 0.35$$

and there is very little to be gained from factors D, E, F and G. The tolerance analysis therefore gives a very clear breakdown of the losses due to the various components.

8.1.11   Effect of reducing the tolerance

The relationship between the variance and standard deviation is simply:

$$variance = [standard\ deviation]^2$$

$$= \sigma^2 \tag{8:1.12}$$

If the tolerance for a factor is reduced from λ to γ, the reduction in standard deviation is γ/λ and the reduction in variance is:

$$variance = [standard\ deviation]^2$$

$$= \left(\frac{\gamma}{\lambda}\ \sigma\right)^2$$

$$= \left(\frac{\gamma}{\lambda}\right)^2 \sigma^2 \qquad (8:1.13)$$

or the variance is reduced by $(\gamma/\lambda)^2$. Therefore, if we reduce the tolerance of a factor from λ to γ then the variance will be reduced by $(\gamma/\lambda)^2$. Since the loss due to an individual component is given by:

$$L_{Factor\ A} = k\left[\frac{\rho_A}{100} \times \sigma^2\right] \qquad (8:1.14)$$

the reduction in loss is:

$$L_{Factor\ A} = k \times \frac{\rho_A}{100} \times \left(\frac{\gamma}{\lambda}\ \sigma\right)^2$$

$$= k \times \frac{\rho_A}{100} \times \left(\frac{\gamma}{\lambda}\right)^2 \times \sigma^2 \qquad (8:1.15)$$

Therefore, if we reduce the tolerance of a resistor from $\pm\ 2\ \%$ to $\pm\ 1\ \%$, then we can expect a reduction of $(1/2)^2$ in the variance. For example, for a factor such as A (Resistor A) the loss at the new tolerance $\pm\ 1\ \%$ is:

$$L_{Resistor\ A} = k \times \frac{\rho_A}{100} \times \left(\frac{\gamma}{\lambda}\right)^2 \times \sigma^2$$

$$= 2000 \times \frac{34.66}{100} \times \left(\frac{1}{2}\right)^2 \times 0.0030 \qquad (8:1.16)$$

$$= \$\ 0.52$$

The gain $\mathfrak{F}$ in changing Resistor A from a tolerance of $\pm\ 2\ \%$ to $\pm\ 1\ \%$ is therefore:

$$\Im = 2.08 - 0.52$$
$$= \$ \ 1.56$$

(8:1.17)

In general, the gain for a factor Ψ can be expressed as $\Im_\Psi$:

$$\Im_\Psi = k \times \frac{\rho_\Psi}{100} \times \sigma^2 - k \times \frac{\rho_\Psi}{100} \times \left(\frac{\gamma}{\lambda}\right)^2 \times \sigma^2$$

$$= k \times \frac{\rho_\Psi}{100} \times \sigma^2 \times \left(1 - \left(\frac{\gamma}{\lambda}\right)^2\right)$$

(8:1.18)

Thus, for component B, the gain $\Im$ can be calculated to be:

$$\Im_B = k \times \frac{\rho_B}{100} \times \sigma^2 \times \left(1 - \left(\frac{\gamma}{\lambda}\right)^2\right)$$

$$= 2000 \times \frac{37.64}{100} \times 0.0030 \times \left(1 - \left(\frac{1}{2}\right)^2\right)$$

$$= \$ \ 1.70$$

And for component C, the gain $\Im$ can be calculated to be:

$$\Im_C = k \times \frac{\rho_C}{100} \times \sigma^2 \times \left(1 - \left(\frac{\gamma}{\lambda}\right)^2\right)$$

$$= 2000 \times \frac{21.78}{100} \times 0.0030 \times \left(1 - \left(\frac{1}{2}\right)^2\right)$$

$$= \$ \ 0.98$$

On this basis, we could replace all components which have a high percent contribution. Of course it is not advisable to upgrade all components since we have to balance the gain with the increased cost of the higher grade component. In other words, we need to establish the contribution ratio (ρ %) above which a component may be upgraded. The next section shows how to calculate the percent contribution above which a component must be upgraded.

8.1.12   Establishing which components to upgrade
Recall that the cost increase to upgrade the tolerance of a resistor from $\pm$ 2 % to $\pm$ 1 % is \$ 0.50. We shall denote this cost as C. Unless the gain $\Im$ in changing from $\pm$ 2 % to $\pm$ 1 % is greater than the cost C, it is not wise to upgrade that factor. Therefore, we need to calculate the percent contribution of the factor above which it is advisable to upgrade it. Since, the gain in upgrading a factor from λ to γ is $\Im$, we can calculate ρ for which $\Im > C$. That is:

$$\Im > C$$

$$k \times \frac{\rho}{100} \times \sigma^2 \times \left(1 - \left(\frac{\gamma}{\lambda} \right)^2 \right) > C \qquad (8:1.19)$$

Simplifying,

$$k \times \frac{\rho}{100} \times \sigma^2 \times \left(1 - \left(\frac{\gamma}{\lambda} \right)^2 \right) > C$$

$$\therefore k \times \rho \times \sigma^2 \times \left(1 - \left(\frac{\gamma}{\lambda} \right)^2 \right) > C \times 100$$

$$\rho > \frac{C \times 100}{k \times \sigma^2 \times \left(1 - \left(\frac{\gamma}{\lambda} \right)^2 \right)} \qquad (8:1.20)$$

For a resistor, $C = $ \$ 0.50. Substituting $C = $ \$ 0.50, $k = 2000$, $\sigma^2 = 0.0030$, $\lambda = $ 2 %, $\gamma = 1$ %;

$$\rho = \frac{C \times 100}{k \times \sigma^2 \times \left(1 - \left(\frac{\gamma}{\lambda} \right)^2 \right)}$$

$$= \frac{0.50 \times 100}{2000 \times 0.0030 \times \left(1 - \left(\frac{1}{2} \right)^2 \right)} \qquad (8:1.21)$$

$$= 11.11 \text{ \%}$$

Thus, any resistor with a percent contribution greater than 11.11 may be upgraded. From the analysis of variance, only resistors A, B and C have $\rho > 11.11$ %. Thus, we only need to replace components A, B and C.

8.1.13   Table of tolerance design

From the foregoing calculations, we can upgrade resistors A, B and C. Figure 8:1.13 shows the cost analysis for the tolerance design. A base price of $ 0.00 is assumed at grade 2 since we only need to analyze the increase in cost. The quality loss is the loss due to variation at the respective grades. The total loss is the loss due to variation and, for grade 1, this includes the cost of upgrading. The Gain column shows the gain in loss-to-society.

From Figure 8:1.13 we can proceed to upgrade resistors A, B and C resulting in a net gain in loss-to-society of $ 2.74 per control circuit. The gain in loss to society can also be calculated from the new variance after upgrading.

$$L(y) = k \left[\frac{\rho_A}{100} + \frac{\rho_B}{100} + \frac{\rho_C}{100} + \frac{\rho_D}{100} + \frac{\rho_E}{100} + \frac{\rho_F}{100} + \frac{\rho_G}{100} \right] \sigma^2 \quad (8:1.22)$$

Since factors D, E, F and G are insignificant, we may rewrite:

$$L(y) = k \left[\frac{\rho_A}{100} + \frac{\rho_B}{100} + \frac{\rho_C}{100} + \frac{\rho_{Others}}{100} \right] \sigma^2 \quad (8:1.23)$$

where $\rho_A + \rho_B + \rho_C + \rho_{Others} = 100$ %. Upon upgrading factors A, B and C, the new loss, $L_{Upgraded}$ will be:

$$L_{Upgraded} = k \left[\frac{\rho_A}{100} \left(\frac{\gamma}{\lambda} \right)^2 + \frac{\rho_B}{100} \left(\frac{\gamma}{\lambda} \right)^2 + \frac{\rho_C}{100} \left(\frac{\gamma}{\lambda} \right)^2 + \frac{\rho_{Others}}{100} \right] \sigma^2$$

$$(8:1.24)$$

$$= 2000 \left[\frac{34.66}{100} \left(\frac{1}{2} \right)^2 + \frac{37.64}{100} \left(\frac{1}{2} \right)^2 + \frac{21.78}{100} \left(\frac{1}{2} \right)^2 + \frac{5.92}{100} \right] \sigma^2$$

$$= \$ 1.76$$

and the cost of this upgrading, L_{Cost} is $ 0.5 $\times$ 3 = $ 1.50.

Factor	Grade	Price $	Quality Loss $	Total $	Gain $	Decision
A	2	0.00	2.08	2.08	1.06	Change to grade 1
	1	+0.50	0.52	1.02		
B	2	0.00	2.26	2.26	1.20	Change to grade 1
	1	+0.50	0.56	1.06		
C	2	0.00	1.31	1.31	0.48	Change to grade 1
	1	+0.50	0.33	0.83		
D	2	0.00				Retain grade 2
	1	+0.50				
E	2	0.00				Retain grade 2
	1	+0.50				
F	2	0.00				Retain grade 2
	1	+2.00				
G	2	0.00				Retain grade 2
	1	+2.00				
Total gain in loss-to-society					2.74	

Figure 8:1.13   Tolerance design cost analysis.

The gain in loss-to-society is therefore:

$$\$ = L_{Original} - L_{Upgraded} - L_{Cost}$$
$$= \$\ 6.00 - \$\ 1.76 - \$\ 1.50$$
$$= \$\ 2.74$$

The tolerance design calculations therefore gives us a clear method of selective component upgrading.

8.1.14   Self-assessment questions

1.   The current y in a tuning circuit is given by the equation:

$$y = \frac{V}{\sqrt{R^2 + (2\,\pi\,f\,L)^2}}$$

where V is the voltage, R is the resistance, π is a constant, f is the frequency and L is the inductance as shown in Figure 8:1.14.

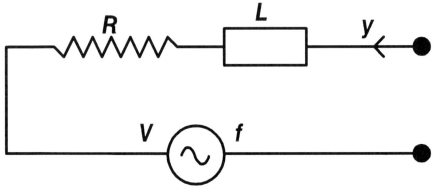

Figure 8:1.14   Example of a tuning circuit.

A parameter design experiment resulted in the nominal values for factor levels shown in Figure 8:1.15:

Factor	Nominal value	Lower Tolerance	Upper Tolerance	Unit
R	9.92	−10 %	+10 %	Ω
L	0.004	−5 %	+5 %	H
V	100.0	−5 %	+5 %	V
f	50	-	-	Hz

Figure 8:1.15   Optimum factor levels.

The specification for the tuning circuit is: $y = 10.0 \pm 4.0$ A. When the circuit goes out of specification, the cost of repairing or replacing it is on average $ 32. The cost of upgrading the current 10 % resistor to 5 % resistors is $ 0.50. The cost of upgrading the current inductor from 5 % to 1 % is $ 2.00 and the cost of upgrading the current voltage regulator from 5 % to 1 % is $ 5.00.

For engineering reasons it is necessary to use a frequency of 50 Hz. This factor is therefore kept constant. The tolerances around the remaining factors, V, R and L, are based on degradation and deterioration over an average life of the product. These are also included in Figure 8:1.15. The objective is to reduce the variability of y around the target value by upgrading the resistor, inductor or voltage regulator. Show how this tolerance design can be conducted.

8.1.15   Answers to self-assessment questions

1.   The current y in a tuning circuit is given by the equation:

$$y = \frac{V}{\sqrt{R^2 + (2 \pi f L)^2}}$$

where V is the voltage, R is the resistance, π is a constant, f is the frequency and L is the inductance as shown in Figure 8:1.14.

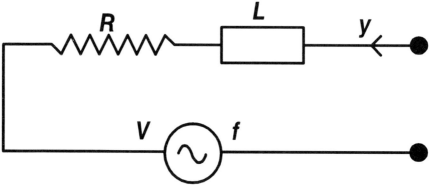

Figure 8:1.14   Example of a tuning circuit.

A parameter design experiment resulted in the nominal values for factor levels shown in Figure 8:1.15:

Factor	Nominal value	Lower Tolerance	Upper Tolerance	Unit
R	9.92	−10 %	+10 %	Ω
L	0.004	−5 %	+5 %	H
V	100.0	−5 %	+5 %	V
f	50	-	-	Hz

Figure 8:1.15   Optimum factor levels.

The specification for the tuning circuit is: $y = 10.0 \pm 4.0$ A. When the circuit goes out of specification, the cost of repairing or replacing it is on average $ 32. The cost of upgrading the current 10 % resistor to 5 % resistors is $ 0.50. The cost of upgrading the current inductor from 5 % to 1 % is $ 2.00 and the cost of upgrading the current voltage regulator from 5 % to 1 % is $ 5.00.

For engineering reasons it is necessary to use a frequency of 50 Hz. This factor is therefore kept constant. The tolerances around the remaining factors, *V*, *R* and *L*, are based on degradation and deterioration over an average life of the product. These are also included in Figure 8:1.15. The objective is to reduce the variability of *y* around the target value by upgrading the resistor, inductor or voltage regulator. Show how this tolerance design can be conducted.

Answer

We shall answer this question in a step-by-step method:

●   assignment of factor levels,
●   conducting the experiment,
●   confidence interval for factors,
●   establishing the quality loss function,
●   tolerance calculations,
●   establishing the loss due to a factor (component),
●   effect of reducing the tolerance,
●   establishing which components to upgrade,
●   table of tolerance design.

1.   Assignment of factor levels

We assign factor A (resistor) to column 1 and factor B (inductor) to column 2 leaving the A×B interaction column (3) unassigned since there many other columns available. Factor D (voltage regulator) is then assigned to column 4 as shown in Figure 8:1.16. Thus, columns 3, 5, 6 and 7 are left unassigned as shown in Figure 8:1.17.

	Factor	Nominal value	Lower Tolerance	Upper Tolerance	Unit
A	*R*	9.92	−10 %	+10 %	Ω
B	*L*	0.004	−5 %	+5 %	H
D	*V*	100.0	−5 %	+5 %	V
-	*f*	50	-	-	Hz

Figure 8:1.16   Optimum factor levels.

2.   Conducting the experiment

The results of the tolerance design calculation are also given in Figure 8:1.17. The corresponding response table of factor effects is shown in Figure 8:1.18. The analysis of variance is shown in Figure 8:1.19.

Exp	1 A	2 B	3 C	4 D	5 E	6 F	7 G	Results
1	1	1	1	1	1	1	1	10.5468
2	1	1	1	2	2	2	2	11.6570
3	1	2	2	1	1	2	2	10.5263
4	1	2	2	2	2	1	1	11.6344
5	2	1	2	1	2	1	2	8.6544
6	2	1	2	2	1	2	1	9.5654
7	2	2	1	1	2	2	1	8.6431
8	2	2	1	2	1	1	2	9.5528

Figure 8:1.17   The $L_8(2^7)$ orthogonal array with simulation data.

	A	B	C	D	E	F	G
Level 1	11.09	10.11	10.10	9.59	10.05	10.10	10.10
Level 2	9.10	10.09	10.10	10.60	10.15	10.10	10.10
Difference	1.99	0.02	0.00	1.01	0.10	0.00	0.00
Rank	1	3	-	2	-	-	-

Figure 8:1.18   Response table of simulation results.

Source	Pool	Sq	ν	Mq	F-ratio	Sq'	rho %
A		7.898	1	7.898	1940.29	7.894	79.28
B		0.001	1	0.001	-	-	-
C	Y	0.000	1	0.000	-	-	-
D		2.039	1	2.039	500.96	2.035	20.43
E	Y	0.020	1	0.020	-	-	-
F	Y	0.000	1	0.000	-	-	-
G	Y	0.000	1	0.000	-	-	-
Pooled e		0.020	5	0.004	1.00	0.028	0.29
St		9.958	7	1.423	-	9.958	100.00
Mean		815.680	1	-	-	-	-
ST		825.637	8	-	-	-	-

Figure 8:1.19   Analysis of variance for the simulation experiment.

3.   Confidence interval for factors

The confidence interval for the factors is calculated from the total number of SN ratio data for the whole experiment. Since there are a total of eight experiments with four experiments at each factor level, $n = 4$. For the calculation of a confidence interval we use $\nu 1 = 1$ degree of freedom. The confidence interval for a factor is:

$$CI = \sqrt{F_{0.05,1,5} \times V_e \times \left[\frac{1}{n}\right]}$$

$$= \sqrt{6.61 \times 0.004 \times \left[\frac{1}{4}\right]}$$

$$= \pm\ 0.08$$

The response graph of factor effects is shown in Figure 8:1.20.

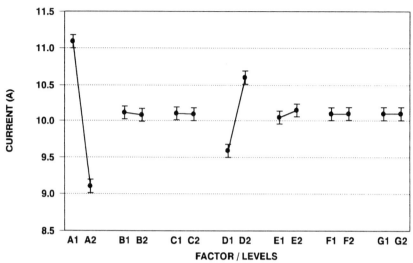

Figure 8:1.20 Response graph of factor effects.

4. Establishing the quality loss function
The cost coefficient for the quality loss function is:

$$k = \frac{A_0}{\Delta_0^2}$$

$$= \frac{32.00 \ \$}{(4.00 \ A)^2}$$

$$= 2.00 \ \$ \ A^{-2}$$

The loss-per-piece is therefore:

$$L(y) = k \ \sigma^2$$

$$= 2.00 \times 1.42$$

$$= \$ \ 2.85$$

5.   Tolerance calculations
The specification for this circuit is $y = 10.00 \pm 4.0$ A. If y exceeds this tolerance, the circuit will be discarded by the customer with a loss of $ 32.00. The cost of upgrading a resistor from ± 10 % to ± 5 % is $ 0.50 per resistor. The cost of upgrading an inductor from ± 5 % to ± 1 % is $ 2.00 per inductor and the cost of upgrading a voltage regulator from ± 5 % to ± 1 % is $ 5.00 per voltage regualator.

The mathematical equation for the loss-per-piece averaged over many pieces is:

$$L(y) = k \, \sigma^2$$

However, σ^2 is the variation due to all the components and the error variance with contribution ratios ρ. Therefore;

$$L(y) = k \left[\frac{\rho_A}{100} + \frac{\rho_B}{100} + \frac{\rho_C}{100} + \frac{\rho_D}{100} + \frac{\rho_E}{100} + \frac{\rho_F}{100} + \frac{\rho_G}{100} \right] \sigma^2$$

and the loss due to an individual component can be calculated from its percent contribution to the variation. Further, since factors B, C, E, F and G are insignificant:

$$L(y) = k \left[\frac{\rho_A}{100} + \frac{\rho_D}{100} + \frac{\rho_B + \rho_C + \rho_E + \rho_F + \rho_G}{100} \right] \sigma^2$$

$$= k \left[\frac{\rho_A}{100} + \frac{\rho_D}{100} + \frac{\rho_{Pool}}{100} \right] \sigma^2$$

6.   Establishing the loss due to a factor (component)
The loss due to factor A is:

$$L_{Resistor\ A} = k \left[\frac{\rho}{100} \right] \times \sigma^2$$

$$= 2.00 \times \left[\frac{79.28}{100} \right] \times 1.42$$

$$= \$ \ 2.26$$

The loss due to factor D is:

$$L_{Voltage \ regulator \ D} = k \left[\frac{\rho}{100} \right] \times \sigma^2$$

$$= 2.00 \times \left[\frac{20.44}{100} \right] \times 1.42$$

$$= \$ \ 0.58$$

The loss due to the remaining factors (inductor and other noise effects) is:

$$L_{Others} = k \left[\frac{\rho}{100} \right] \times \sigma^2$$

$$= 2.00 \times \left[\frac{0.29}{100} \right] \times 1.42$$

$$= \$ \ 0.01$$

7.   Effect of reducing the tolerance

Recall that the cost increase to upgrade the tolerance of a resistor from $\pm$ 10 % to $\pm$ 5 % is $ 0.50. We shall denote this cost as $\mathbb{C}$. Unless the gain $\mathfrak{I}$ in changing from $\pm$ λ % to $\pm$ γ % is greater than the cost $\mathbb{C}$, it is not wise to upgrade that factor. Therefore, we need to calculate the percent contribution of a factor above which it is advisable to upgrade that factor. Since, the gain in upgrading a factor from λ to γ is $\mathfrak{I}$, we can calculate ρ for which $\mathfrak{I} > \mathbb{C}$. That is:

$$\mathfrak{I} > \mathbb{C}$$

$$k \times \frac{\rho}{100} \times \sigma^2 \times \left(1 - \left(\frac{\gamma}{\lambda} \right)^2 \right) > \mathbb{C}$$

Simplifying,

$$k \times \frac{\rho}{100} \times \sigma^2 \times \left(1 - \left(\frac{\gamma}{\lambda}\right)^2\right) > C$$

$$\therefore k \times \rho \times \sigma^2 \times \left(1 - \left(\frac{\gamma}{\lambda}\right)^2\right) > C \times 100$$

$$\rho > \frac{C \times 100}{k \times \sigma^2 \times \left(1 - \left(\frac{\gamma}{\lambda}\right)^2\right)}$$

8.   Establishing which components to upgrade

For a resistor, $C = \$ 0.50$. Substituting $C = \$ 0.50$, $k = 2.00$, $\sigma^2 = 1.42$, $\lambda = 10$ %, $\gamma = 5$ %;

$$\rho = \frac{C \times 100}{k \times \sigma^2 \times \left(1 - \left(\frac{\gamma}{\lambda}\right)^2\right)}$$

$$= \frac{0.50 \times 100}{2.00 \times 1.42 \times \left(1 - \left(\frac{5}{10}\right)^2\right)}$$

$$= 23.43 \text{ %}$$

Thus, any resistor with a percent contribution greater than 23.43 may be upgraded. From the analysis of variance, resistor A has $\rho > 23.43$ %. Thus, we may upgrade component A.

For a voltage regulator, $C = \$ 5.00$. Substituting $C = \$ 5.00$, $k = 2.00$, $\sigma^2 = 1.42$, $\lambda = 5$ %, $\gamma = 1$ %;

$$\rho = \frac{C \times 100}{k \times \sigma^2 \times \left(1 - \left(\frac{\gamma}{\lambda}\right)^2\right)}$$

$$= \frac{5.00 \times 100}{2.00 \times 1.42 \times \left(1 - \left(\frac{1}{5}\right)^2\right)}$$

$$> 100.00 \%$$

Thus, it is unwise to upgrade a voltage regulator. Similarly, the loss due to an inductor (and other noise effects) is only $ 0.01 and there is no need to even consider it.

9.   Table of tolerance design
From the foregoing calculations, we should only upgrade the resistor. Figure 8:1.21 shows the cost analysis for the tolerance design.

Factor	Grade	Price $	Quality Loss $	Total $	Gain $	Decision
A	2	0.00	2.26	2.26	1.19	Change to grade 1
	1	+0.50	0.56	1.06		
B	2	0.00	0.58	0.58		Retain grade 2
	1	+2.00	0.02	2.02		
Total gain in loss-to-society					1.19	

Figure 8:1.21   Tolerance design cost analysis.

A base price of $ 0.00 is assumed at grade 2 (current grade) since we only need to analyze the increase in cost. The quality loss is the loss due to variation at the respective grades. The total loss is the loss due to variation and, for grade 1, this includes the cost of upgrading. The Gain column shows the gain in loss-to-society. Thus, it is advisable to upgrade only the resistor.

CHAPTER 9

MANAGING THE DESIGN OF EXPERIMENTS

AIMS:
To provide a management overview on implementing and managing robust design within an organization.

OBJECTIVES:
When you have completed studying this chapter you should be able to:
- explain the key principles of total quality management,
- explain the role of problem solving,
- implement and manage robust design in an organization.

OVERVIEW:
This chapter introduces a few principles of total quality management, including quality awareness of a company, quality perception and the cost of quality. A number of management issues concerning the implementation of robust design within a company and guidelines for successful experimentation are also given. Some common criticisms of robust design are also addressed, with explanations of how to overcome these criticisms.

9.1 Total Quality Management

9.1.1 Introduction

Quality is still seen by many business managers as a problem to be solved by the *Quality Inspector* or the Quality Assurance Department. Business is becoming increasingly difficult as competitive pressures increase. Essentially, many managers have failed to recognize quality management as a strategy for success in every sector and for every business. For these managers, Total Quality Management (TQM) is a sophisticated version of Quality Assurance, something that is to be added at the end of the process and which increases costs. However, quality cannot be added on as something that is included in the *packaging box*. It must be built into the culture of the business and the design of the process and product. Thus, it includes everyone and is reflected in every activity at all levels of management. It belongs to us all. Poor quality is not *what we get from others, it is the result of the culture of which we are a part.*

9.1.2 Quality awareness – where do you stand?

The quality awareness of an organization is an indication of the commitment of the organization to the customer requirements. The levels of awareness of an organization are illustrated in Figure 9:1.1.

Figure 9:1.1 Levels of quality awareness.

At the Product oriented phase, inspections are made after production. There are many audits of products and processes. Problem solving activities are widespread and many problems recur. An organization at this level usually has a large Quality Control department, and quality control is seen to be the job of the Quality Assurance Department. A characteristic of this phase is endless *firefighting*.

At the Process oriented phase, quality assurance is conducted during production. Statistical Process Control is frequently used to monitor products and processes. Such methods emphasize finding the causes of problems and resetting the process to its nominal condition. The attempt in this phase is to reduce the incidence of rejects and to minimize scrap. Quality Assurance is now seen to be the job of the Quality Assurance *and* the Manufacturing Departments.

At the Systems oriented phase, quality assurance begins to involve not only the Quality Assurance and Manufacturing Departments but *also* the Design, Purchasing, Marketing and Sales and Service Departments. The emphasis moves from largely inspection to largely monitoring.

At the Personnel oriented phase, the emphasis moves from departmental involvement to individual involvement. There is a significant change in the thinking of employees from *their problem* to *our problem*. Each employee recognizes how his/her actions affect the customer. Training is crucial at this stage. Indeed, this is the phase of investing in people.

At the Society oriented phase, the organization can realize a unit of operation whose goal is set on the customer. The organization has sufficient internal control and can begin to look at the society. In particular, the organization must reduce the loss-to-society. That is, the organization should now attempt to produce the same product at a lower cost.

At the Consumer oriented phase, the aim of an organization is to provide customer delight. At this phase, product optimization of basic functions occurs in the design stage before the customer is even aware of the problem. Where the customer sees a problem, we define the *Voice of the Customer* and translate it into *engineering specifications*. At this phase, customer is king or queen.

9.1.3 Total Quality Management – a new concept

Quality is the priority in achieving industrial competitiveness. It is the backbone of excellence; a never-ending journey that seeks to improve continuously. For the customer, it means good design, reliable performance, prompt delivery and, in short, providing goods or services that meet the customer's expectations in full, at a competitive price. For a company, this means ensuring that inefficient and wasteful efforts are kept to a minimum. This requires, above all, a commitment by the organization to a Total Quality Management programme.

Total Quality Management sets out with a mission. This is a definition of what the organization exists to do and an aim which it works towards. It is the statement of the fundamental objectives of the company.

Total Quality Management requires a philosophy. A definition of what the company wants to do; the way the company wants to do it and the commitment of the company to the customer. Inevitably this shapes the corporate culture of the company.

The implementation of a Total Quality Management programme however, is not as simple as it may seem. Various factors, such as

- organization,
- top management,
- customer orientation,
- communication,
- ownership,
- prioritization and timescale,

need to be checked in what we may well call a *health check*. Unless an organization is healthy, it would be futile to implement Total Quality Management. These factors are discussed below.

1.   Organization

The organization needs to define the responsibility, authority and the interrelationships among all personnel who manage, perform and verify work affecting quality. See Figure 9:1.2. This is particularly important for personnel who need the organizational freedom and authority to control further processing, delivery or installation of any non-conforming product (process or service) until any deficiencies have been corrected.

TQM - A New Concept

Figure 9:1.2 Total Quality Management: a new concept.

A decentralized organization can be particularly beneficial in making employees feel close to the customer, by making the employee aware of how his/her non-conformance ultimately affects the customer. As with the introduction of any new technique or procedure it is essential to start in the right way: the selection of the first project and team is vital. The team needs to be very clear about the objectives, goals, targets, etc. Equally, if not more important, it needs to consider the limitations that need to be placed on the project; the time, cost and area to be considered, the requirements of safety, weight, material and performance for a particular market sector. With these clearly documented the team is less likely to stray and can focus on the areas of high priority and required customer satisfaction. There is a learning period for the team and the need to obtain more detailed information across a broad base. Small projects can be undertaken relatively quickly, but major projects will probably require 50 – 60 hours of meetings to take the project through the whole process as well as individual work by each team member. This time needs to be accommodated in the project and company manpower plans.

2. Top management

Top management must make a committed decision to have a Total Quality Management programme, the quality policy, objectives and commitment must be expressly stated, understood and maintained at all levels in the organization. This would mean that not only managers, supervisors and factory operators understand the policy but also the security personnel at the gate, the secretary and the cleaners.

A quality system such as BS/EN/ISO 9000 may be adopted. Such a system does not itself ensure quality but it is a business system that heralds *we do what we say and we say what we do*. The quality system adopted to satisfy customer requirements must be reviewed at periodic intervals to ensure its continuing suitability and effectiveness. The management must also adopt a system of management by facts and its activities must be highly visible.

3. Customer orientation

Again, all levels of the organization must understand that customer is Number 1 and right first time is the target. Everyone needs to understand customers want their product (or service) to be right every time. This requires that everyone constantly thinks of quality and improving quality. See Figure 9:1.3. In other words, the company defines the customer requirements and conforms to all these requirements without exception.

Another problem that frequently arises in most companies is the parochial approach of various departments, where departments tend to work independently and often non-cooperatively in what is frequently called the *fortress mentality*. Such an approach must be demolished and each department must be familiar with other departments and their contribution, so that the entire organization becomes just one team. Taken a step further, the organization must realize not only the final customer, but also the internal customer, where the previous department, process or person is the supplier to the current department, process or person which is in turn the supplier to the next customer.

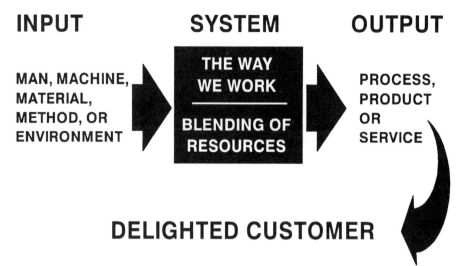

Figure 9:1.3 Managing resources for customer satisfaction.

4.   Communication

The importance of communication cannot be overemphasized. The mechanism of the flow of communication must be bilateral at all levels of the organization. That is, from the top management downwards to the shop floor, and vice-versa. Communication must also develop the right attitude for inter-personnel communication. The basics of supervision are important to cultivate good morale and *together we stand* attitude is of prime importance. Management itself is no exception to this aspect of communication. Thus, all levels of the organization require good bilateral communication *where problems and not people are criticized*.

5.   Ownership

Ownership creates the feeling and reality of contributing and hence winning. This sense of belonging comes with the objectives and function of a department and the psychological ownership of individuals in that department. To achieve psychological ownership, every individual must know what he/she is supposed to do, what he/she is doing and have the means to control what he/she is doing, i.e. a state of *self-control*.

  Taken further, individuals from an area could form a quality circle – a voluntary means of tackling their own work-related problems. This provides the team ownership of and responsibility for their work area and gives them the chance to put things right. This is not only rewarding to the team but is also beneficial to the organization. Top management sometimes sees this as a threat to their authority because there is a change from *command and control* to *empower and approve.* However, this initial fear can to be dispelled by way of good bilateral communication.

6.   Prioritization and timescale

Having discussed the outline of the activities required for Total Quality Management, a priority and implementation timescale is needed. It may be noted that the timescale may be determined by the company size with respect to its people count. For a typical company, this would be projected over a period of twelve months for a start-off and another year or two to develop the new culture. Indeed, it is not how, but how fast we change in an enterprise where *Quality is a never-ending journey to excellence* for industrial competitiveness.

9.1.4 'Flavour-of-the-month'

There are a myriad of quality improvement methods. Many are based on persuasion and motivation while others are based on instillation of fear. Also there is a general confusion of what constitutes a quality improvement technique. A good example of this is the British Standard 5750 (BS 5750) and subsequently BS/EN/ISO 9000. Many companies believed that compliance with BS 5750 would improve their process or product quality. When it did not, BS 5750 was construed as an unnecessary burden with hugh amounts of paperwork. However, BS 5750 is a quality system that must be achieved for ensuring a system of conducting business. BS 5750 does not show anyone how to establish the correct aperture size for a camera, nor does it show how we may literally improve the signal-to-noise ratio in the earpiece of a telephone. That is an engineering function. Nevertheless, BS 5750 addresses a very important question: *Do you have a system for improving quality?*

Similarly, there is a widespread belief (particularly in the West) that Statistical Process Control (SPC) is a cure-all for manufacturing problems. Great gurus like Deming and Juran advocated it. Many companies are proud to display a large number or control charts displaying all kinds of information. But SPC is a *reactive method* of ensuring quality. Shigeo Shingo referred to it as a mirror that can only reflect the system. Indeed in robust design, we regard it as a downstream quality improvement technique. Of course when quality has been improved, by any method, we need to ensure that these levels are maintained. In that sense, SPC may be the ideal tool. But to expect to improve the sensitivity of a temperature controller that switches on or off a control circuit is beyond the domain of SPC.

Many companies believe that *their company* has the best 'flavour'. A company may adamantly believe that nothing but Just-In-Time (JIT) is the only technique they need. Unfortunately, JIT is an inventory control system that must be part of a company's way of working. On its own, JIT is an ideal technique to identify bottlenecks in a production environment, and focuses on the resources required to overcome a constraint. Additionally, JIT enables problem areas to be easily identified by virtue of low work-in-progress. JIT is geared towards minimizing all forms of waste – a philosophy that is not far removed from minimizing the loss-to-society.

Admittedly, some techniques (or at least the gurus) do seem to contradict each other. Poka-yoke, for example, would suggest that any form of SPC is a waste and that there must be a system that automatically indicates a defective product and prevents it from moving further down the assembly line. This is a complete reversal from the much hailed statistical sampling techniques to 100 % inspection at the of source of the defect.

For the manager, it is imperative to understand that (most) quality techniques actually complement each other. It is the role of top managers to foster the right attitude in their operations so that quality improvement is not restricted to any one method.

9.1.5 Cost of quality

BS 4788 defines quality as *The totality of features and characteristics of a product or service that bear on its ability to satisfy stated or implied needs.* This definition is ideal for a quality system in which quality is taken to be *fitness for purpose.* From an engineering viewpoint, such a definition is inadequate for two reasons: first, it does not take account of loss due to unintended side effects, and second, there is no unit of measure for quality or the lack of it.

Taguchi defines quality as *The quality of a product is the (minimum) loss imparted by the product to society from the time the product is shipped, other than losses caused by its intrinsic functions.* In the present context, the phrase *after the product is shipped* has a great significance in that it distinguishes between *loss* and *cost.*

Loss is associated with waste that can and should be prevented. Loss can be divided in two categories:
- loss due to functional variability,
- loss due to harmful side effects.

1.   Loss due to functional variability
Loss due to functional variability is the loss caused by the deviation of the product characteristic from the intended ideal. This is the price paid for variability in the product function. This loss is borne by the producer (if the product is within warranty) or the consumer (if the product is without warranty). An unscrupulous manufacturer can reduce his/her loss while actually imparting a larger and unfair loss to the consumer. The major emphasis of Taguchi's philosophy is to reduce this loss by designing products and processes that are robust. See Figure 9:1.4.

2.   Loss due to harmful side effects
Loss due to harmful side effects is the loss caused by an unintended side effect. Thus, by definition, *loss imparted by the product to society* would include the losses due to automobile exhausts, toxic gases from incineration, noisy machinery and non-biodegradable packaging. Hence, although a manufacturer may produce a product at a price that a customer sees as good value, if in the process, the manufacturer generates toxic substances that cause pollution, then that is still a quality problem. In that case, the society (third party) has to pay

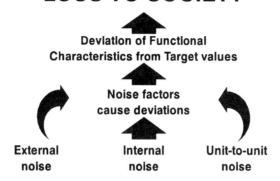

LOSS TO SOCIETY

Any departure of the Quality Characteristic from the intended ideal is a loss. Because we cannot achieve perfect Quality, we cannot eliminate loss.

Figure 9:1.4 Noise factors and loss-to-society.

for this side effect. In Taguchi's definition, that does not constitute quality improvement because it does not reduce the loss-to-society i.e. manufacturer, the customer and society. Thus, Taguchi's definition insists that quality improvement must reduce not just the loss to the manufacturer or the consumer but also the loss to society as a whole.

Taguchi's definition *other than losses caused by its intrinsic function* excludes losses caused by a product's intrinsic function. Most people drink intoxicating liquor because of the feeling it causes. However, a drunk person may drive his/her car into a busy street and cause great damage. Such a loss is a personality problem of the drunk driver and is not a quality issue. To stop the production of liquor because it causes intoxication is unreasonable, as that is the function of liquor.

Costs are associated with the price of manufacturing a product. These can be classified into:
- prevention cost,
- appraisal cost,
- internal failure cost,
- external failure cost.

1.   Prevention cost
Prevention cost is the price of any action taken to investigate, prevent or reduce the risk of a nonconformity or defect.

2.   Appraisal cost

Appraisal cost is the price of measuring, process control, testing, etc., before a product is shipped. It is the cost of evaluating the achievement of quality requirements including, for example, the cost of verification and control performed at any stage of the quality spiral.

3.   Internal failure cost

Internal failure cost is the price of making right a defective product before it is shipped. It is the cost arising within an organization due to nonconformities or defects at any stage of the quality spiral, including the costs of scrapping, reworking, retesting, reinspection and redesigning.

4.   External failure cost

External failure cost is the cost arising after delivery to a customer of products with nonconformities or defects which may include the cost of claims against warranty, replacement and consequential losses, and evaluation on penalties incurred. By our definition of loss and cost, external failure cost may be regarded as a loss since this is the price the customer has to pay. Of course, we may claim that when a product is under warranty, the producer will pay for the repair, and when the product is without warranty, the customer will have to pay. In either case, this is loss-to-society and we should endeavour to minimize external failure cost.

9.1.6 Benefits of robust design

Traditional quality control has effected little change in the last 50 years. Robust design, however, is a powerful tool to effect change.

The first major benefit introduced by robust design is the definition of quality itself. Quality is seen not only as the loss due the poor functional performance but also the loss due to harmful side effects. The central theme of parameter design is focused around reducing variability, which is itself intimately tied to cost, so that quality is measured in monetary units. Hence, conformance (or non-conformance) is no longer an acceptable measure of quality.

The second benefit introduced by robust design is the system of financial control. Conventional producers do not usually make positive attempts to improve product characteristics that are already within specification because they see no benefit in doing so. The quality loss function, however, measures the loss-to-society for product characteristics that are not only outwith specification but also within specification. This provides a perspective for continuous improvement, by reducing variability and thus reducing costs.

The third benefit introduced by robust design is tolerancing. This refers to the use of the quality loss function to establish realistic manufacturing tolerances based on economic and functional considerations. Here, the quality loss function is used to balance the loss between the producer and the customer and forms the basis for a contract between the producer and the customer.

The fourth benefit introduced by robust design is the tolerance design. Tolerance design provides a trade-off between quality and cost. Additionally, tolerance design can be used to identify critical factors for process control. This greatly reduces the need for extensive use of statistical process control charts all over the plant and highlights those processes that do need statistical process control charts.

9.1.7 Company implementation of robust design

The key factors involved in the successfully introduction of robust design into a company are:

- senior management commitment,
- internal steering group,
- structured training,
- communication.

1.   Senior management commitment

Senior management commitment is a fundamental requirement for success in implementing robust design. This first step can be accomplished by an overview of robust design. Once senior management commitment is secured the company may proceed to identify potential areas of application.

2.   Internal steering group

An internal steering group is necessary to coordinate the many activities involved in implementing robust design. Ideally, the steering group must consist of at least one senior manager and an engineer who has completed a course on robust design.

3.   Structured training

Structured training is a crucial step in the process of implementing robust design. See Figure 9:1.5 This does not mean to say that all employees must be trained in robust design. Various levels of training must be established within the company, e.g. some employees may only need a little training, while others may need an awareness training to be able to participate in brainstorming sessions, others may need to be trained to understand robust design at a basic level, and a few may wish to become specialists in data analysis.

Also, consultancy services may be necessary in the initial stages as it may be nearly impossible for engineers to complete a robust design by themselves. Additionally, it is recommended that in-house workshops be set up to complement training.

Communication is perhaps the most important factor in the success of any quality programme. Many quality programmes have failed because of poor communication. It is essential that there is good communication of successfully completed experiments or projects as well as proposed experiments or projects. This will also highlight the teams working on areas of continuous improvement, and will spur others to do the same. Recognition awards to teams that make significant improvements to a process or product may be made to encourage others to form teams to improve quality in their work environment.

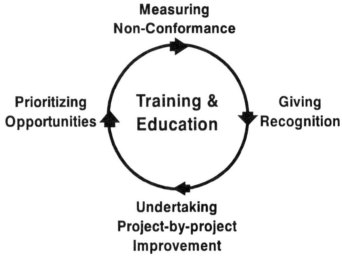

Figure 9:1.5 Role of training and education.

The results and experience gained from experiments may be held centrally to build up a database. Such a database will form a pool of information on the reasons why certain engineering or management decisions were made. The gradual accumulation of knowledge will establish a record of quality progress from one level to another.

9.1.8 Self-assessment questions

1. Explain how you would identify where a company stands in terms of quality awareness.

2. What are the key principles of Total Quality Management.

3. Distinguish between the cost of quality and loss due to poor quality.

4. What are the benefits of robust design?

5. How would you implement robust design within a company.

9.1.9 Answers to self-assessment questions

1.   Explain how you would identify where a company stands in terms of quality awareness.

Answer

At the Product oriented phase, inspections are made after production. At the Process oriented phase, quality assurance is conducted during production. At the Systems oriented phase, quality assurance begins to involve not only the Quality Assurance and Manufacturing Departments but also the Design, Purchasing, Marketing and Sales and Service Departments. At the Personnel oriented phase, the emphasis moves from departmental to individual involvement. At the Society oriented phase, the organization can set up a unit of operation whose goal is set on customer satisfaction. At the Consumer oriented level, the aim of an organization is to provide customer delight.

2.   What are the key principles of Total Quality Management.

Answer

Total Quality Management is the priority for achieving industrial competitiveness. It is the backbone of excellence; a never-ending journey that seeks to improve continuously. The key factors in the successful introduction of a Total Quality Management programme are:
- organization,
- top management,
- customer orientation,
- communication,
- ownership,
- prioritization and timescale.

3.   Distinguish between the cost of quality and loss due to poor quality.

Answer

Cost is associated with the price of manufacturing a product. When this product has been shipped to the customer and the customer (or producer) incurs an expenditure, then that is a loss. Prevention, appraisal and internal failure are costs. External failure cost is a loss.

4.   What are the benefits of robust design?

Answer

The first and major benefit introduced by robust design is that quality is seen not only as the loss due to poor functional performance but also the loss due to harmful side effects. The second benefit introduced by robust design is the system of financial control. The quality loss function measures the loss-to-society for product characteristics that are not only outwith specification but also within specification. The third benefit introduced by robust design is the use of the quality loss function to establish realistic manufacturing tolerance based on economic and functional considerations to balance the loss between the producer and the customer and so form the basis for contract between the producer and the customer. The fourth benefit introduced by robust design is tolerance design. Tolerance design provides a trade-off between quality and cost. Additionally, tolerance design can be used to identify critical factors for process control. This greatly reduces the need for the use of statistical process control charts all over the plant.

5.   How would you implement robust design within a company.

Answer

I would implement robust design within a company by starting with senior management. Senior management commitment is a fundamental requirement for success in implementing robust design. An internal steering group is necessary to coordinate the many activities involved in implementing robust design. Structured training is a crucial step in the process of implementing robust design. Also, consultancy services may be necessary in the initial stages as it would be nearly impossible for engineers to complete a robust design by themselves. Additionally, it is recommended that in-house workshops be set up to complement training. Communication is perhaps the most important factor of the success of any quality programme. It is essential that there is good communication of completed experiments or projects as well as proposed experiments or projects.

9.2   Problem-solving Process

9.2.1 Introduction

If the only tool we have is a hammer, then all problems are nails. But if not all problems are nails then we need more than a hammer. Whatever approach a company takes to achieve Total Quality Management, the company will require a problem-solving process together with a selection of tools and techniques. A problem-solving process establishes a systematic approach to continuous improvement using various tools and techniques. The problem-solving process avoids jumping to the first possible solution and the likely adverse side-effects of such a quick-fix solution. Indeed, the problem-solving process provides a structured approach to problem-solving by carefully considering various activities necessary for a permanent solution. Of course, no single tool or technique is more important than another. Each has a role to play and contributes in a specific way to the quality improvement process. Most tools and techniques are simple to use and a TQM company usually encourages every employee to know at least seven basic problem-solving tools.

9.2.2 Quality improvement

Total quality management requires continuous quality improvement. In many organizations, ad hoc problem solving results in problem recurrence. In such organizations, there is always a large number of chronic problems. When a problem occurs it occurs on a shutdown scale. Management do not understand why, except that they are likely to miss production schedules. Engineers are expected to fix the problem immediately, if they need their jobs. Manufacturing do not understand why all the control charts did not help and neither do Quality Assurance. Often, many solutions are suggested and implemented. The problem goes away, everybody believes he/she has helped, but nobody knows what solved the problem and indeed what caused it. Until another day, and another problem.

For successful problem solving, a solution to a problem must ensure that the problem does not recur. See Figure 9:1.6. Moreover, the solution must be robust. Unless that is the case, firefighting will be part of the company culture.

Only when problems are solved permanently, can we expect continuous improvement where an initial quality level is taken to a new quality level permanently. The lessons learned and the experienced gained must be fed into the next stage of quality planning and control. See Figure 9:1.7.

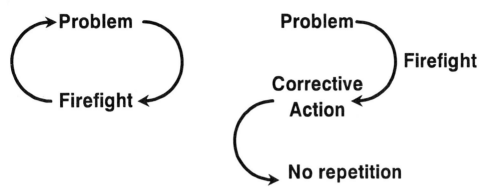

Figure 9:1.6 Breaking the loop.

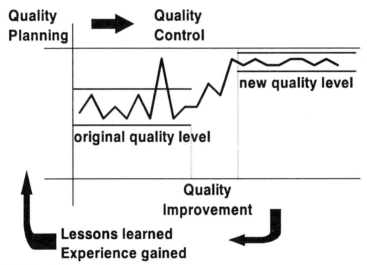

Figure 9:1.7 Breakthrough improvement.

9.2.3 Tools and techniques – 7 basic tools

Facilitating permanent quality improvements requires many tools and techniques. All of these tools and techniques may be referred from standard textbooks on quality. Conducting designed experiments frequently requires the use of some of these tools and techniques as well. A diagrammatic representation of these tools and techniques are shown in Figure 9:1.8.

- process flow chart,
- check / tally chart,
- histogram,

- cause-effect diagram,
- pareto analysis,
- scatter diagram,
- control charts,

1.   Process flow chart
In the planning or examination of any process it is frequently necessary to record the series of stages, events, activities and decisions in a form which can be easily understood and communicated to all. The process flow chart provides a team with a clear overview of the process. It is particularly useful to answer the question *where is it wrong?*

2.   Check / tally chart
The check / tally chart is a simple tool which must not be under-estimated. It is used to count the frequency of incidences. It thus makes data collection easier and more systematic. Frequently, it is used to provide data for Pareto Analysis. It is particularly useful to answer the question *how often is it wrong?*

3.   Histogram
A histogram is a column graph that can be used to display the distribution of values obtained whenever numerical data is collected for analysis. A histogram suggests the possible distribution of the population from which the sample was taken. It shows the mean and the dispersion (standard deviation) of the data. It is particularly useful to answer the question *what is the variation?*

4.   Cause-effect diagram
The cause-effect diagram is a technique for identifying the possible causes affecting a problem. It helps to break problems into manageable pieces and thus to focus on the most likely or root cause(s) of a problem. It is particularly useful to answer the question *what caused the problem?*

5.   Pareto analysis
The Pareto analysis is a technique for recording and analyzing information relating to a problem or cause and thus enable the most significant aspect to be identified. The Pareto analysis separates the *vital few* from the *useful many*. It is particularly useful to answer the question *what is the biggest problem?*

6.   Scatter diagram
A scatter diagram shows the relationship between a problem and a likely cause. It is particularly useful to answer the question *is there a relationship?*

7. Control charts

A control chart is a tool that can be utilised at any stage of a process, or when a product can be sampled and quality measured either as attribute or variable data. When the differences in quality between the sampled items are due only to inherent variation, then the process is in a state of *statistical control* with regard to that particular quality characteristic. When differences are due to *assignable causes*, then the process is not in statistical control. An assignable cause may result from special causes such as changes to important factors like machinery, materials and operating environment. A control chart is used to determine the state of control of the process, to sustain the process in a state of control and to help achieve the optimum state of control. A control chart is a time series plot of a sample statistic that reflects the state of the process in the recent past. It is particularly useful to answer the question *can we control and reduce the variation?*

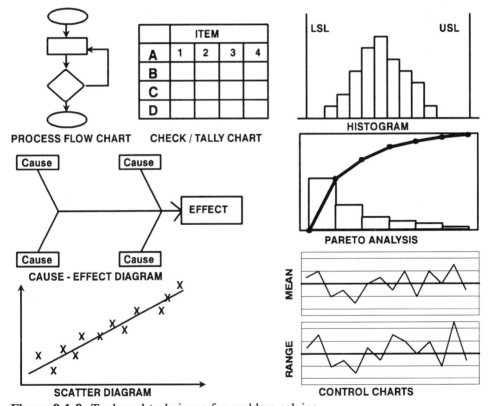

Figure 9:1.8 Tools and techniques for problem solving.

9.2.4 Team formation

The team approach has been successfully used by Quality Circles. This approach is recommended for planning and evaluating robust design. There are several reasons why this approach is superior to one that involves just one or two people. Typically, both salaried and non-salaried personnel should be valued members of the team. This is especially true if non-salaried personnel are a normal part of the process that is being investigated. In a typical manufacturing environment the following (but not limited) suggested contributors should be considered as members of the planning team:

- manufacturing supervisor,
- process engineer,
- manufacturing engineer,
- quality assurance engineer,
- machine operator,
- process inspector,
- line foreman,
- line technician,
- statistical consultant.

The statistical consultant may not be a separate individual but one of the other team members. Ideally, the person providing statistical support should be knowledgeable in analysis of variance techniques, and in particular, the Taguchi approach to robust design. It is also important not to omit the non-salaried personnel from the team for two important reasons:

- the machine operator and inspector can usually help to identify some of the significant process variables that may be overlooked by the supervisor or engineers,
- they are usually the ones who will be performing the experiment and evaluating the results.

If the operator and inspector are a part of the planning team, they are more likely to follow the exact steps or procedures that are specified for the experiment. Psychologically, they share in the success or failure of the experiment or project.

Once a team has been selected to plan and conduct an experiment, one of the members should be chosen or designated as team leader. It is important that someone guides the team during the *brainstorming* sessions and helps to minimize controversy or conflict between team members as well as motivate the team. The manufacturing supervisor or process engineer would be a good choice as team leader.

9.2.5 Brainstorming

A fundamental component of experiment planning is to brainstorm all factors that might have an effect on the quality characteristic of interest.

A successful brainstorming session lets people be as creative as possible and does not restrict their ideas in any way. This freeform approach can generate excitement in the team, equalize involvement in the process, generate more complete lists of possible factors and often results in original solutions to problems. The basic rules for brainstorming sessions are:

1.   Go in rotation with one idea per person per turn.
2.   Encourage everyone to freewheel. Do not hold back on any ideas. Strive for quantity rather than quality of ideas at this point. Ideas breed other ideas.
3.   No discussion or criticism during the actual brainstorming. That will come later. No judgment. No one is allowed to criticize another's ideas, not even with groans or grimaces.
4.   'Pass' if you have no idea this turn.
5.   Exaggerate! Stretch your imagination. Let people build upon ideas generated by others in the team.
6.   Write ALL ideas on a flipchart, or similar tool, so that the whole team can easily scan them.
7.   Laugh! Good natured laughter is great!
8.   Use Pareto analysis (multiple-voting or a series of voting), to select the most important items from the generated list.

Brainstorming should be approached with an open mind. The team leader should serve as arbitrator during these brainstorming discussions. Within the brainstorming phase, it is important to encourage all the team members to freely submit process variables for discussion. All variables that might influence the effect must be investigated. These would include all potential sources of variation and could be qualitative as well as quantitative. Following the brainstorming, there must be a screening process. The purpose of screening the list generated during the brainstorming is to ensure an efficient study of the potentially key factors that may affect the process or product of concern.

9.2.6 Cause-effect diagram

The cause-effect (also known as the Ishikawa or Fishbone) diagram is a very useful tool for identifying and listing process or product factors.

When the principal problem has been selected by Pareto analysis, the next stage will be to classify the most probable causes. To do this the team may construct a cause-effect diagram. The team leader can write the ideas on the cause-effect diagram as they are suggested.

The problem or effect is written in a box on the right-hand side of a large sheet of paper fixed to the wall or a flip-chart where it can be seen by everyone in the team. An arrow is then drawn pointing towards the box.

The team must then decide the most appropriate headings under which the probable causes can be listed. In most cases there are four headings and these are:

●   man,
●   machine,

- material,
- method.

1.   Man
The people doing the work.

2.   Machine
Equipment or tooling used to do the work.

3.   Material
Materials supplied or required to do the work.

4.   Method
Specifications or job instructions.

Occasionally other headings such as vendor, supplier, environment, etc. may be more applicable, or may be included in addition to those mentioned above. These headings form the ends of further arrows pointing towards the main arrow already drawn on the sheet of paper. Once this has been prepared, the team is ready to commence the brainstorming session to identify what the team thinks are likely to be the most probable causes. See Figure 9:1.9.

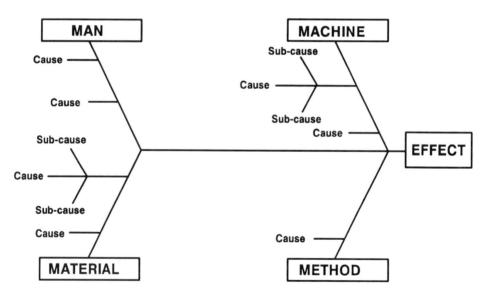

Figure 9:1.9 Cause-effect diagram.

When the brainstorming session is completed, it is necessary to evaluate or screen the suggestions. This evaluation is carried out in much the same way as the initial brainstorming, except that this time the team is attempting to identify the most likely major causes from amongst the many potential causes that have been listed.

The team leader will take each idea in turn and ask the members whether they think the idea in question is likely to be a major cause of the problem. If not, a mark is made against the item to signify it has been evaluated and the team leader will move on to the next idea. If the team, by consensus, believes that the idea may be a major cause, then the leader will circle that idea and move on to the next. This process continues until all the ideas have been evaluated. At this stage, it is likely that there will be a smaller number of ideas, usually less than ten. The next step is to rank these ideas in order of priority and again, the typical approach is to vote and select the most important ones.

Sometimes an idea that the team believes to be the cause of a problem may need to be explored. Such an assessment is necessary to home-in on the most possible causes rather than to spread attention over the less likely. The evaluation will depend on the nature of the cause. If the assessment is right, the team may proceed to the next stage. If wrong, the team must return to the cause-effect diagram and the next most likely cause can be evaluated and verified in the same way.

9.2.7 Steps in robust design of experiments
Robust design involves eight steps that can be grouped into four major categories:
1.   Planning the experiment
        Step 1.   Define the problem,
        Step 2.   Determine the objective,
        Step 3.   Define the quality characteristic,
        Step 4.   Design the experiment,
2.   Performing the experiment
        Step 5.   Perform the experiment,
3.   Analyzing the experimental results
        Step 6.   Analyze and interpret the results,
        Step 7.   Predict the process average,
4.   Verifying the experimental results
        Step 8.   Conduct a confirmation experiment.

9.2.8 Planning the experiment
Planning the experiment is an important process that will determine whether the experiment will be a success or a failure. Four key points need to be addressed:
- define the problem,
- determine the objective,
- define the quality characteristic,
- design the experiment.

1. Define the problem

Define what the problem is and ensure that the problem is a consistent one. It is also a good idea to obtain a record of the current process either in terms of the mean and standard deviations or process control charts.

2. Determine the objective

Defining the objective is the most important step in the experiment. The objective should be carefully and precisely defined to eliminate any possibility of misunderstanding. This should be reached by using some quantifiable measurement if possible. The team should reach an agreement on the objective before proceeding to the next step.

3. Define the quality characteristic

Identify the quality characteristic to be observed and the objective function to be optimized. Defining the quality characteristic is extremely important to the success of the experiment. Whenever possible, the quality characteristic should be defined in terms of quantifiable units (measurable characteristics) as opposed to classifications (attribute characteristics). A quality characteristic should also be defined so that potential interactions between factors tend to be small and additivity is assured. In developing an appropriate quality characteristic, the experimentation team should ensure that all members have a clear understanding of the definition of the quality characteristic. They should also address all pertinent issues associated with obtaining accurate and reliable quality characteristic data. These issues include:

- units of measure,
- measurement environment, e.g. temperature and time,
- instruments used for measurements,
- accuracy of instruments,
- readout accuracy required,
- who will take the measurements,
- whether it is variable or attribute data,
- classification criteria for success or failure,
- criteria for success or failure in meeting the objective.

4. Design the experiment

Identify the control and noise factors and their alternative levels. Identify the testing conditions for evaluating the quality characteristic. Control factors are placed in the control factor (inner) array. Noise factors are placed in the noise factor (outer) array. Factors designated as noise factors must be controlled during experimentation. Design the matrix experiment. Select the appropriate orthogonal array and assign the control factors. Select an outer array and assign noise factors.

9.7.9 Performing the experiment

Conduct the matrix experiment and collect the data. Note that experiments can also be simulated. When dealing with an unfamiliar quality characteristic, it is often helpful to measure trial samples prior to conducting the experiment. This will help prevent the use of a quality characteristic which has little relation to the process or product performance. It can also give an indication of the repeatability of the process or product. When data collection is complete, it may be necessary to reset the process variables to their original settings prior to experimentation until a decision regarding a permanent change in the factor level setting (optimum condition) is made. Engineering change specifications must also be initiated where necessary.

9.7.10   Analyzing the experimental results

Once data has been collected from either a practical experiment or a computer simulation, the next steps are to:

- analyze and interpret the results,
- predict the process average.

1.   **Analyze and interpret the results**
      We analyze the data using a response table, response graphs and in particular, analysis of variance. It is always good engineering practice to perform an analysis of variance to ensure that experimental errors are acceptable. Sometimes we may only need to analyze the mean data. Frequently, however, we may also need to analyze the signal-to-noise ratio or Omega transformation. In any case, the analysis result should identify the optimum levels for the control factors.

2.   **Predict the process average**
      Once the optimum levels have been identified, we then predict the performance of the quality characteristic under these levels and establish the confidence intervals. This is a critical step that allows us to test whether the experimental results are additive with respect to the control factors studied. If the predicted performance can be verified then additivity must be present and we may expect reproducibility of the experiment.

9.7.11   Verifying the experiment results

Following an analysis of the experimental results, we need to conduct a confirmation experiment. Since robust design experiments are largely saturated designs without interaction assignment, it is vital to perform a confirmation experiment. A confirmation experiment is necessary to verify the predicted performance. If the results of the confirmation experiment verify the predicted performance, we may implement the optimum condition into the process. Even at this stage it may be necessary to make a pilot-run under production conditions. If there are no adverse problems, the optimum condition can be implemented full scale. Process data may then be generated to show diagrammatically what improvement has been accomplished. This may done by using the improvement in the process trend chart or appropriate data distributions. If the results of the confirmation experiment do not correspond to the predicted results, or are otherwise unsatisfactory, then the experimental design should be re-evaluated and additional experiments may be required. It may be that one or more key variable(s) has been left out or the team may have failed to identify a good quality characteristic.

9.2.12   Caution on factor level settings

The experimental level settings may be critical to personnel or plant safety. The experimenting team should accept all responsibility for the choice of factors, test extremes and verify that all experimental test combinations will result in safe operating conditions if applied to actual system level of use.

9.2.13   Self-assessment questions

1.   Explain objectively what quality improvement is.

2.   What are the eight steps in robust design?

3.   Describe the tools and techniques that may be used to identify control and noise factors.

9.2.14   **Answers to self-assessment questions**

1.   Explain objectively what quality improvement is.

Answer

Any quality improvement can be objectively regarded as an improvement only if it reduces the net loss-to-society.

2.   What are the eight steps in robust design?

Answer

Step 1.	Define the problem.
Step 2.	Determine the objective.
Step 3.	Define the quality characteristic.
Step 4.	Design the experiment.
Step 5.	Perform the experiment.
Step 6.	Analyze and interpret the results.
Step 7.	Predict the process average.
Step 8.	Conduct a confirmation experiment.

3.   Describe the tools and techniques that may be used to identify control and noise factors.

Answer

The cause-effect diagram is the most useful tool for identifying control and noise factors. These can be used during a brainstorming session to systematically look for causes and sub-causes. The most critical of these can be determined by Pareto analysis. Other techniques such as the scattergraph may be used (before or after a brainstorming session) to establish any specific relationships between given factors. Check / tally sheets may be used to determine the frequency of occurrence of a problem over a period of time, machine or operator. Histograms may be used to quantify the distribution of the mean and dispersion of the problem. Flowcharts may also be used to scrutinize processes and product movements.

9.3   A Critique of Robust Design

9.3.1 Introduction
Robust design has been heavily criticized in recent years. This has been due to the lack of understanding, particularly at management levels. Because robust design is rather mathematical, it is seldom evaluated correctly by beginners and is often concluded to be difficult, inefficient, full of interactions, etc. This is a misconception, however, which the author hopes to overcome through education and knowledge, such as through this book.

9.3.2 Comparison of traditional and robust design
In comparing traditional design and robust design methods we must admit that some traditional designs may have some characteristics of robust designs and some robust designs may have some characteristics of traditional designs. However, we discuss two major differences:
- mean modelling versus two-step optimization,
- significance of the F-test.

   1.   Mean modelling versus two-step optimization
   In traditional methods of experimental design the model is based on the mean response. Robust design is primarily concerned with finding control factor settings that minimise variation while attaining the mean on target. As a consequence, if the target value is changed, traditional design of experiments will require a new problem solution. The robust design approach however, uses the two-step optimization. The two-step optimization identifies the factor levels with high SN ratios and uses these levels to reduce variation. Another factor that changes the mean without affecting the SN ratio is used to adjust to the target. Thus, in robust design, when the target value is changed, attaining the new target is merely an adjustment problem that usually does not require re-experimentation.

   2.   Significance of the F-test
   In traditional methods of experimental design, F-testing plays an important (if not paramount) role. This is because it is necessary to know whether that factor is significant and if it should be included in the model, i.e. the emphasis is on *factor significance*. In robust design, the contribution ratio (ρ %) of a factor to the total sum of squares determines the *factor contribution* to the total variation. Frequently, the better levels of *insignificant factors* are also included in optimization due to the concept of factor contribution as opposed to factor significance.

9.3.3 Interactions

Perhaps the greatest common misunderstanding in traditional design of experiments is that the more interactions studied the better. But recall that the purpose of design of experiments is to ensure that the results obtained from an experiment in the laboratory at the design stage can be reproduced downstream in manufacturing and in the customer's environment. In other words, the response effects must be additive over a wide spectrum of uses and a control factor that is important must exert its influence strongly. The only way to determine that is to ensure that the influence of that factor will stand out clearly, irrespective of noise conditions and interaction effects.

The presence of interactions causes inconsistencies in control factor effects. Thus, control factor effects are not predictable because the effect of one factor varies when another factor changes. In scientific research where it is necessary to understand and establish a model of natural phenomena, it is reasonable to include interactions. Here, the principal objective is to understand the underlying mechanism. In engineering research, however, it is necessary to reproduce laboratory scale experiments in a downstream environment. The principal objective here is reproducibility.

In robust design, interactions are seen to exist. However, we purposefully treat interactions as noise and therefore do not assign them to orthogonal arrays. This allows interactions to be confounded with control factors. However, we expect to identify control factors that can overcome this noise. We do this by predicting the optimum condition based on the additivity of factors and proceed to validate the result with a confirmation experiment. If our conclusion is confirmed then the design is stable enough to overcome the effects of noise. That is robust design.

In this respect, the use of the orthogonal array is largely to establish a predicted process average and test this prediction with a confirmatory experiment primarily because we do not know whether interactions exist. If the results are not upheld then we know that the additivity of factor effects does not exist, or, in other words, interactions do exist. But even then, why should we seek these interactions? To study interactions because they exist is entirely wasteful. If we know that a particular quality characteristic is additive, a one-factor-at-a-time experiment is enough. But the problem is that no one knows whether a quality characteristic has additivity. It is for this reason we use the orthogonal array to test for additivity.

Indeed, unlike most subjective techniques, robust design does not depend on feelings. Robust design is an engineering method that considers many types of factors (e.g. control, noise, signal) that cause variability in a characteristic. It starts with an assumption, conducts research, tests the results, and then makes a prediction. If the prediction can be verified then we proceed to the conclusion. As engineers we have the option of what model we choose. But how well the model fits the data must be evaluated strictly and correctly without hypothesis, as is done in robust design.

In any case, an unsuccessful confirmation experiment warns the engineer that the experiment has failed to demonstrate a reproducible conclusion. Of course, it is still possible to learn something from an unsuccessful experiment, but certainly we cannot deny the need for further investigation.

9.3.4 Violation of the assumption of a normal distribution

In conventional statistical experiments, we assume that factors have equal variance. Such statistical assumptions are seldom used in robust design. In most cases, the distribution of data does not follow the normal distribution. Additionally, there is usually no independence between mean and variance. Often when the mean is reduced, the variance is also reduced simultaneously. In parameter design, our objective is to identify control factor levels that reduce variability. Therefore, to hope to reduce variability in parameter design while assuming there is equal variance from a statistical viewpoint and unequal variance from an engineering viewpoint is totally inadmissible. Therefore, this is a fundamental change from the concept of equal variability.

9.3.5 Calculations are difficult

Perhaps the second most common criticism of robust design of experiments is that it is mathematically complex. No doubt there is a great deal of calculation involved in the analysis of robust design experiments, but, mathematical complexity cannot be regarded as a problem, just as quantum mechanics cannot be disregarded simply because it is complex. Indeed, many of the calculations involved in design of experiments are repetitive, even tedious. Nevertheless, with the availability of programmable calculators and computers, mathematical complexity cannot be regarded as a constraint. Additionally, many spreadsheet programs can be easily set up to perform relatively complex calculations using a variety of built-in functions. Finally, there are also many software packages that support the design of experiments. Therefore, the difficulty of calculations, be it for traditional or robust design, is not an acceptable criticism.

9.3.6 Randomization

Randomization is a technique applied to the order of conducting the experimental runs and to the division or allocation of raw materials to be used in experiments. In traditional design of experiments, this important technique is used to nullify the effects of extraneous variables that are not directly included in the experiment and guards against systematic biases in the data. This is particularly important since noise factors are not included in experimentation the way they are included in robust design. Therefore, randomization provides some assurance against drawing invalid conclusions because of the confounding of unknown variables with control variables.

However, robust design does not place very much emphasis on randomization as does classical design of experimentation. If the failure to randomize affects the results, the results will tend to be useless anyway. The logic is that such effects are actually generated by noise factors which have not been incorporated into the experiment. Unless these noise factors are included in the experiment, key information is lacking for making the process or product robust anyway.

Also, because the array used in robust design is an orthogonal array, randomizing the experiments will simply result in one factor being randomized while another gets 'unrandomized'. For example, in the standard $L_8(2^7)$ orthogonal array shown in Figure 9:1.10, it would appear that factor A has a column of four 1's followed by four 2's. Randomizing with respect to factor A could cause factor G to become unrandomized. Hence, randomization does not scramble the order of all factors effectively. Even after arranging the experiments in a random order, some factors would appear in a nearly systematic order for one or more factors. This is shown clearly in Figure 9:1.10.

	Standard $L_8(2^7)$ orthogonal array									Randomized $L_8(2^7)$ orthogonal array						
	A	B	C	D	E	F	G			A	B	C	D	E	F	G
1	1	1	1	1	1	1	1		1	1	1	1	1	1	1	1
2	1	1	1	2	2	2	2		4	1	2	2	2	2	1	1
3	1	2	2	1	1	2	2		7	2	2	1	1	2	2	1
4	1	2	2	2	2	1	1		6	2	1	2	2	1	2	1
5	2	1	2	1	2	1	2		2	1	1	1	2	2	2	2
6	2	1	2	2	1	2	1		8	2	2	1	2	1	1	2
7	2	2	1	1	2	2	1		5	2	1	2	1	2	1	2
8	2	2	1	2	1	1	2		3	1	2	2	1	1	2	2

Figure 9:1.10   Effect of randomization in $L_8(2^7)$ orthogonal array.

9.3.7 Self-assessment questions

1. Differentiate between traditional design and robust design.

2. Explain the importance of interactions in relation to robust design.

3. Explain the approach of assumption of normal distributions.

4. Suggest ways of performing difficult calculations.

5. Explain the unimportance of randomization.

9.3.8 Answers to self-assessment questions

1.   Differentiate between traditional design and robust design.

Answer

The fundamental difference between traditional design and robust design is that traditional design methods emphasize modelling the mean value, whereas robust design methods emphasize identifying factors with the highest signal-to-noise ratio.

2.   Explain the importance of interactions in relation to robust design.

Answer

Interactions are a nuisance in any experimental design. The presence of large interactions is an indication of likely downstream problems. In robust design interactions are treated as noise. This is to enable the identification of control factors that can overcome the effect of interactions. Of course we cannot tell whether interaction effects exist until we perform a confirmation experiment.

3.   Explain the approach of assumption of normal distributions.

Answer

Robust design does not assume that factors have equal variance.

4.   Suggest ways of performing difficult calculations.

Answer

Difficult calculations can be performed using programmable calculators or computers. A large number of spreadsheets may also be used. Specialized software packages are also available.

5.   Explain the unimportance of randomization.

Answer

Randomization is not an important issue in robust design where orthogonal arrays are used. This is because randomizing with respect to one factor would unrandomize another factor so that the second factor now appears to need randomization. In any case, since noise factors are usually included in robust design this has the effect of randomization.

CHAPTER 10

CONDUCTING AN EXPERIMENT

AIMS:
To provide an example of conducting, analyzing and reporting a parameter design experiment.

OBJECTIVES:
When you have completed this chapter you should be able to:
* conduct a parameter design experiment,
* analyze a parameter design experiment,
* report the experiment clearly.

OVERVIEW:
This chapter introduces a method of reporting an experiment. The chapter gives a method of approaching experimentation particularly with respect to forming a multi-disciplinary team for brainstorming and team activities. The relevant data that need to be collected before and after experimentation are highlighted. The verification of quality improvement through quantitative graphics for production personnel and cost-savings calculations for management are demonstrated.

10.1 A Robust Design Experiment

10.1.1   Introduction

An engineering problem has to be approached in a systematic way. Often, engineers perform experiments in an unsystematic way that not only is difficult to understand but also very difficult to follow. This section aims to provide a standard approach. Although this example is not intended to be a perfect method, it nevertheless shows some fundamental points that must be addressed in a good engineering report. In particular, we focus on the basic techniques needed in robust design and on quality loss calculations.

10.1.2   Example of an orthogonal array experiment

Robustization of the final camera tester calibration for improved light exposure distribution in the Supersnap[1] camera. This experiment is based on the $L_8(2^7)$ orthogonal array. Brainstorming techniques were used to identify key factors and factor levels. Seven 2-level control factors were studied in the experiment. Analysis of variance has been performed on both the mean and SN ratio data. The quality characteristic type is nominal-the-best. Quality loss function evaluation has also been performed to establish the cost saving from the experiment.

10.2 An Industrial Experiment
10.2.1   Objective

Robustization of the shutter speed for improved light exposure time in the Supersnap camera.

10.2.2   Abstract

Supersnap cameras have the special advantage that picture prints are available soon after a picture is taken. This immediate print feature has benefited both amateurs and professionals alike. One problem related to the contrast in the pictures is caused by variations in the light exposure time. This light exposure time is controlled by a battery-powered electronic circuit that measures the light energy falling on a photocell, which accumulates the energy over the time during which another photoreceptor (in place of a negative film) is flooded. The latter indicates the amount of light energy that strikes the negative film over the period of exposure.

---

[1]   This experiment is adapted from an industrial experiment conducted in Polaroid Camera Division, United Kingdom. Security reasons do not allow the true results or values to be disclosed here. Hence, the product type, the Supersnap camera, is fictitious and all other data, particularly monetary values have been coded for academic purpose.

10.2.3   Statement of the problem

Figure 10:2.1 shows schematic diagrams of the shutter assembly in which the aperture is normally closed (A) by the spring release shutter (SRS). When the camera is triggered, the SRS moves down, the aperture is opened and magnet hold started. Simultaneously, the printed circuit board (PCB), battery and photocell assembly measures the amount of light falling on the film. Depending on ambient light conditions the PCB assembly switches off the magnet hold. This releases the magnet release shutter (MRS) which closes the aperture again. The period during which the shutter is open (between Figure 10:2.1 (B) and (C)) is the shutter time and is measured in camera shutter stops. The problem was that the shutter stops variation tested at the Final Camera Test (FCT) after assembly was much larger than that before the shutter assembly was built into the camera.

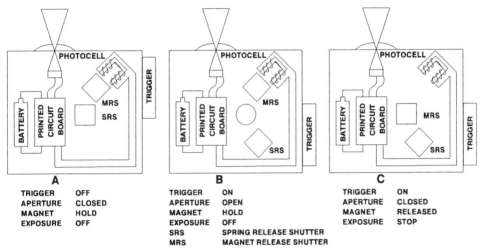

A			B			C	
TRIGGER	OFF		TRIGGER	ON		TRIGGER	ON
APERTURE	CLOSED		APERTURE	OPEN		APERTURE	CLOSED
MAGNET	HOLD		MAGNET	HOLD		MAGNET	RELEASED
EXPOSURE	OFF		EXPOSURE	OFF		EXPOSURE	STOP
			SRS	SPRING RELEASE SHUTTER			
			MRS	MAGNET RELEASE SHUTTER			

Figure 10:2.1   Schematic of camera showing basic operations.

Figure 10:2.2 shows the current process variability of shutter stops for a sample of 50 cameras. The specification for this process is 0.0 ± 0.3 stops. From Figure 10:2.2 it is obvious that the entire specification range is used up.

10.2.4   Cause-effect diagram

The factors thought most likely to affect the shutter stops were brainstormed by a multi-functional team and the result is shown in the cause-effect diagram in Figure 10:2.3.

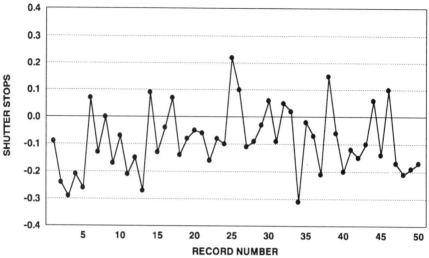

Figure 10:2.2   Current process records of shutter stops.

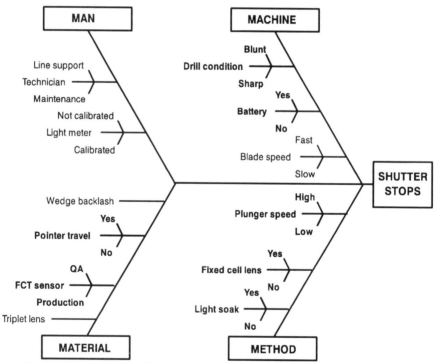

Figure 10:2.3   Cause-effect diagram showing factors affecting shutter stops. Factors chosen for experimentation are shown bolded.

10.2.5   Establishing factors for experiment

The brainstorming revealed many factors that may affect the quality characteristic. However, since engineering resources are limited, it was necessary to identify those factors that were regarded as most important. A tally chart was used to identify the factors that were regarded as most important by the technical or engineering experts. From the factors identified in the brainstorming session, the control factors (current conditions shown shaded) identified for this experiment are shown in Figure 10:2.4.

Factor	Parameter	Level 1	Level 2
A	Drill condition	Sharp	Blunt
B	Battery	Yes	No
C	Plunger speed	High	Low
D	Light soak	No	Yes
E	Pointer travel	Yes	No
F	FCT sensor	Prod	QA
G	Fixed cell lens	No	Yes

Figure 10:2.4   Current control factor conditions are shown shaded.

10.2.6   Selection of orthogonal array

Since seven 2-level factors were selected for experimentation, an orthogonal array with seven degrees of freedom was required. The orthogonal array selected was the $L_8(2^7)$ shown in Figure 10:2.5. The orthogonal array was saturated with factors without interaction assignment. This is necessary to identify factors which overcome the effect of interactions. Additionally, a noise factor array was set up to simulate noise conditions in the use of films. Since a primary factor in the picture contrast is ambient light, it was decided to include five levels of light conditions.

      Normally, the assignment of control factors would require the use of linear graphs. Referring to an $L_8(2^7)$ orthogonal array (Figure 10:2.5) and linear graph (Figure 10:2.6), if factor A is assigned to column 1 and factor B is assigned to column 2, then the interaction of factors A and B is found in column 3. Assuming that the interaction of A and B is negligible, then one may assign another factor, say factor C, to column 3. Of course if there is any interaction between factors A and B then the effect of this interaction will be confounded with the effect of factor C. Similarly, if factor D is assigned to column 4, then the interaction of factors A and D is represented in column 5. If this interaction is assumed to be insignificant, we may assign factor E. Likewise, the interaction of factors B and D is confounded with factor F (column 6). Finally, factor G may be assigned to the 3-factor (ABC) interaction in column 7. In this experiment, since the interactions are treated as noise, all columns have been assigned to control factors. If the assumption above is invalid, then a confirmation experiment will show poor results.

	1	2	3	4	5	6	7
1	1	1	1	1	1	1	1
2	1	1	1	2	2	2	2
3	1	2	2	1	1	2	2
4	1	2	2	2	2	1	1
5	2	1	2	1	2	1	2
6	2	1	2	2	1	2	1
7	2	2	1	1	2	2	1
8	2	2	1	2	1	1	2

Figure 10:2.5 $L_8(2^7)$ orthogonal array.

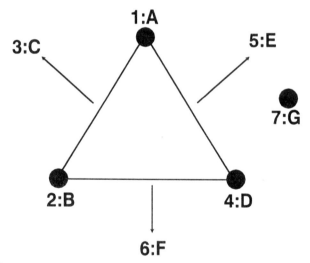

Figure 10:2.6 Factor assignment using a linear graph.

10.2.7   Planning the experiments

From Figure 10:2.5 the experimental factor settings are substituted accordingly to obtain the set-up for the experimental runs. For example, experiment 1 would have factor A – Drill Condition set at level 1 – Sharp; factor B – Battery set to level 1 – Yes; factor C – Plunger Speed set to level 1 – High, etc., for all the factors. Five cameras made with this setting were then marked clearly for testing on the Final Camera Tester. Other similar experimental runs were also prepared and labelled for calibration. The results of the experimental runs are then entered as shown in Figure 10:2.7.

	A	B	C	D	E	F	G	Results				
1	1	1	1	1	1	1	1	−0.20	−0.30	−0.30	−0.32	−0.18
2	1	1	1	2	2	2	2	−0.23	−0.22	−0.14	−0.15	−0.23
3	1	2	2	1	1	2	2	−0.18	−0.30	−0.27	−0.33	−0.32
4	1	2	2	2	2	1	1	−0.33	−0.20	−0.17	−0.25	−0.19
5	2	1	2	1	2	1	2	−0.13	−0.13	−0.09	−0.16	+0.04
6	2	1	2	2	1	2	1	−0.31	−0.32	−0.23	−0.33	−0.27
7	2	2	1	1	2	2	1	−0.05	−0.10	−0.12	−0.03	−0.10
8	2	2	1	2	1	1	2	−0.17	−0.29	−0.16	−0.15	−0.28

Figure 10:2.7   Results of the parameter design experiment.

10.2.8   Analysis of means

The traditional analysis which is performed with data from a designed experiment is the regular analysis of the mean response. If y_1 to y_8 represents the totals for each of the five replications in experimental runs 1 through 8, then for factor A at level 1 the sum of responses is denoted by A1, where A1 = $y_1 + y_2 + y_3 + y_4$ as observed from the orthogonal array of factor levels. The average response Ā1 is then A1/(5×4) since A1 is a total of 20 observations. A summary of the average responses is given in Figure 10:2.8. The *Rank* is the magnitude of differences between levels 1 and 2 for the factors and represents an arithmetic method of identifying the most important factors.

An alternative method would be to use the analysis of variance. An analysis of variance is performed on the data as shown in Figure 10:2.9. To make allowance for error, the factors with smaller sums of squares are pooled into the error sum of squares. From Figure 10:2.9 the factors E (35.90 %), A (14.16 %) and D (7.17 %) are significant.

	A	B	C	D	E	F	G
Level 1	−0.241	−0.210	−0.186	−0.179	−0.261	−0.198	−0.215
Level 2	−0.169	−0.200	−0.224	−0.231	−0.149	−0.212	−0.195
Difference	0.072	0.010	0.038	0.052	0.112	0.014	0.020
Rank	2	7	4	3	1	6	5

Figure 10:2.8   Response table of factor effects (mean analysis).

Source	Pool	Sq	ν	Ms	F-ratio	Sq'	rho %
A		0.051	1	0.051	14.94	0.048	14.16
B	Y	0.001	1	0.001	-	-	-
C		0.014	1	0.014	4.11	0.011	3.16
D		0.028	1	0.028	8.06	0.024	7.17
E		0.124	1	0.124	36.34	0.121	35.90
F	Y	0.002	1	0.002	-	-	-
G	Y	0.004	1	0.004	-	-	-
e	Y	0.113	32	0.004	-	-	-
Pooled e		0.120	35	0.003	1.00	0.133	39.61
St		0.337	39	0.009	-	0.337	100.00
Mean		1.677	1	-	-	-	-
ST		2.014	40	-	-	-	-

Figure 10:2.9   Analysis of variance (mean) for the camera experiment.

The confidence interval (CI) for the factor levels in Figure 10:2.8 is:

$$CI = \pm \sqrt{F_{0.05,1,35} \times V_e \times \frac{1}{n}}$$

$$= \pm \sqrt{4.12 \times 0.0034 \times \frac{1}{20}}$$

$$= \pm 0.03 \text{ stops}$$

10.2.9   Analysis of the signal-to-noise ratios

In addition to the analysis of means, it is important to study the variation of the response using the signal-to-noise ratio. In its simplest form, the SN ratio is the ratio of the mean (signal) to the standard deviation (noise) and abbreviated SN ratio. Regardless of the characteristic, the transformations are such that the SN ratio is always interpreted the same way; the larger the SN ratio, the better. A unique feature of this experiment was that of the choice of the quality characteristic. Although it was established that the quality characteristic was nominal-the-best, it was decided to use the SN ratio for a signed-target characteristic, since it was felt that it was more critical to reduce the variation. Hence, we use $\eta = -10 \log_{10} \sigma^2$ where σ^2 is the variance in the five data for each of the eight experiments. The unit of the SN ratio is the decibel (dB). Denoting the SN ratio for experimental conditions one through eight as η_1 to η_8, the sum of SN ratios for factor A at level 1 is A1, where $A1 = \eta_1 + \eta_2 + \eta_3 + \eta_4$. The average SN ratio for factor A at level 1 is A1/4. From these values of the SN ratio responses, it is possible to determine the factors that contribute most to the variation by calculating the difference between the factor levels, as shown in Figure 10:2.10.

	A	B	C	D	E	F	G
Level 1	24.730	25.098	25.578	24.647	24.756	23.227	25.912
Level 2	25.330	24.962	24.482	25.413	25.304	26.833	24.148
Difference	0.600	0.136	1.093	0.766	0.548	3.606	1.764
Rank	5	7	3	4	6	1	2

Figure 10:2.10   Response table of factor effects (SN ratio analysis).

Mathematically, a better method would be to calculate the analysis of variance as shown in Figure 10:2.11. To make allowance for error, the factors with smaller sums of squares are pooled into the error sum of squares.

Source	Pool	Sq	ν	Ms	F-ratio	Sq'	rho %
A	Y	0.72	1	0.72	1.59	-	-
B	Y	0.04	1	0.04	0.08	-	-
C		2.40	1	2.40	5.30	1.95	5.24
D		1.18	1	1.18	2.60	0.72	1.95
E	Y	0.60	1	0.60	1.33	-	-
F		26.02	1	26.02	57.45	25.56	68.75
G		6.23	1	6.23	13.76	5.78	15.54
Pooled e		1.36	3	0.45	1.00	3.17	8.53
St		37.18	7	5.31	-	37.18	100.00
Mean		5012.00	1	-	-	-	-
ST		5049.18	8	-	-	-	-

Figure 10:2.11   Analysis of variance (SN ratio) for the camera experiment.

The confidence interval for the factor levels is calculated using the formula:

$$CI = \pm \sqrt{F_{0.05,1,3} \times V_e \times \frac{1}{n}}$$

$$= \pm \sqrt{10.13 \times 0.4528 \times \frac{1}{4}}$$

$$= \pm 1.07 \text{ dB}$$

The effects of the factors and levels can now be summarized for both the mean and SN ratio analyses in the form of response graphs, as shown in Figures 10:2.12 and 10:2.13, respectively. In this experiment the mean response is used to optimize the factors that affect the mean and the SN ratio response is used to optimize the factors that affect the variance. The response graph has the practical importance of graphical representations, particularly for management and line personnel who may not comprehend analysis of variance. The disadvantage of a graphical representation alone, however, is that they do not enable the engineer to know how much of the effects are attributable to error and whether or not the current set of factors can explain all the variance.

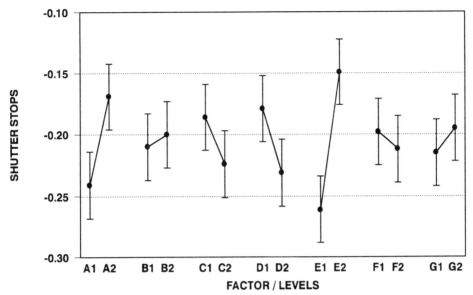

Figure 10:2.12   Response graph of mean analysis.

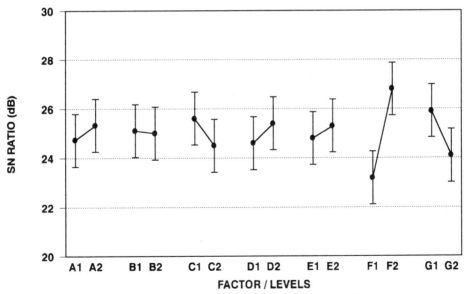

Figure 10:2.13   Response graph of signal-to-noise ratio analysis.

10.2.10   Recommendation of optimal levels

In optimizing the quality characteristics it was important to use the two-step optimization process:

1.   to reduce the variance,
2.   to adjust to target.

To accomplish this, a table of comparison for factor effects can be made, as shown in Figure 10:2.14.

	$\bar{y}$	σ	Affects	Comment	Use
A	✓2	X	mean only	use A to adjust to target	A2
B	X	X	neither	choose on other criteria	B1
C	✓4	✓3	mean and variance	increasing $\bar{y}$ increases SN ratio	C1
D	✓3	✓4	mean and variance	decreasing $\bar{y}$ increases SN ratio	D1
E	✓1	X	mean only	use E to adjust to target	E2
F	X	✓1	variance only	use F to attain high SN ratio	F2
G	X	✓2	variance only	use G to attain high SN ratio	G1

Figure 10:2.14   Comparison of factor effects where ✓ indicates the factor is important and X indicates the factor is unimportant. The number beside the ✓ indicates the rank.

From Figure 10:2.14 it is clear that:

1.   Factors F and G only affect the variance and hence these should be set at the higher SN ratio levels, namely F2 and G1.
2.   Factors E and A only affect the mean and hence these should be used adjust the mean to target. Initially, these were set at E1 and A2. However, these may be changed to adjust the mean to target.
3.   Interestingly, factors C and D affect both the mean and the variance. Therefore, care should be exercised in using these factors. If the mean value is smaller than the target, then it is recommended to use factor C to assist in adjusting to target. If the mean value is larger than the target then it is recommended to use factor D to assist in adjusting to target. This is because increasing C increases the SN ratio and decreasing D increases the SN ratio.
4.   Factor B does not affect either the mean or the variance. This suggests that factor B may not be related to the shutter stops characteristic at all. It may be set at a nominal value and maintained at that value. The choice

of which level to use may be made on the basis of whichever level is cheaper, easier or more convenient.

From Figure 10:2.14 it was decided to use the level combination A2, B1, C1, D1, E2, F2 and G1. Coincidently, this factor combination is similar to experiment 7 with the exception of the insignificant factor B. In such a case, we may proceed to use SN ratio for this experiment number to continue with the quality loss function calculations. In any case, it is imperative to conduct a confirmation experiment at the optimum condition before going into full scale implementation.

10.2.11   Prediction of response

In establishing to what extent the data of the experiment is successful, it is necessary to predict the mean and expected SN ratio at the optimum condition and then compare them against a confirmation experiment. For the mean response the overall average of the data $\bar{y} = -0.205$. The predicted mean response μ is:

$$\mu_{Predicted} = \text{Estimate of the process mean at optimum condition}$$

$$= \bar{y} + (\overline{E2} - \bar{y}) + (\overline{A2} - \bar{y}) + (\overline{D1} - \bar{y}) + (\overline{C1} - \bar{y})$$

$$= \overline{E2} + \overline{A2} + \overline{D1} + \overline{C1} - 3 \times \bar{y}$$

$$= -0.149 - 0.169 - 0.179 - 0.186 + 3 \times 0.205$$

$$= -0.068 \text{ stops}$$

The confidence interval of the predicted mean CI_{Mean} is:

$$CI_{Mean} = \pm \sqrt{F_{\alpha,v1,v2} \times V_e \times \left[\frac{1}{n_{eff}}\right]}$$

where n_{eff} is:

$$n_{eff} = \frac{\text{total number of experiments}}{\text{sum of degrees of freedom used in estimate of mean}}$$

In this case $n = 40$ (total number of experiments) and a total of four factors are used in the estimate of the predicted mean. Hence,

$$n_{eff} = \frac{total\ number\ of\ experiments}{sum\ of\ degrees\ of\ freedom\ used\ in\ estimate\ of\ mean}$$

$$= \frac{8 \times 5}{\nu_\mu + \nu_A + \nu_C + \nu_D + \nu_E}$$

$$= \frac{40}{1 + 1 + 1 + 1 + 1}$$

$$= 8$$

The confidence interval for the predicted mean is therefore:

$$CI_{Mean} = \pm \sqrt{F_{0.05,1,35} \times V_e \times \left[\frac{1}{n_{eff}}\right]}$$

$$= \pm \sqrt{4.12 \times 0.0034 \times \left[\frac{1}{8}\right]}$$

$$= \pm 0.04\ stops$$

Similarly, the overall SN ratio can be denoted $\bar{\eta}$, where

$$\bar{\eta} = Overall\ SN\ ratio\ of\ the\ experimental\ data$$

$$= 25.030\ dB$$

The SN ratio at the predicted optimum condition is:

$$\eta_{Predicted} = Estimate\ of\ the\ process\ SN\ ratio\ at\ optimum\ condition$$

$$= \bar{\eta} + (\overline{F2} - \bar{\eta}) + (\overline{G1} - \bar{\eta}) + (\overline{C1} - \bar{\eta}) + (\overline{D1} - \bar{\eta})$$

$$= \overline{F2} + \overline{G1} + \overline{C1} + \overline{D1} - \bar{\eta}$$

$$= 26.83 + 25.91 + 25.58 + 24.65 - 3 \times 25.03$$

$$= 27.88\ dB$$

The confidence interval of the predicted SN ratio CI_{SN} is:

$$CI_{SN} = \pm \sqrt{F_{\alpha,v1,v2} \times V_e \times \left[\frac{1}{n_{\text{eff}}}\right]}$$

where the total number of SN ratio experiments is 8, and a total of four factors are used in the estimate of the predicted SN ratio ($\bar{\eta}$). Hence,

$$n_{\text{eff}} = \frac{total\ number\ of\ experiments}{sum\ of\ degrees\ of\ freedom\ used\ in\ estimate\ of\ mean}$$

$$= \frac{8}{v_\mu + v_F + v_G + v_C + v_D}$$

$$= \frac{8}{1 + 1 + 1 + 1 + 1}$$

$$= 1.60$$

The confidence interval is therefore:

$$CI_{SN} = \pm \sqrt{F_{0.05,1,3} \times V_e \times \left[\frac{1}{n_{\text{eff}}}\right]}$$

$$= \pm \sqrt{10.13 \times 0.453 \times \left[\frac{1}{1.6}\right]}$$

$$= \pm\ 1.69\ dB$$

In order to allow for the possibility of an over-estimate due to error variance, only the strong effects have been used in calculating the predicted mean and SN ratio at the optimum process.

10.2.12   Confirmation experiment

A confirmation experiment was conducted at the optimum conditions for the SN ratio response. The confidence interval for the confirmation experiment is:

$$CI_{SN} = \pm \sqrt{F_{0.05,1,3} \times V_e \times \left[\frac{1}{n_{\mathit{eff}}} + \frac{1}{r}\right]}$$

$$= \pm \sqrt{10.13 \times 0.453 \times \left[\frac{1}{1.6} + \frac{1}{5}\right]}$$

$$= \pm\ 1.95\ \mathbf{dB}$$

where $r = 5$ is the number of confirmation trials. The result of a confirmation experiment using five trials gave a mean of -0.04 and a SN ratio of 28.39 dB. Both the values are well within the predicted confidence intervals.

Figure 10:2.15 shows the process improvement over 50 cameras. Figure 10:2.16 shows the process improvement in the form of normal distributions before and after the process improvement.

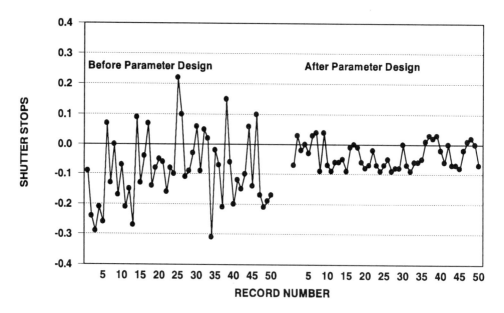

Figure 10:2.15   Process record before and after optimization.

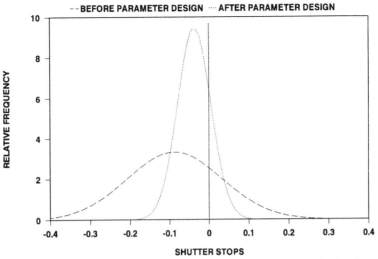

Figure 10:2.16   Comparison of processes before and after optimization.

10.2.13   Cost savings calculations

Supersnap cameras are sold with a 16-day warranty period. If, during this period, the customer perceives that the variation in picture contrast is not acceptable then the customer is likely to return the camera causing a loss of $90 to the producer. On this basis,

$$L(y) = k \ (y - m)^2$$

$$\therefore k = \frac{A_0}{\Delta_0^2}$$

$$= \frac{\$ \ 90.00}{(0.30 \ stops)^2}$$

$$= 1000.00 \ \$ \ stops^{-2}$$

The quality loss function may thus be written:

$$L(y) = 1000.00 \ (y - m)^2$$

The gain $\mathcal{G}$ in loss-per-piece in going from the existing condition to the optimum condition is determined by taking the difference between the loss-per-piece at the existing and optimum conditions.

$$\mathfrak{G} = k\left([MSD]_{Existing} - [MSD]_{Optimum}\right)$$

At this stage, there were two alternatives for calculating the gain in loss-to-society:

- experimental data,
- production data.

In this experiment the current production condition is represented in experiment 5. The predicted optimum condition also appears as experiment 8. Therefore, the process improvement could be calculated purely from this information alone. However, production data for the current condition abound on the factory floor. The results of the test run at the predicted condition gave the optimum production condition.

Of course, where experimentation is costly or data is scarce, it is reasonable to use the experimental data. In the present experiment, it would be better to use production floor data (with equal sample size) given in Figure 10:2.17.

	Before optimization	After optimization
mean	−0.0876	−0.0374
standard deviation	0.1197	0.0424
mean squared deviation [MSD]	0.0220	0.0032

Figure 10:2.17   Comparison of process before and after optimization.

Proceeding with production data, the gain in loss-per-piece is:

$$\mathfrak{G} = k\left([MSD]_{Existing} - [MSD]_{Optimum}\right)$$

$$= k\left([\sigma^2 + (\bar{y} - m)^2]_{Existing} - [\sigma^2 + (\bar{y} - m)^2]_{Optimum}\right)$$

$$= k\left([0.1197^2 + (-0.0876 - 0.0000)^2] - [0.0424^2 + (-0.0374 - 0.0000)^2]\right)$$

$$= 1000.00 \times (0.0220 - 0.0032)$$

$$= \$\ 18.80$$

If the annual sales is 25,000 cameras, the gain in loss-to-society is:

$$Gain = \$ \ 18.80 \times 25,000$$

$$= \$ \ 470,000$$

This saving is realized in the form of increased customer satisfaction, reduced warranty costs, better reputation and long run gains in market share.

10.2.14   Summary of the confirmation experiment
Referring to the graph of the production run before and after implementing the optimum condition, the results verify the reduced variation in the confirmation experiment.

10.2.15   Conclusion of experiment
Using the parameter design of experiment it was possible to identify the key factors and levels to reduce the variation in the Supersnap camera shutter stops calibration, resulting in a potential cost saving of $ 470,000 per annum.

REFERENCES

1   Alan Wu, *Introduction to Dynamic Characteristics*. First Symposium On Taguchi Methods, Birmingham. 6–7 December 1988.

2   Anand M. Joglekar and Raghu N. Kackar, *Graphical and Computer-Aided Approaches to Plan Experiments*. Quality And Reliability Engineering International. 1989, vol. 5, pp. 113–23.

3   Bandurek, G. R., Hughes, H. L. and Crouch, D., *The Use of Taguchi Methods in Performance Demonstrations*. Quality And Reliability Engineering International. 1990, vol. 6, pp. 121–31.

4   Bandurek, G. R., Disney, J. and Bendell, A., *Application of Taguchi Methods to Surface Mount Processes*. Quality Reliability Engineering International, April–June 1988, vol. 4, no. 2, pp. 171–81.

5   Barker, Thomas B., *Quality Engineering by Design: Taguchi's Philosophy*. Quality Progress (BY ASQC), December 1986, pp. 32–42.

6   Barker, Thomas B., *Quality Engineering by Design: Taguchi's Philosophy*. Quality Assurance September, 1987, vol. 13, no. 3, pp. 72–80.

7   Bendell, *The Quality Gurus. What Can They Do For Your Company?* Department of Trade and Industry Enterprise Initiative. October 1991.

8   Bendell, Wilson, G. and Millar, R. M. G., *Taguchi Methodology within Total Quality*, IFS Publications, 1990.

9   Bendell, A., Disney, J. and Pridmore, W. A., *Taguchi Methods: Application in World Industries*. 1989, IFS Publications, UK.

519

10 Box, G., Bisgaard, S. and Fung, C., *An Explanation and Critique of Taguchi's Contribution to Quality Engineering.* Quality Reliability Engineering International, April–June 1988, vol. 4, no. 2, pp. 123-31.

11 Box, G. and Draper, N., *Evolutionary Operation, A Statistical Method for Process Improvement* 1969, John Wiley & Sons, Inc.

12 Box and Müller, *A Note on the Generation of Random Normal Deviates.* Annals of Mathematics and Statistics, 1958, vol. 29, pp. 610-1.

13 Box, G. E. P., Hunter, William G. and Hunter, J. Stuart, *Statistics for Experimenters: An Introduction to Design, Data Analysis and Model Building.* John Wiley & Sons, 1978.

14 Byrne, Diane M. and Taguchi, Shin, *The Taguchi Approach to Parameter Design.* ASQC Quality Congress Transaction, Anaheim. 1986.

15 Dale, B. G. and Plunkett, J. J., *Managing Quality.* 1990, Philip Allan.

16 David Hutchins, *In Pursuit of Quality: Participative Techniques for Quality Improvement* by Pitman Publishing (UK) 1990.

17 Draper and Smith, *Applied Regression Analysis*, 2nd edition, John Wiley & Sons, Inc., 1966.

18 Duncan, A. J., *Quality Control and Industrial Statistics*, 3rd edition, Richard D. Irwin, Inc. Homewood, Illinois, 1965.

19 Eugene Grant and Richard Leavenworth, *Statistical Quality Control*, 4th edition, McGraw-Hill Book Company, 1972.

20 Feigenbaum, A. V., *Total Quality Control: Engineering and Management*, McGraw-Hill Book Company, 1961.

21 Fisher, R. A., *A System of Confounding for Factors with more than Two Alternatives, giving Completely Orthogonal Cubes and Higher Powers.* Annals Of Eugenics, 1945, vol. 12, pp. 283-90.

22 Fisher, R. A., *The Design of Experiments*, 7th edition, Oliver and Boyd, 1960.

23 Fisher R. A. *Statistical Methods for Research Workers*, 7th edition. Oliver and Boyd, 1938.

24   Fisher, Ronald A., *The Analysis of Variance. Statistical Methods For Research Workers*. 1938, pp. 256–306.

25   Ganter, William A., *Quality in Design*. Quality Reliability Engineering International. January–March 1988; vol. 4, no. 1, pp. 4–6.

26   George Bandurek, *Tolerance Design*. Taguchi Club News. June 1990, pp. 1–13.

27   Gupta, V. K. and Parmar, R. S., *Fractional Factorial Technique to Predict Dimensions of the Weld Bead in Automatic Submerged Arc Welding*. Journal of the Inst. Of Engineers (India), Mechanical Engineering. November 1989, vol. 70, no. 4, pp. 67–75.

28   Hamada, M. and Wu, C. F. J., *A Critical Look at Accumulation Analysis and Related Methods*, with discussion. Technometrics, May 1990, Vol. 32, No. 2, pp. 119–62.

29   Hammersley, J. M. and Handscomb, D. C., *Monte Carlo Methods, Monographs on Statistics and Applied Probability*. Chapman and Hall, 1964.

30   Hicks, Charles R., *Fundamental Concepts in the Design of Experiment*, 2nd edition, Holt, Rinehart and Winston Incorporation, USA, 1964.

31   Ishikawa, Kaoru, *Quality and Standardization: Program for Economic Success*, Quality Progress. January 1984, vol. 1, pp. 16–20.

32   Shingo, S., *Zero Quality Control: Source Inspection and the Poka-yoke System*, Productivity Press, 1986.

33   John MacDonald and John Piggott, *Global Quality. The new management culture*. Mercury Books, Gold Arrow Pub. Ltd. 1990.

34   Juran, J. M., *Quality Control Handbook*, 2nd edition, McGraw-Hill Book Company Inc., 1962.

35   Kackar, R. N., *Taguchi's Quality Philosophy: Analysis and Commentary*. Quality Progress (BY ASQC) December 1986, pp. 21–9.

36   Kackar, Raghu N., *Off-Line Quality Control, Parameter Design and Taguchi Method*. George E. P. Box pp. 189–90. Robert G. and John S. Ranberg. Easterling pp. 191–2. Richard A. Freund pp. 193–4. James M. Lucas pp. 195–7. Joseph J. Pignatiello, Jr. pp. 198–206. Journal Quality Technology, October 1985, vol. 17, no. 4, pp. 176–88.

37   Kackar, R. N., *Taguchi's Quality Philosophy: Analysis and Commentary*. Quality Assurance September 1987, vol. 13, no. 3, pp. 65–71.

38   Kapur, Kailash C. and Chen, Guangming, *Signal-to-Noise Ratio Development for Quality Engineering*. Quality Reliability Engineering International, April–June 1988, vol. 4, no. 2, pp. 133–41.

39   Karabatsos, N. A., *In memoriam. Dr. Kaoru Ishikawa: Quality Organiser*. 1915–1989. Quality Progress. June 1989, p. 20.

40   Khosrow Dehnad, *A Geometric interpretation of Taguchi's Signal-to-Noise ratio. Quality Control, Robust Design and the Taguchi Method*, Wadsworth and Brooks/Cole, California, USA, 1989.

41   Logothetis, N. and Wynn H. P., *Quality through Design – Experimental Design, Off-line Quality Control and Taguchi's Contributions*, Oxford Series on Advanced Manufacturing, Clarendon Press, Oxford, 1989.

42   Logothetis, N., *The Role of Data Transformations in Taguchi Analysis*. Quality Reliability Engineering International, January–March 1988, vol. 4, no. 1, pp. 49–61.

43   McEwan, W., Belavendram, N. and Abou-Ali, M. *Improving Quality through Robustisation*, Journal of the Institute of Quality Assurance, June 92, vol 18, no. 2, pp 56–61.

44   Montgomery, D. C., *Design and Analysis of Experiments*, John Wiley and Sons, 1976.

45   Myron Tribus and Geza Szonyi, *An Alternative View of the Taguchi Approach*. Quality Progress (BY ASQC) May 1989, pp. 46–52.

46   Neave, H. R., *The Deming Philosophy*, The Department of Trade and Industry, HMSO, UK, 1989.

47   Phadke, M. S. and Taguchi, G., *Selection of Quality Characteristics and SN Ratios for Robust Design*. Globecom 1987, January–December 1987, pp. 1002–7.

48   Phadke, M. S. and Dehnad, K., *Optimization of Product and Process Design for Quality and Cost*. Quality Reliability Engineering International, April–June 1988, vol. 4, no. 2, pp. 105–12.

49   Philip J. Ross, *Taguchi Techniques for Quality Engineering*, McGraw-Hill Book Company, 1988.

50   Philip B. Crosby, *Quality is Free – the art of making quality certain*, McGraw-Hill Book Company, 1979.

51   Plackett, R. L. and Burman, J. P., *The Design of Optimum Multifactorial Experiments*. Biometrika. 1946, vol. 33, pp. 305–25.

52   Rao, C. R., *Factorial Experiments Derivable from Combinatorial Arrangements of Arrays*. Journal Of The Royal Statistical Society. 1947, vol. 9, no. 1, p. 128–39.

53   Raveendra, J. and Parmar, R. S., *Mathematical Models to Predict Weld Bead Geometry for Flux Cored Arc Welding*. Metal Construction. January 1987, pp. 31R–5R.

54   Ross, Philip J., *The Role of Taguchi Method and Design of Experiments in QFD*. Quality Progress (BY ASQC) June 1988, pp. 41–7.

55   Ryan, Thomas P., *Taguchi's Approach to Experimental Design: Some Concerns*. Quality Progress (BY ASQC) May 1988, pp. 34–6.

56   Schonberger, Richard J., *World Class Manufacturing*, The Free Press, Collier Macmillan Publishers, 1986.

57   Shainin, D. and Shainin, P., *Better than Taguchi Orthogonal Tables*. Quality Reliability Engineering International, April–June 1988, vol. 4, no. 2, pp. 143–9.

58   Shewhart, W. A., *Economic Control of Quality of Manufactured Product*, D.Van Nostrand Company Inc., 1931. pp. 37–40, 173.

59   Shingo, S., *A Revolution in Manufacturing: The SMED System*, Productivity Press, 1985.

60   Shingo, S., *Study of "Toyota" Production System*, Japan Management Association, Tokyo, Japan, 1981.

61   Shoemaker, Anne C. and Kackar, Raghu N., *A Methodology for Planning Experiments in Robust Product and Process Design*. Quality Reliability Engineering International, April–June 1988, vol. 4, no. 2, pp. 95–103.

62   Sullivan, L. P., *The Power of Taguchi Methods*. Quality Progress (BY ASQC), June 1987, pp. 76-9.

63   Sullivan, L. P., *The Power of Taguchi Methods to Impact Change in U.S. Companies*. Springs. October 1987, vol. 26, no. 2, pp. 67-73.

64   Sullivan, L. P., *The Power of Taguchi Methods*. Quality Assurance, September 1987, vol. 13, no. 3, pp. 88-90.

65   Taguchi, G. and Konishi, S., *Orthogonal Arrays and Linear Graphs - Tools for Quality Engineering*, ASI Press, 1987.

66   *Taguchi Methods in Simultaneous Engineering, A Pre-Conference Technical Session with Professor Yuin Wu*. July 16 1991, Dearborne, MI.

67   Taguchi, G. and Wu, Y., *Introduction to Off-line Quality Control*. Central Japan Quality Control Association, 1985.

68   Taguchi, G., *System of Experimental Design*, Unipub Kraus International Publications, New York, 1987, Volumes 1 and 2.

69   Taguchi, G., [Japan Standards Association, Tokyo] *Quality Engineering in Japan*. Bulletin of Japan Society Of Precision Engineering. December 1985, vol. 19, no. 4, pp. 237-42.

70   Taguchi, G., Elsayed, Elsayed A. and Thomas Hsiang. *Quality Engineering in Production Systems*, McGraw-Hill Book Company, 1989.

71   Taguchi, G., *Introduction to Quality Engineering, Course Manual*, American Supplier Institute Inc., 1987.

72   Taguchi, G., *How Japan Defines Quality*. Design News, July–August 1985, vol. 41, no.13, pp. 99-100, 102, 104.

73   Taguchi, G., *Introduction to Quality Engineering; Designing Quality into Products and Processes*. 1987. American Supplier Institute Incorporated.

74   Taguchi, G. and Don Clausing, *Robust Quality*. Harvard Business Review, January–February 1990, vol. 90, no. 1, pp. 65-75.

75   Taguchi, G. and Phadke, M. S., *Quality Engineering Through Design Optimization*. Global Telecommunications Conference.

76   *Taguchi's Parameter Design: A Panel Discussion*, edited by Vijayan N. Nair. Technometrics, May 1992, vol. 34, no. 2, pp. 127-61.

77   The Distribution of Colour Television Sets. April 17, 1979. The Asahi.

78   Thomas Pepper, *The Japanese Challenge*, Thomas Y Crowell Publishers, New York, 1979.

79   Tippett, L. H. C., *Application of Statistical Methods to the Control of Quality in Industrial Production*, Transactions of the Manchester Statistical Society, (1935–36).

80   William Mendenhall, *Introduction to linear methods and the Design and Analysis of Experiments*, Duxbury Press, Wadsworth Publishing Company, Belmont, California, 1986.

81   Wu, C. F. J., Mao, S. S. and Ma, F. S., *An Investigation of OA Based Methods for Parameter Design Optimization*. Miscellaneous Reports. Centre For Quality And Productivity Improvement, University Of Wisconsin, Madison. April 1987, Report No. 24.

82   Yates, F., *Experimental Design*, Charles Griffin & Company Ltd., 1970.

83   Yuin Wu, *Evaluation and Improvement of Measuring Systems: A New Methodology Developed by Dr. Genichi Taguchi*. First Symposium On Taguchi Methods, Birmingham, 6–7 December 1988.

84   Yuin Wu, *Quality Engineering, Process and Product Optimization*, American Supplier Institute Inc., 1985.

GLOSSARY

Accumulation analysis

To attain additivity, one of the approaches is to replace irrational characteristics, such as yield or percent defective within a certain range, with ordered categorical data. Categorical data often have technical meaning in terms of the sequence of classes, such as first, second or third. Thus, instead of using percent yield as a quality characteristic, Taguchi suggests, it may be more appropriate to use the accumulation analysis. This technique is however a subject of much controversy.

Alpha error

The error of rejecting a true hypothesis. Also referred as Type I or producer's risk.

Analysis of variance

The separation of the total variation displayed by a set of observations, as measured by the sums of squares of deviations from the mean, into components associated with defined sources of variations such as control factors.

Anova

Abbreviation of analysis of variance. See analysis of variance.

Assignment of interaction

Assignment of interaction uses a linear graph to assign interaction effects to particular columns in an orthogonal array. Such a method facilitates the estimation of the interaction effect.

Attribute characteristic

Attribute characteristics are not continuous variables, but can be classified on a discretely graded scale. They are often based on subjective judgments such as None/Some/Severe or Good/Fair/Bad.

Beta error

> The error of accepting a false hypothesis. Also referred as Type II or consumer's risk.

Characteristic types

> There are mainly five types of quality characteristics, depending on what target value is required in the quality characteristic.

Classified attribute

> A classified attribute characteristic has at least three classes and provides more information than the fundamental Go/No-go characteristic. Typical levels are grades (e.g. A, B, C or D).

Combined column

> This is a method of assigning two 2-level factors to a 3-level column in an orthogonal array.

Compounding noise

> The most effective way to create meaningful noise factors is to compound noise factors. Compounding noise factors can be done with engineering knowledge and preliminary experiments on noise factors such that the resultant noise is representative of opposite conditions. An examples of compounding noise is one where a part is machined with N1 = (higher grade material, no run-out, new tooling) and N2 = (lower grade material, some run-out, worn tooling). Note that N1 and N2 are opposite conditions. Without compounding there would be eight combinations which however, could be assigned to an $L_4(2^3)$.

Confirmation experiment

> Confirmation experiments are the trials that are conducted to verify the validity (addivity) of the experimental results. Usually, such an experiment is run with the optimum factor level settings for all the control factors corresponding to the number of experimental samples in a replicate. The results of a confirmation experiment must be compared with the predicted optimum condition.

Contribution ratio

> Also called rho (ρ). An estimate of the sum of squares contributed by a source to the total sum of squares of the experimental results.

Control factors

> These are factors whose values can be selected and controlled by the design or manufacturing engineer, e.g. welding current, material type and preheat temperature. These are the product parameter specifications whose values are the responsibility of the designer. Each of the control factors can take more than

one value which is referred to as a level. It is the objective of the design activity to determine the best levels of these factors.

Conventional approach to quality

The conventional approach to quantify loss has been the make-to-specifications approach. In this method all products within the specifications are considered as being equally good and of having no loss. Any product outside the specification, however trivial, would be considered as bad and having a fixed loss.

Data transformation

This is a process of data manipulation to create more additivity in a set of data. For example, data such as percent yield are not additive in the neighbourhood of 0 % or 100 %. The omega transformation however improves this additivity. Many other transformations are also available.

Degrees of freedom

The degrees of freedom refers to the number of independent measurements available to estimate pieces of information or the number of independent (fair) comparisons that may be made within a set of data. For a factor, the degree of freedom is one less than the number of levels.

Degrees of freedom of interaction

The number of degrees of freedom of the interaction of two factors is the product of the degrees of freedom of each factor.

Degrees of freedom of an orthogonal array

The number of degrees of freedom in an orthogonal array is one less than the number of experiments.

Design process

This consists of System Design, Parameter Design and Tolerance Design. See the respective designs.

Distribution of interactions

In some orthogonal arrays (e.g. $L_{12}(2^{11})$, $L_{18}(2^1 \times 3^7)$, etc.) interaction effects between columns are distributed throughout the other columns in the array. In the $L_{12}(2^{11})$ for example, the interaction between any two columns is confounded with the remaining nine columns. In the $L_{18}(2^1 \times 3^7)$, the interactions between 3-level columns are partially confounded with the remaining 3-level columns. These arrays are therefore recommended when the experimenter is only interested in the main effects.

Dummy treatment

Dummy treatment is a technique that can be used to incorporate a 2-level factor into a 3-level series orthogonal array. It is recommended that this method be used when there are only a few 2-level factors and many 3-level factors. In order to assign a 2-level factor into a 3-level series orthogonal array, the factor is formally treated as a 2-level factor. One of the two levels of the factor is then assigned in place of the third level in the column. It is usual to duplicate the level which seems to be of greatest importance.

Dynamic characteristic

A dynamic characteristic is one whose target value changes with the signal factor. Dynamic characteristics can be studied using a parameter design which studies the characteristic over a range of signal factors. Often this is set at a number of levels. Optimization is therefore over a range of the quality characteristic.

Error factor

Also called noise factor. See noise factor.

Factors

These are parameters or variables that have an impact on the process or product performance. A factor is any experimental condition which may be assigned at will from one experimental run to another. Examples of experimental factors include temperature, time, pressure, voltage, equipment, operator, work shift, raw material, etc. There are four important types of factors.
1.   Noise factors.
2.   Control factors.
3.   Signal factors.
4.   Scaling factors.
For explanations of these factors see the appropriate entries in this glossary.

Factor levels

These are levels of values or attributes assigned to factors such as force, temperature or current, which may be represented as:

Level 1	Level 2	Level 3.	
10	20	30	(N, °C, A, etc.)
low	medium	high	(attribute type)

Failure of a confirmation experiment

If a confirmation run does not confirm the experimental results, this could be due to: Poor additivity, (equivalent to the existence of interactions) or not enough control factors were selected to ensure reproducibility. A significant factor may have been missed or factor levels may have been set too narrowly to detect effects of changes in factor levels.

Full factorial design

A full factorial design permits the effects of many factors to be investigated at the same time. While full factorial designs certainly cover all possible test conditions, they are not efficient when the number of factors and levels become even moderately large. For seven 2-level factors, the number of experiments is 2^7 or 128 experiments.

Go/No-go characteristic

The simplest type of attribute characteristic is the Go/No-go characteristic which has only two levels. Typical levels are Pass/Fail and Good/Bad.

Harmonic mean

The reciprocal of the average of the reciprocals of a set of observations.

Indicative factors

An indicative factor is one that possesses levels like those of a control factor but which are determined by the conditions of use. An engineer may not select the optimum condition of an indicative factor.

Inner array

The inner array is also called the main factor array or control factor array.

Interaction

When the effect of one factor depends on the level of another factor, an interaction is said to exist. In other words, an interaction occurs when two or more factors together have an effect on the quality characteristic(s) that is different from those of the factors individually.

Interactions, Study of

Rather than to study interactions among control factors, it is recommended to study the following:

1. Select energy related quality characteristics.
2. Use a sliding level for obvious interactions.
3. Take one factor level as a ratio to the other factor level.

Larger-the-better

A larger-the-better characteristic is a non-negative measurable characteristic that has an ideal state or target value of infinity. Examples of larger-the-better characteristics are: fuel efficiency, life, mean time between failures and strength.

Latin square

> An experimental design in which the effects of one factor are grouped according to levels of two (or more) factors, the levels of the first factor being assigned at random, with the restriction that no one level of the first will appear more than once with any given level of the other two (or more) factors.

Levels

> See factor levels.

Linear graph

> This is a graphic tool to facilitate the assignment of complicated factors and interactions to an orthogonal array. The linear graph is a series of numbered lines and dots which have a one-to-one correspondence to the columns of the related orthogonal array. Each linear graph is associated with one orthogonal array. However, a given orthogonal array can have several linear graphs. The linear graph facilitates the assignment of factors to specific columns of the orthogonal array.

LD-50

> The term LD-50 is borrowed from the medical profession where it is used to designate the lethal dose of a drug at which the live/die ratio is 50 : 50 %. For the purpose of manufacturing the LD-50 can be defined as the point where 50 % of the customers will be unhappy with the performance of a product. Once this is established (through market research) one can then define the customer tolerance as being the difference between the target and the LD-50 point. In other words this would correspond to the 3σ point.

Loss reduction

> Quality is associated with a certain loss-to-society. When quality is improved, the loss-to-society must be reduced. The reduction in loss-to-society is called loss reduction.

Loss-to-society

> The factors which cause a quality characteristic to deviate from its target value are called noise factors. Noise factors cause variability and loss of quality. Any departure (or deviation) of a characteristic from the intended ideal is a loss. Because we cannot achieve perfect quality, we cannot eliminate loss. The total loss to the producer and the consumer is called loss-to-society.

Main effects

> Taguchi stresses the strategy of focusing on main effects and characteristics with good additivity which eliminate or reduce the impact of potential interactions. The effect of interactions can be minimized by redefining the factors associated with the interaction, by the use of SN ratios or the use of sliding level factors.

Manufacturer's loss
>   For the manufacturer loss becomes apparent as:
>   1.   Customer dissatisfaction.
>   2.   Customer's time and money.
>   3.   Added warranty cost.
>   4.   Bad reputation.
>   5.   Long run loss of market share.

Noise factors
>   Any uncontrollable factor that causes the deviation of a quality characteristic from the target value is called a noise factor. Noise factors are impossible, difficult or expensive to control. They influence the output and their levels change from environment to environment, time to time and unit to unit. Only statistical characteristics can be known or specified, but not their actual values. There are three types of noise factors.
>   1.   External noise (outer noise) such as the variation in operating environment due to temperature, humidity, etc.
>   2.   Internal noise (inner noise) such as deterioration, wear of parts, etc.
>   3.   Unit-to-unit noise (between-product noise or part-to-part noise) such as manufacturing imperfections.
>   Noise factors are usually difficult to control, expensive to control or one does not intend to control. However, for experimental purposes, they may need to be controlled on a small scale.

Nominal-the-best
>   A nominal-the-best characteristic is a measurable characteristic with a specific user-defined target value. Values may either be positive or negative. Examples of nominal-the-best characteristics are: length, pressure, viscosity and weight.

Omega transformation
>   The omega transformation transforms the proportion p into decibels through the formula $\Omega = -10 \log (1/p - 1)$. Also known as the logit transformation.

One-factor-at-a-time
>   This method of experimentation loses information with multiple factors and is less realistic with large number of factors. It cannot guarantee reproducibility of results under actual manufacturing conditions.

Orthogonal array
>   An orthogonal array is a matrix of numbers arranged in columns and rows. Each column represents a specific factor or condition that can be changed from experiment to experiment. Each row represents the state of the factors in a given experiment. The array is called orthogonal because the effects of the various factors are balanced and can be separated from the effects of the other factors

within the experiment. That is, an orthogonal array is a balanced matrix of factors and levels, such that the effect of any factor or level is not confounded with any other factor or level. The orthogonal array is a fractional factorial array which assures a balanced, fair comparison of levels of any factor or interaction of factors whereby all columns can be evaluated independently of one another.

Outer array

This is also called the noise factor array or the error factor array.

Panic mode

This is a problem-solving mode when a problem is identified and it is essential to fix it immediately at all costs. It usually ends up with many fixes and not knowing which fix really solved the problem. It does not allow a cost-effective solution and is frequently called firefighting.

Number of levels

If a new process is being investigated, it may be desirable to run three levels for some of the variables to evaluate nonlinearity over the range of the variables. If more is known about the effects of certain variables, then two levels may be sufficient to derive the desired information from the analysis results. When uncertainty exists about the number of levels to choose for a given variable, then three levels might provide sufficient information. A lot depends on the cost of experimentation and how much the scope of the experiment is increased by going from two to three levels.

Off-line quality control

Off-line quality control is the design activities at the product design stage through research and development of a prototype product. This stage involves quality improvement through identifying parameter settings that make the process or product insensitive to noise.

On-line quality control

On-line quality control refers to the quality activities at the production stage and includes both the process and product. This leads to the customer service stage where the quality activity is the care and service of the product by the producer.

Parameter design

Parameter design is used to identify the parameter settings of a process or product that will make the process or product robust.

Pooling down

Pooling down strategy entails pooling all but the largest factor effect and F-testing the factor effect against the remaining factor effects pooled together. If that factor effect is significant, then the next largest factor effect is removed from the pool and those two factor effects are F-tested against the remaining factor effects pooled together until some insignificant F-ratio is obtained. This strategy tends to minimize the number of significant factors.

Quadratic loss function

Taguchi has found the quadratic representation of the Quality Loss Function (QLF) to be an efficient and effective way to assess the loss due to deviation of a quality characteristic from its target value.

Quality characteristic

The quality characteristic is the objective measure of a particular feature that one requires in a process or product. It is also called the functional characteristic. Examples of quality characteristics are room temperature, mileage and tyre wear. While all three are representative quality characteristics, room temperature is a quality characteristic that one would like to be nominal, say 20 °C. Such a quality characteristic is referred to as nominal-the-best. On a similar argument, mileage is a quality characteristic that one would like to be as large as possible. Such a quality characteristic would be referred to as larger-the-better. Finally, tyre wear is a quality characteristic that one would like to be as small as possible. Such a quality characteristic is referred to as smaller-the-better.

Quality characteristic and quality loss function

It may be implied that the quality characteristic determines the type of quality loss function that may be used. If the quality characteristic is nominal-the-best, then the quality loss function must also be nominal-the-best. If the quality characteristic is smaller-the-better, the quality loss function must also be smaller-the-better. Similarly, if the quality characteristic is larger-the-better, the quality loss function must also be larger-the-better.

Quality loss function evaluation

Taguchi realizes that loss is a continuous function. Loss does not occur suddenly. The quality loss function evaluation is a new concept and the objective of Taguchi's evaluation is the quantitative evaluation of quality loss due to function variation. In reality there exists some function which uniquely defines the relationship between economic loss and the deviation of a quality characteristic from its target. However, it is usually uneconomical to determine this exact relationship. Therefore, depending on the target characteristic, Taguchi proposes the nominal-the-best, larger-the-better or smaller-the-better quality loss functions for evaluating quality loss.

Quality ideas
Quality ideas based on the quality perceived of customers are:
1.   Conformance to specifications limits is an inadequate measure of quality or of loss due to poor quality.
2.   Quality loss is caused by customer dissatisfaction.
3.   Quality loss can be related to product characteristics.
4.   Quality loss is a financial loss.
5.   The quality loss function is an excellent tool for evaluating loss at the earliest stage of product or process development.

Randomization
Randomization is a technique applied to the order of conducting the experimental runs and to the division or allocation of raw materials to be used in the experiments. This is an important technique used to neutralize the effects of extraneous variables that are not directly included in the experiment and guards against systematic biases in the data. Randomization provides some insurance against drawing invalid conclusions because of the confounding of unknown variables with control variables. However, the Taguchi methodology does not place as great an emphasis on randomization as classical design of experimentation. If failure to randomize affects the results, the results will tend to be useless anyway. The logic is that such effects are actually generated by noise factors which have not been incorporated into the experiment. Unless noise factors are included in the experiment, key information is lacking for making the process robust.

Range of factor levels
Assuming a linear relationship, the wider the range used in the experiment, the better is the chance of discovering the real effect of that variable on the quality (functional) characteristic. However, the wider the range, the less reasonable is the assumption of a linear effect for the variable. The selection of the range depends in part on whether the purpose of the experiment is exploration over a broad region or fine tuning to achieve optimum conditions. In either case, three levels are preferable to two levels.

Repetition
In repetition, the order of experimental runs is randomized while individual units or observations within the run are kept grouped together. This allows for an estimate of experimental error known as within experimental error, sampling error or secondary error. Repetition requires less setup than replication and may be used when variation between process setup is small and the cost of setup change is high.

Replication

Replication is used when significant variation exists or is suspected within the process setup. Within replication, all experimental units or observations are fully randomized. This allows for an estimate of experimental error, known as between experimental error or primary error. Replication will yield better results than repetition if sampling error is small. However, replication can be costly if the initial setup cost is high.

Rho (ρ)

This the percent contribution attributable to a source to the total sum of squares in an analysis of variance.

Robustization

Robust product and process design is an important technique for achieving high quality at low cost. It involves making the product's function much less sensitive to various sources of noise such as manufacturing variation, environmental variation and deterioration. This is a problem in optimization involving minimization of the mean square loss resulting from the deviation of the product's function from its target. The purpose of selecting noise factor levels is to simulate current variability and to ensure robustness to the variability. In most cases, a lower and upper boundary setting (two levels) will be sufficient for ensuring robustness. However, in the case of discrete noise factors, the number of levels may be larger. For example, the noise factor 'Shift' could include first shift, second shift and third shift (three levels).

Scaling factors

These are factors which are used to shift the mean level of a characteristic to achieve the functional relationship between a signal factor and the response.

Signal factors

These are factors which change the true values of the object to be measured.

Signal-to-noise ratio

The signal-to-noise ratio (SN ratio) is used in evaluating the quality of the product. The signal-to-noise ratio measures the level of performance and the effect of noise factors on performance and is an evaluation of the stability of performance of an output characteristic. Higher performance as measured by a higher SN ratio implies a smaller loss as measured by the corresponding quality loss function. Like the quality loss function, the signal-to-noise ratio is an objective measure of quality that takes both the mean and the variance into account.

Signed target

This a characteristic whose mean value is easily changed. The objective in this characteristic is to reduce the standard deviation about the mean value.

Sliding levels

This is a method of overcoming interaction effects in certain types of data. Suppose, for example, that the yield of a reaction depends on the temperature and time. Suppose the reaction is studied at three temperature levels, say 500 °C, 600 °C and 700 °C. Since high temperature and long time may be known to cause an over-reaction, it may be necessary to study times of 6, 7 and 8 hours for 500 °C and times of 5, 6 and 7 hours for 600 °C and times of 4, 5 and 6 hours for 700 °C. The utilization of such changing levels constitutes the sliding level method.

SN ratio

See signal-to-noise ratio.

Smaller-the-better

A smaller-the-better characteristic is a non-negative measurable characteristic that has an ideal state or target value of zero. Examples of smaller-the-better characteristics are: deterioration, shrinkage and wear.

Social loss

When a product (or service) is competitive, the customer accepts the loss. When a product (or service) is not competitive, the customer no longer accepts the loss and the producer must accept the loss. Both customer losses and producer losses are called social loss.

Suggested arrays

The $L_{12}(2^{11})$, $L_{18}(2^1 \times 3^7)$ and $L_{36}(2^3 \times 3^{13})$ are highly recommended orthogonal arrays. Since any interactions are uniformly distributed among all remaining columns, main effects can be studied without much concern for confounding.

System design

System design is the genius of creating something new, often the first feasible design. Examples of system design are the thermionic diode, thermionic valve, transistor, silicon chip, integrated circuits and very large scale integrated circuits.

Taguchi's meaning of quality

Taguchi's meaning of quality is to provide products and services that meet the customers needs and expectations over the life of the product or service, at a cost that represents customer value.

Taguchi's definition of quality

Taguchi defines quality as 'The quality of a product is the (minimum) loss imparted by the product to the society from the time the product is shipped'.

Target value

There are three types of target values:

1. Smaller-the-better where the target is to get as small as possible (i.e. 0, zero).
2. Nominal-the-best, where the target is to be centred at a given value.
3. Larger-the-better where the target is to be as large as possible (i.e. ∞, infinity).

Tolerance design

Tolerance design is used when quality can no longer be improved by parameter design. It involves investing money to reduce tolerances, e.g. more precise components, higher grade material, etc. Tolerance design is used to identify which among many components need to be upgraded.

Traditional quality loss evaluation

Traditionally, quality control has had the objective of controlling functional variation and its related problems. However, since no method of quantitatively evaluating quality and/or loss was established, the problems of quality control and their resolutions were centred on the removal of the cause with little or no connection to the economic loss in monetary terms.

Type I error

The error of rejecting a true hypothesis. Also referred to as α-error or producer's risk.

Type II error

The error of accepting a false hypothesis. Also referred to as β-error or consumer's risk.

INDEX